Sustainable Development of Food Waste Biorefineries

Sustainable Development of Food Waste Biorefineries

Editors

Jose Luis García-Morales
Francisco Jesús Fernández Morales

MDPI

Basel • Beijing • Wuhan • Barcelona • Belgrade • Novi Sad • Cluj • Manchester

Editors
Jose Luis García-Morales
Department of Environmental Technologies, Faculty of Sea and Environmental Sciences
University of Cádiz
Puerto Real
Spain

Francisco Jesús Fernández Morales
Department of Chemical Engineering
University Castilla-La Mancha
Ciudad Real
Spain

Editorial Office
MDPI AG
Grosspeteranlage 5
4052 Basel, Switzerland

This is a reprint of articles from the Special Issue published online in the open access journal *Fermentation* (ISSN 2311-5637) (available at: www.mdpi.com/journal/fermentation/special_issues/448KD102G4).

For citation purposes, cite each article independently as indicated on the article page online and as indicated below:

Lastname, A.A.; Lastname, B.B. Article Title. *Journal Name* **Year**, *Volume Number*, Page Range.

ISBN 978-3-7258-1978-2 (Hbk)
ISBN 978-3-7258-1977-5 (PDF)
doi.org/10.3390/books978-3-7258-1977-5

Contents

About the Editors

Jose Luis García-Morales

Dr. García-Morales has held various academic positions at the University of Cádiz, progressing from Associate Professor (1998–2002), to Tenured Lecturer (2002–2018), and since 2018, Professor in Environmental Technologies. He obtained his Bachelor's degree in Chemical Sciences (1992) and Ph.D. in Chemical Engineering (1997) from the same university, specializing in the anaerobic digestion of effluents from the food industry. He completed a postdoctoral fellowship at INRA-INSTITUT NATIONAL AGRONOMIQUE PARIS-GRIGNON in France in 1998, funded by a Technologist Enhancement Grant from the Spanish Ministry of Science. In 2018, he expanded his international experience as a Visiting Research Scholar at the University of Florida's Institute of Food and Agricultural Sciences (IFAS), Department of Soil and Water Sciences, and in 2024 as a Visiting Professor at University College Dublin (UCD), College of Engineering and Architecture, School of Biosystems and Food Technology. His research focuses on biological treatment within the biorefinery concept, particularly anaerobic digestion and composting of biowaste and industrial wastewater, especially from the agri-food sector and urban environments. Currently, he is investigating advanced oxidation processes, mainly ozone-based, for treating various wastes and effluents and their applications in materials science.

Francisco Jesús Fernández Morales

Holding a Ph.D. in Chemical Engineering, Professor Fernández-Morales specializes in the areas of sustainable process engineering, biorefineries, and waste valorisation. His research focuses on developing innovative solutions for the transformation of waste materials into valuable products, emphasizing the principles of sustainability and circular economy. With a rich academic background and extensive experience in the field, Professor Fernández Morales has significantly contributed to the advancement of waste valorization from the chemical engineering perspective.

Throughout his career, Professor Fernández-Morales has published numerous articles in leading scientific journals and has presented his work at many international conferences. His research has been instrumental in advancing the understanding of bioprocesses and the integration of environmentally friendly technologies in industrial practices. In addition to his research contributions, Professor Fernández-Morales has been involved in developing and delivering a range of undergraduate and postgraduate courses, aiming to equip the next generation of engineers with the knowledge and skills required to address contemporary challenges in the environmental and chemical engineering fields.

Professor Fernández-Morales is also actively engaged in collaborative projects with industry and academic institutions, fostering innovation and knowledge transfer. His work has earned him recognition and awards, reflecting his commitment to excellence in research and education.

In his current position as Full Professor of Chemical Engineering at the University of Castilla-La Mancha, Professor Fernández-Morales continues to inspire students and colleagues alike with his passion for environmental and chemical engineering and his unwavering dedication to sustainable development.

Preface

The global challenge of sustainable development has become more pressing than ever due to environmental degradation and resource scarcity. Among the myriad strategies to address these issues, the valorization of food waste through biorefineries stands out as a promising and multifaceted approach. This Reprint entitled "Sustainable Development of Food Waste Biorefineries" delves into the innovative methodologies, technological advancements, and holistic frameworks that are shaping the future of food waste management.

Food waste is a growing issue, as various stages of the supply chain discard significant portions of produced food. This not only leads to substantial economic losses but also contributes to severe environmental impacts and resource depletion. Biorefineries, which convert biomass into valuable products, present a compelling solution by transforming food waste into biofuels, biochemicals, and other bioproducts, thereby closing the loop in the food production and consumption cycle.

This Reprint brings together a diverse array of research articles, reviews, and case studies that highlight the state-of-the-art in food waste biorefineries. The contributions cover a wide range of topics, including but not limited to novel pretreatment technologies, enzymatic processes, microbial fermentation techniques, and the integration of biorefineries within circular economy frameworks. The insights provided by these works not only underscore the technological feasibility of food waste biorefineries but also address the economic, social, and regulatory dimensions essential for their sustainable development.

These studies offer innovative approaches to enhancing the efficiency and yield of bioconversion processes, thereby maximizing the recovery of valuable compounds and reducing waste. Additionally, the issue explores the environmental benefits of food waste biorefineries, such as reductions in carbon footprint and improvements in waste management practices, reinforcing their role in achieving sustainability goals.

The editorial team extends its deepest gratitude to all the authors, reviewers, and contributors whose dedicated efforts have made this Reprint possible. Their expertise and insights are invaluable in advancing our understanding of how biorefineries can contribute to a more sustainable future.

We hope that this Reprint not only informs but also inspires readers, researchers, policymakers, and industry practitioners to further explore and implement sustainable practices in food waste management. By fostering collaboration and innovation, we can advance towards a more resilient and sustainable bioeconomy.

Jose Luis García-Morales and Francisco Jesús Fernández Morales
Editors

fermentation

Editorial

Sustainable Development of Food Waste Biorefineries

José Luis García-Morales [1] and Francisco Jesús Fernández-Morales [2,*]

1 Department of Environmental Technologies, Faculty of Marine and Environmental Sciences, IVAGRO-Wine and Agrifood Research Institute, University of Cadiz, 11510 Puerto Real, Spain; joseluis.garcia@uca.es
2 Department of Chemical Engineering, University of Castilla-La Mancha, Avda. Camilo José Cela S/N, 13071 Ciudad Real, Spain
* Correspondence: fcojesus.fmorales@uclm.es

The sustainable development of food waste biorefineries is crucial for a number of reasons, and these reasons have environmental, economic, and social dimensions. These biorefineries help mitigate the significant problem of food waste, which is a major environmental challenge, by turning this waste into valuable products, thereby reducing the volume of the waste discarded into landfills and decreasing emissions of carbon dioxide and methane, potent greenhouse gases. In addition, recycling food waste through biorefineries conserves natural resources and minimizes the environmental impact associated with the production and processing of raw materials, thus contributing to a more sustainable and circular economy (CE) [1].

Processing food waste in biorefineries can significantly reduce its environmental impact compared to traditional waste management methods such as landfilling and incineration, while the production of biochemicals, biofuels, biogas, etc., from food waste provides renewable energy sources that can replace fossil fuels, further reducing carbon emissions [2].

The biorefinery concept refers to the sustainable processing of biomass (in this Special Issue, food wastes) into a wide spectrum of bio-based products (such as chemicals and materials) and bioenergy (biofuels, power, heat) [3]. Similar to conventional fossil refineries converting crude oil into multiple products like fuels, lubricants, and chemicals, biorefineries aim to maximize the value derived from biowastes by using integrated and sustainable process configurations to produce multiple outputs [4]. In this context, there are two key components to the biorefinery concept: biomass and sustainable conversion technologies [3].

Resource consumption could be optimized by integrating both of these biorefinery concepts, reducing waste generation rates and improving economic viability. Because of that, this integration involves not only the efficient use of all components of biomass, but also minimizes its environmental impact and contributes to a circular economy.

Regarding biomass feedstocks, the raw material for biorefineries can include agricultural residues, forest residues, dedicated energy crops, micro and macro-algae, and organic waste from various sources such as food wastes. These feedstocks are processes using sustainable conversion technologies able to transform the biomass into the desired products. There are two main types of technologies used here: biochemical and thermochemical processes. Biochemical processes use microorganisms or enzymes to convert biomass into products. This includes fermentation (to produce ethanol or other chemicals), anaerobic digestion (to produce biogas), and enzymatic hydrolysis. Thermochemical processes use heat and chemical reactions to convert raw materials into desired products. This includes processes like pyrolysis (producing bio-oil), gasification (producing syngas), hydrothermal liquefaction, etc.

A wide spectrum of different products can be obtained by means of biorefinery transformation processes, such as biochemicals derived from biomass, which can be used as building blocks for producing polymers, solvents, and other industrial chemicals and bio-based materials including bioplastics, biofibers, and other renewable materials. Also,

Citation: García-Morales, J.L.; Fernández-Morales, F.J. Sustainable Development of Food Waste Biorefineries. *Fermentation* **2024**, *10*, 301. https://doi.org/10.3390/fermentation10060301

Received: 30 May 2024
Accepted: 4 June 2024
Published: 5 June 2024

a wide spectrum of high-energy-density chemicals can be obtained, including ethanol, biodiesel, biogas, and advanced biofuels (e.g., biobutanol, biohydrogen) [5].

From an economic perspective, food waste biorefineries transform a stream of low-value waste into valuable products such as biofuels, bioplastics, chemicals, fertilizers, etc., creating new revenue streams and economic opportunities and stimulating local economies by creating jobs in the collection, processing, and conversion of food waste, as well as fostering the innovation and growth of new industries focused on sustainable technologies. In addition, biorefineries can reduce the waste management costs for municipalities and businesses, allowing those savings to be redirected towards other sustainability initiatives. In this sense, the biorefinery concept also allows for a more efficient use of resources, simultaneously creating a sustainable economy. This is because, in using all of the biomass and producing multiple products, biorefineries enhance the overall efficiency of their raw material and the value derived from it. Moreover, this higher efficiency in the use of raw material also leads to environmental benefits due to the reduction of the exploitation of natural resources and their associated energy consumption, greenhouse gas emissions, effluent generation, etc., promoting the sustainability of raw materials and energy consumption [5]. These changes could create new economies by creating new jobs and promoting the development of local industries around biomass supply chains, reducing transport and logistics costs. Finally, this concept could lead to interesting energy scenarios due to the more diversified and secure energy supply created. In terms of the actual structure of global energy, it is attractive to make countries less dependent on energy imports and diversify their energy supply sources.

From a social point of view, these biorefineries improve waste management practices, fostering awareness and education on the importance of reducing, reusing, and recycling food waste, which can lead to broader societal shifts towards more sustainable behaviors. These biorefineries not only increase energy security by diversifying energy sources and reducing our dependence on imported fossil fuels, but they also reduce our dependence on the by-products from biorefineries, such as compost and biofertilizers, which are used to enrich soils and improve agricultural productivity, closing the nutrient cycle and supporting sustainable agricultural practices. This approach also aligns with several United Nations Sustainable Development Goals (SDGs), such as Goal 2 (Zero Hunger), Goal 7 (Affordable and Clean Energy), Goal 12 (Responsible Consumption and Production) and Goal 13 (Climate Action), contributing to more efficient resource management and fostering a circular economy, making it a vital component of global sustainability efforts.

Additionally, this new biorefinery scenario presents significant challenges and calls for considerations related to the technological development required to ensure a proper processing of the substrates in order to reach adequate efficiencies at achievable economical costs. In this sense, biorefineries need to be economically competitive with traditional petrochemical refineries and other industrial processes. Another important aspect is the availability of a reliable, sustainable, and cost-effective supply of biomass feedstock, which is critical to achieve the desired results. Finally, in a mainly competitive economic environment, supportive policies should also be implemented to promote the development and adoption of biorefinery technologies that, in some cases, could be burdened by higher economical costs [6].

In summary, the biorefinery concept represents a holistic approach to biomass utilization, aiming to produce a wide array of products and energy in an efficient and sustainable way. The potential integration various bio-technologies and bio-processes may play a significant role in advancing the bioeconomy and contributing to environmental sustainability.

This Editorial is part of the Special Issue "Sustainable Development of Food Waste Biorefineries", which highlights new opportunities and challenges in advancing assessments of the performance of food waste biorefineries, focusing on technological advancements and management initiatives, including the recovery of material and energy from these wastes.

Fifteen manuscripts were submitted to this Special Issue, and all of them were subject to *Fermentation*'s rigorous review process. Ten research papers were then accepted for publication and inclusion in this Special Issue.

As shown in Table 1, the contributions contained herein cover both material and energy valorization technologies. The majority of these contributions relate to experimental work in the laboratory and industry, but modelling and sustainability studies were also included in this Special Issue. Nine contributions were research articles and only one was a brief report (Table 1).

Table 1. Analysis of the contributions published in this Special Issue.

N# Contribution	Contribution Type	Research Area	Focus	Type of Biomass	Approach
1	Research Article	Acidogenic fermentation	Fatty acids generation	Food waste	Laboratory research
2	Research Article	Sustainability assessment	Rural areas	Food waste	Full-scale study
3	Research Article	Autolysis enzymes	Nutritional compounds	Spent yeast residue	Case study
4	Research Article	Modelling amino acids fermentation and flash distillation	Ammonium bicarbonate extraction	Anaerobic digestion digestate	Laboratory research and modeling
5	Research Article	Anaerobic digestion	Energy and CO_2	Craft brewery wastes	Bench-scale research
6	Brief Report	Modeling	Biomethane	Whey and sugar cane	Laboratory research
7	Research Article	Holistic valorization	Antioxidants, phenolic compounds, PHA, and ethanol	Cocoa by-products	Laboratory research
8	Research Article	Acidogenic fermentation and modeling	H_2 and VFA	Corn bioethanol effluent	Laboratory research and modeling
9	Research Article	Anaerobic digestion	VFA, caproic acid	Winery effluent	Laboratory research
10	Research Article	Anaerobic digestion	Carboxylic acid	Synthetic amino acid medium	Laboratory research

The article by Pooja Radadiya et al. (contribution 1) discusses the effects of one of the most relevant variables, pH, on the production of short- and medium-chain fatty acids when fermenting food wastes under acidogenic conditions in a leachate bed reactor. To do so, they evaluated the hydrolysis and acidogenesis of a simulated food waste at pH values ranging from 5.5 to 8.5, as well as under uncontrolled conditions. In this research, the authors observed that the optimum pH was 6.5. Butyrate was the predominant fermentative product when operating at 5.5–6.5, whereas acetate was dominant when operating at pH 7.5–8.5. Under uncontrolled pH conditions, lactate was the predominant fermentation product.

The article by Carlos Ariel Cardona et al. (contribution 2) focuses on a sustainability assessment of food waste biorefineries in rural areas of Colombia. In this work, the authors analyze the sustainability of food wastes biorefineries in boosting the rural economy in Colombia, paying special attention to six food wastes (acai, annatto, sugarcane bagasse, rejected plantain and avocado, and organic kitchen food waste) from three different rural areas (Chocó, Caldas, and Sucre). In this study, it was observed that biogas production was the most convenient for the complete use of these residues, while levulinic acid was the most feasible and sustainable by-product to generate.

Joana F. Fundo et al.'s article (contribution 3) explores the optimizing of autolysis conditions for genetically engineered yeast residues to enhance their downstream processing and valorization. This research identifies that increasing the pH to 8 and conducting autolysis at 50 °C for 2 h significantly improves the release of valuable compounds like amino acids and minerals. This method efficiently reduces processing times and costs, yielding higher concentrations of leucine, aspartic acid, and potassium. The autolysis process also generated higher dry weight yields and a higher protein content in the supernatants compared to the untreated samples. Additionally, the autolyzed yeast residues were increased in essential amino acids, making them a rich source of nutrients. This innovative approach presents a sustainable and cost-effective method for converting yeast residues into valuable products which are particularly useful in animal feed supplementation and have other potential commercial applications.

The fourth article, by Alejandro Moure Abelenda et al., focuses on the modelling of amino acid fermentations (arginine, glycine, and histidine) and on the study of the subsequent flash distillation of the digestate to recover ammonium bicarbonate. The most adequate flash distillation conditions included a high moisture content, and the process was enhanced by adding hydrochloric acid or sodium hydroxide to maximize the stabilization of the digestates.

The article by Dhanashree Rawalgaonkar et al. (contribution 5) investigates the feasibility of anaerobic digestion (AD) and CO_2 recovery systems in small craft breweries. Utilizing biochemical methane potential tests and anaerobic sequencing batch reactor studies, their research assesses the production of biomethane from high-strength brewery waste, including hops. Their results indicate that the co-digestion of yeast waste with 20% hops slightly reduces the methane yield but remains economically feasible. The developed spreadsheet tool evaluates economic feasibility based on production volume, waste management, and energy costs, revealing that AD and CO_2 recovery are viable for breweries producing over 50,000 barrels annually. Additionally, their study highlights the environmental and economic benefits of implementing these systems, such as reduced waste surcharges and lower energy costs, thus promoting sustainability in the craft brewing industry. Future recommendations include pilot-scale AD studies with varying hop dosages and the further exploration of resource recovery pathways, including compressed natural gas and liquefied CO_2.

The article by Huaita Pacari Arotingo Guandinango et al. (contribution 6) deals with the modeling of biomethane generation from the anaerobic digestion of whey and sugar cane mixtures made in Ecuador. The experimental results were fitted to six different kinetic models, five of which were previously studied by other authors while the sixth one was developed in this work by modifying an existing first-order model. From the experimental results, the authors observed that the model developed in their work and the modified two-stage Gompertz model were those with the best fitting. Additionally, their model depended on five parameters, one less than the modified two-stage Gompertz model, making it more robust and straightforward.

In the seventh article of this Special Issue, Licelander Hennessey Ramos et al. focus on the valorization of cocoa pod husks (CPHs) and cocoa bean shells (CBSs), major by-products of the cocoa industry. Their research characterizes the chemical composition of these residues, highlighting the richness of their protein, lipid, and bioactive compound contents. CBSs exhibited higher protein and lipid contents compared to CPHs. The study demonstrated that both CPHs and CBSs contain significant amounts of phenolic compounds, particularly pyrogallol, which grants them antioxidant properties. The alkaline pretreatment and then enzymatic hydrolysis of CPH efficiently released glucose, which supported the growth of Cupriavidus necator DSM 545 and Saccharomyces cerevisiae Fm17 to produce polyhydroxyalkanoates (PHAs) and bioethanol, respectively. Their findings suggest that cocoa residues can be sustainably converted into high-value bioproducts, promoting a circular economy in the cocoa industry. Future research could explore the further optimization of pretreatment conditions and the scalability of their bioconversion processes.

The eighth article, by María Eugenia Ibañez López et al., studied the generation of volatile fatty acids and hydrogen during the acidogenic fermentation of corn bioethanol effluent under uncontrolled pH conditions. The researchers evaluated acidogenic fermentation with starting pHs ranging from 4 to 6. When starting at pH 4, the system develops extreme acidic conditions that stop fermentation, causing uncoupling growth during the fermentative process. However, when starting the fermentation at pH 5 and 6, the initial substrate was completely fermented, and significant amounts of fermentation products were obtained. The optimum initial pH for uncontrolled pH fermentation was found to be pH 5, which yielded the highest biomass growth rate and highest hydrogen and butyric yields.

The ninth text published in this Special Issue, by M. Eugenia Ibáñez-López et al., explores the optimization of the anaerobic digestion of winery wastewater (WW) using an upflow anaerobic sludge blanket (UASB) reactor. Their research focuses on producing volatile fatty acids (VFAs), with an emphasis on caproic acid (HCa), by varying the hydraulic retention time (HRT) to 8, 5, and 2.5 h. Their results indicate that a 5 h HRT optimizes VFA production, yielding a maximum HCa concentration of 0.9 gCOD/L and enhancing the overall production of VFAs by approximately 20%. Microbial analysis revealed the dominance of Firmicutes, particularly Clostridium species, which are known to produce HCa. This study demonstrates the potential of UASB reactors in biorefineries to efficiently convert WW into valuable bioproducts, promoting sustainability in the wine industry. Further research is recommended to explore other operational conditions and to scale up the process for industrial application.

The tenth article, which is by Leandro Conrado et al., focuses on the biotransformation of proteinogenic amino acids into volatile fatty acids (VFAs) under anaerobic conditions with methanogenesis inhibition. Through batch experiments on a microbiome from an anaerobic digester, this study found that lysine, glutamate, and serine primarily produced butyrate, while other amino acids generated lesser amounts of propionate, iso-butyrate, and iso-valerate. 16S rRNA gene amplicon sequencing identified *Anaerostignum*, *Intestimonas*, *Aminipila*, and *Oscillibacter* as the key microbes in butyrate production. The findings suggest that these amino acids could be significant feedstocks for VFA production in non-methanogenic conditions, highlighting the potential of optimizing biorefining processes to produce higher-value carboxylic acids.

The contributions published within this Special Issue were the following papers:

1. Radadiya, P.; Latika, A.; Fei, X.; Lee, J.; Mishra, S.; Hussain, A. The Effect of pH on the Production and Composition of Short- and Medium-Chain Fatty Acids from Food Waste in a Leachate Bed Reactor at Room Temperature. Fermentation 2023, 9(6), 518; https://doi.org/10.3390/fermentation9060518.
2. Cardona, C.A.; Ortiz-Sanchez, M.; Salgado, N.; Solarte-Toro, J.C.; Orrego, C.E.; Perez, A.; Acosta, C.D.; Ledezma, E.; Salas, H.; Gonzaga, J.; Delgado, S. Sustainability Assessment of Food Waste Biorefineries as the Base of the Entrepreneurship in Rural Zones of Colombia. Fermentation 2023, 9(7), 609; https://doi.org/10.3390/fermentation9070609.
3. Fundo, J.F.; Deuchande, T.; Rodrigues, D.A.; Pimentel, L.L.; Vidigal, S.S.M.P.; Rodríguez-Alcalá, L.M.; Pintado, M.E.; Amaro, A.L. Induced Autolysis of Engineered Yeast Residue as a Means to Simplify Downstream Processing for Valorization—A Case Study. Fermentation 2023, 9(7), 673; https://doi.org/10.3390/fermentation9070673.
4. Moure Abelenda, A.; Aggidis, G.; Aiouache, F. Modelling of Amino Acid Fermentations and Stabilization of Anaerobic Digestates by Extracting Ammonium Bicarbonate. Fermentation 2023, 9(8), 750; https://doi.org/10.3390/fermentation9080750.
5. Rawalgaonkar, D.; Zhang, Y.; Walker, S.; Kirchman, P.; Zhang, Q.; Ergas, S.J. Recovery of Energy and Carbon Dioxide from Craft Brewery Wastes for Onsite Use. Fermentation 2023, 9(9), 831; https://doi.org/10.3390/fermentation9090831.
6. Arotingo Guandinango, H.P.; Espín Valladares, R.C.; Núñez Pérez, J.; Lara Fiallos, M.V.; Pereda Reyes, I.; Pais-Chanfrau, J.M. Modelisation of the Biomethane Accumulation in Anaerobic Co-Digestion of Whey and Sugarcane Molasse Mixtures. Fermentation 2023, 9(9), 834; https://doi.org/10.3390/fermentation9090834.
7. Ramos, L.H.; Cisneros-Yupanqui, M.; Santisteban Soto, D.V.; Lante, A.; Favaro, L.; Casella, S.; Basaglia, M. Exploitation of Cocoa Pod Residues for the Production of Antioxidants, Polyhydroxyalkanoates, and Ethanol. Fermentation 2023, 9(9), 843; https://doi.org/10.3390/fermentation9090843.

8. Ibañez-López, M.E.; Díaz-Domínguez, E.; Suffo, M.; Makinia, J.; García-Morales, J.L.; Fernández-Morales, F.J. Biorefinery Approach for H_2 and Acids Production Based on Uncontrolled pH Fermentation of an Industrial Effluent. Fermentation 2023, 9(11), 937; https://doi.org/10.3390/fermentation9110937.
9. Ibáñez-López, M.E.; Frison, N.; Bolzonella, D.; García-Morales, J.L. Enhancing Anaerobic Digestion with an UASB Reactor of the Winery Wastewater for Producing Volatile Fatty Acid Effluent Enriched in Caproic Acid. Fermentation 2023, 9(11), 958; https://doi.org/10.3390/fermentation9110958.
10. Conrado, L.; McCoy, J.; Rabinovich, L.; Davoudimehr, M.; Stamatopoulou, P.; Scarborough, M. Anaerobic Conversion of Proteinogenic Amino Acids When Methanogenesis Is Inhibited: Carboxylic Acid Production from Single Amino Acids. Fermentation 2024, 10(5), 237; https://doi.org/10.3390/fermentation10050237.

Analyzing the papers published in this Special Issue reveals several research gaps in this field. It is worth noting that no research or technical studies were carried out using thermochemical processing technologies. Additionally, only one study was carried out at full scale. In this sense, more full-scale studies are required to determine the technical feasibility of the processes developed. Finally, more studies focused on the economic aspects of biorefineries should be developed in order to accurately determine whether the projects and investments proposed are financially viable and able to generate a positive return.

Author Contributions: Conceptualization, F.J.F.-M.; formal analysis, F.J.F.-M. and J.L.G.-M.; writing—original draft preparation, F.J.F.-M. and J.L.G.-M.; writing—review and editing, F.J.F.-M. All authors have read and agreed to the published version of the manuscript.

Funding: This research received no external funding.

Institutional Review Board Statement: Not applicable.

Data Availability Statement: Not applicable.

Acknowledgments: As Guest Editors of the Special Issue "Sustainable Development of Food Waste Biorefineries", we would like to express our deep appreciation to all the authors whose valuable work was published in this Issue and thus contributed to its success.

Conflicts of Interest: The authors declare no conflicts of interest.

References

1. Geissdoerfer, M.; Savaget, P.; Bocken, N.M.; Hultink, E.J. The Circular Economy e A New Sustainability Paradigm? *J. Clean. Prod.* **2017**, *143*, 757–768. [CrossRef]
2. Caldeira, C.; Vlysidis, A.; Fiore, G.; De Laurentiis, V.; Vignali, G.; Sala, S. Sustainability of Food Waste Biorefinery: A Review on Valorisation Pathways, Techno-Economic Constraints, and Environmental Assessment. *Bioresour. Technol.* **2020**, *312*, 123575. [CrossRef]
3. Morais, A.R.; Bogel-Lukasik, R. Green Chemistry and the Biorefinery Concept. *Sustain. Chem. Process.* **2013**, *1*, 1–3. [CrossRef]
4. Magama, P.; Chiyanzu, I.; Mulopo, J. A Parametric Experimental Validation of a Biorefinery Concept Based on Anaerobic Digestion of Fruit and Vegetable Waste. *Biofuels Bioprod. Biorefining* **2022**, *16*, 972–985. [CrossRef]
5. Ahmed, S.F.; Kabir, M.; Mehjabin, A.; Oishi, F.T.Z.; Ahmed, S.; Mannan, S.; Mofijur, M.; Almomani, F.; Badruddin, I.A.; Kamangar, S. Waste Biorefinery to Produce Renewable Energy: Bioconversion Process and Circular Bioeconomy. *Energy Rep.* **2023**, *10*, 3073–3091. [CrossRef]
6. Zetterholm, J.; Pettersson, K.; Leduc, S.; Mesfun, S.; Lundgren, J.; Wetterlund, E. Resource Efficiency or Economy of Scale: Biorefinery Supply Chain Configurations for Co-Gasification of Black Liquor and Pyrolysis Liquids. *Appl. Energy* **2018**, *230*, 912–924. [CrossRef]

fermentation

Article

Anaerobic Conversion of Proteinogenic Amino Acids When Methanogenesis Is Inhibited: Carboxylic Acid Production from Single Amino Acids

Leandro Conrado, Jacob McCoy, Leo Rabinovich, Mona Davoudimehr, Panagiota Stamatopoulou and Matthew Scarborough *

Department of Civil and Environmental Engineering, University of Vermont, Burlington, VT 05405, USA
* Correspondence: matthew.scarborough@uvm.edu

Abstract: Proteins are an abundant biopolymer in organic waste feedstocks for biorefining. When degraded, amino acids are released, but their fate in non-methanogenic microbiomes is not well understood. The ability of a microbiome obtained from an anaerobic digester to produce volatile fatty acids from the twenty proteinogenic amino acids was tested using batch experiments. Batch tests were conducted using an initial concentration of each amino acid of 9000 mg COD L^{-1} along with 9000 mg COD L^{-1} acetate. Butyrate production was observed from lysine, glutamate, and serine fermentation. Lesser amounts of propionate, iso-butyrate, and iso-valerate were also observed from individual amino acids. Based on 16S rRNA gene amplicon sequencing, *Anaerostignum*, *Intestimonas*, *Aminipila*, and *Oscillibacter* all likely play a role in the conversion of amino acids to butyrate. The specific roles of other abundant taxa, including *Coprothermobacter*, *Fervidobacterium*, *Desulfovibrio*, and *Wolinella*, remain unknown, but these genera should be studied for their role in fermentation of amino acids and proteins to VFAs.

Keywords: carboxylic acids; amino acids; proteins; anaerobic digestion; chain elongation

Citation: Conrado, L.; McCoy, J.; Rabinovich, L.; Davoudimehr, M.; Stamatopoulou, P.; Scarborough, M. Anaerobic Conversion of Proteinogenic Amino Acids When Methanogenesis Is Inhibited: Carboxylic Acid Production from Single Amino Acids. *Fermentation* **2024**, *10*, 237. https://doi.org/10.3390/fermentation10050237

Academic Editors: Thaddeus Ezeji, Jose Luis García-Morales and Francisco Jesús Fernández Morales

Received: 18 March 2024
Revised: 15 April 2024
Accepted: 25 April 2024
Published: 29 April 2024

1. Introduction

Organic wastes have been proposed as a renewable source of carbon to produce beneficial chemicals, including methane and volatile fatty acids (VFAs) [1]. While highly variable, these wastes consist of mixtures of biopolymers encompassing carbohydrates, proteins, lipids, and other trace organic and inorganic molecules [2]. Carbohydrates and proteins are abundant in many common organic wastes, including food wastes. The degradation of complex carbohydrates is well-studied, but biological protein degradation in anaerobic systems needs to be better understood. While it is known that proteins and amino acids can be converted to biogas through conventional anaerobic digestion [3–5], little is known about the fermentation of amino acids when methanogenesis is inhibited. To unlock the full potential of proteins as a feedstock for biorefining, more needs to be known about the biotransformation of amino acids under anaerobic conditions.

Volatile fatty acids have been targeted as end products and key intermediates in advanced organic waste fermentation processes, such as the carboxylate platform and chain elongation [6,7]. VFAs, such as acetate, propionate, butyrate, iso-butyrate, valerate, and iso-valerate, serve as fundamental chemical components extensively utilized across various industries, including the food, textile, and pharmaceutical sectors [8]. Amino acids can be fermented into organic acids, including VFAs, through three classes of metabolic pathways: the Stickland reaction, oxidative deamination of individual amino acids, and reductive deamination of individual amino acids [9–11]. Generally, Stickland reactions are common and couple the oxidation of an amino acid with the reduction of another. Oxidation of amino acids typically yields hydrogen as a product, while many amino acid reduction pathways rely on hydrogen as an electron donor. For over five decades, the ability of pure cultures to produce a variety of VFAs from amino acids has been known [12,13], but the fate and role of amino acids in anaerobic microbiome-based processes, either in conventional anaerobic digestion or in alternative fermentations, is largely unexplored.

Only a few studies have reported results on amino acid bioconversion under anaerobic conditions when methanogenesis is inhibited. Recent work by Wang et al. provided a detailed analysis of the fermentation of eight amino acids (alanine, glutamate, glutamine, leucine, lysine, methionine, threonine, and valine) using metagenomic and metaproteomic approaches and found that L-isomers of amino acids promoted VFA production more than D-isomers [14]. Additionally, Regueira et al. performed an extensive modeling analysis for the conversion of amino acids to carboxylic acids [15]. While these studies have provided valuable modeling and experimental insights into the potential to convert amino acids to beneficial products, the fate of amino acids and the microbial communities involved in amino acid fermentations are only beginning to emerge.

To further understand the potential role of amino acids and proteins in biorefining, the anaerobic biotransformation of all twenty proteinogenic amino acids was tested while chemically inhibiting methanogenesis. In addition to measuring the consumption of amino acids, the production of VFAs was assessed. Further, 16S rRNA gene amplicon sequencing was used to assess how microbial communities changed during incubations. In total, this work provides a comprehensive analysis of amino conversion to VFAs and a preliminary assessment of the microbial communities responsible for these conversions. These results should inform future work with bioreactors to convert single or mixed amino acids to VFAs.

2. Materials and Methods

2.1. Materials

Sludge was acquired from a full-scale anaerobic digester at a wastewater treatment facility in South Burlington, VT, USA, and utilized as an inoculum. Sludge was collected from the same location for all experiments. A synthetic medium containing 10 g COD L^{-1} of acetate and 10 g COD L^{-1} of twenty individual L-isomer amino acids was prepared for all experiments. The chemical composition of the synthetic medium was as follows (g/L): one amino acid (Table 1), sodium acetate anhydrous ($NaC_2H_3O_2$) 12.82, potassium phosphate dibasic (K_2HPO_4) 2.0, sodium phosphate monobasic (NaH_2PO_4) 0.4, magnesium sulfate heptahydrate ($MgSO_4{\cdot}7H_2O$) 0.1, calcium chloride ($CaCl_2$) 0.05, and 2-bromoethanesulfanoate (BRES) 3.3. Acetate was added as a supplemental electron acceptor and to provide a substrate that is commonly produced by anaerobic microbiomes. BRES was added to inhibit methanogenesis [16]. Further, 1 mL of trace-mineral supplement (ATCC, Manassas, VA, USA) was added to 1 L of media to provide trace metals. Potassium phosphate dibasic (K_2HPO_4) and sodium phosphate monobasic were added to the media to help maintain a pH greater than 5.5 during fermentation. No additional pH control mechanism was used. The initial pH of the media is provided (Table 1) and varied according to the amino acid added with basic amino acids, resulting in an increased pH and acidic amino acids, which resulted in a decreased pH. Arginine resulted in a high pH of 9.68 due to its highly basic characteristics [17].

Table 1. Concentrations of amino acids used in media preparation.

Amino Acid	Chemical Name (As Added)	Concentration Added (g L^{-1})	Media pH
Alanine	L-alanine	9.34	6.69
Arginine	L-arginine	9.90	9.68
Asparagine	L-asparagine monohydrate	15.63	7.10
Aspartate	L-aspartic acid potassium salt	17.83	6.88
Cysteine	L-cysteine	15.14	6.00
Glutamine	L-glutamine	10.15	6.78
Glutamate	Sodium L-glutamate monohydrate	12.98	6.98
Glycine	Glycine	15.64	6.20

Table 1. *Cont.*

Amino Acid	Chemical Name (As Added)	Concentration Added (g L^{-1})	Media pH
Histidine	L-histidine	9.70	7.36
Isoleucine	L-isoleucine	5.46	6.28
Leucine	L-leucine	5.47	6.34
Lysine	L-lysine monohydrochloride	8.15	6.73
Methionine	L-methionine	8.48	7.07
Phenylalanine	L-phenylalanine	5.15	6.96
Proline	L-proline	6.54	7.34
Serine	L-serine	13.14	6.88
Threonine	L-threonine	9.31	6.90
Tryptophan	L-tryptophan	5.67	5.67
Tyrosine	L-tyrosine	5.96	5.96
Valine	L-valine	6.10	6.39

2.2. Experimental Setup

Biological triplicates were used for each amino acid along with control experiments without any amino acid provided to the media. The medium for the control experiments included acetate as the supplemental carbon source and BRES to inhibit methanogenesis, but no amino acids were added. Serum bottles with a 125 mL total volume were inoculated with 10 mL of sludge and 90 mL of medium, meaning that the inoculum sludge biomass and chemical constituent concentration were diluted 1:10 with the synthetic medium. After combining sludge and media, nitrogen gas was sparged into the bottles for at least 1 min, and the bottles were sealed with butyl rubber stoppers to maintain anaerobic conditions. The serum bottles were incubated in an incubator shaker at 35 °C while being agitated at a setting of 100 rpm. The bottles were incubated for a total of 14 days. In total, six rounds of batch experiments were performed with different batches of inoculum sludge with 2 to 4 amino acids being tested for each round.

2.3. Analytical Methods

The inoculum sludge was tested for pH, total solids (TS), volatile solids (VS), soluble chemical oxygen demand (COD), and soluble ammoniacal nitrogen before the start of experiments. These tests were performed according to Standard Methods for the Examination of Water and Wastewater [18]. TS and VS tests were conducted according to standard methods 2540B and 2540E [18]. For soluble COD and ammoniacal nitrogen analyses, samples were filtered through a 0.20 µm syringe filter prior to testing. COD testing was performed according to standard method 5220 with Hach test kits for high-range COD (Hach Method 8000, Hach, Loveland, CO, USA). Ammoniacal nitrogen analyses were performed according to standard method 4500 using Hach test kits (Hach Method 8155, Hach, Loveland, CO, USA).

On sampling days, headspace gas was collected from the serum bottles and analyzed with a gas chromatogram with a thermal conductivity detector (GC-TCD). An SGE 25 mL gas-tight fitted syringe was used to collect headspace samples. The samples were directly injected into a GC-TCD Nexis GC-2030 system (Shimadzu, Kyoto, Japan) to quantify the methane. After de-capping the serum bottles, liquid samples were centrifuged at 9600 rcf for 10 min, and the supernatant and the biomass pellet were stored separately for further analyses. Liquid samples were stored in the −20 °C freezer and used to measure amino acids, VFAs, and ammoniacal nitrogen. Samples for amino acid analyses were delivered to the Dana Farber Cancer Institute at Harvard University for analysis by high-performance liquid chromatography (HPLC) with derivatization of amino acids with fluorescent tags. Cysteine and tryptophan were excluded from the amino acid analyses because they could not be measured using the methods employed. Cysteine is challenging to quantify, given its instability in air [19], and tryptophan degrades substantially during sample preparation to derivatize amino acids before analyzing with chromatography [20]. VFAs, including acetic acid, propionic acid, iso-propionic acid, butyric acid, iso-butyric acid, valeric acid, iso-valeric acid, caproic acid, heptanoic acid, and caprylic acid, were analyzed using gas chromatography–mass spectrometry (GC-MS) with a GC-MS TQ8040NX system (Shimadzu, Kyoto, Japan). For the GC-MS method, a 2.5 mL smart syringe at 150 °C was used to collect 1 mL of the headspace gas in 10 mL vials. The incubation temperature was 95 °C, the incubation time was 40 min, and the agitator's speed was 250 rpm. The column used was the DB-FATWAX UI (Agilent Technologies, CA, USA) with 30 m length, 0.25 μm thickness, and 0.23 mm diameter. The column temperature started at 80 °C for 1 min and increased by 20 °C/minute until it reached 240 °C for 1.5 min. The MS acquisition mode was a Q3 scan, with the ion source temperature and the interface temperature at 280 °C and 250 °C, respectively.

2.4. Microbial Community Analyses

The biomass pellets were stored at −80 °C for DNA extraction using the Qiagen DNAEasy PowerSoil extraction kit (Qiagen, Hilden, Germany). After extraction, DNA quality and quantity were assessed using a Nanodrop 2000 spectrophotometer (Thermo Fisher Scientific, Waltham, MA, USA) and a Qubit 3.0 fluorometer (Thermo Fisher Scientific, Waltham, MA, USA). Samples were diluted to provide a DNA concentration of 1 ng ml^{-1} for library preparation. The Vermont Integrative Genomics Resource performed 16S rRNA gene amplification using the standard V3/V4 primer set [21] and sequenced the resulting amplified libraries on the Illumina MiSeq platform to generate 2 × 300 bp reads. The reads were analyzed using the Qiime2 [22] pipeline with the following steps: (1) raw reads and sample metadata were imported as a Qiime artifact; (2) reads were denoised and quality trimmed with dada2 [23]; (3) taxonomy was assigned using the Silva database [24]; and (4) abundance tables were created using default Qiime commands. Relative abundance data was visualized using the superheat package in R [25]. Raw DNA-sequencing data are available from NCBI under Bioproject number PRJNA1087599.

3. Results and Discussion

3.1. Characterization of Inoculum Sludge

In total, six batches of sludge were used for seeding the batch experiments (Table 2). The sludge TS was 2.3 ± 0.6% with 69 ± 14% of the TS being vs. Further, the soluble COD was 809 ± 192 mg COD L^{-1}, and the soluble ammonia was 451 ± 125. The pH varied from 6.95 to 7.56 with an average of 7.36 ± 0.26. In total, there was some fluctuation between sludge characteristics, but the overall characteristics align with what is expected in a methanogenic anaerobic digester consuming waste-activated sludge at a municipal wastewater treatment facility. In total, the results also suggest that, when diluted 1:10 with the synthetic media, the initial ammoniacal nitrogen concentration should be low, between 25 and 57 mg N L^{-1}. Further, when mixed with the media, the soluble COD remaining in the sludge would be only a minor fraction (e.g., <1 percent) of the soluble COD provided to the incubations.

Table 2. Characteristics of inoculum sludge.

	TS (%)	VS (% of TS)	Soluble COD (mg L^{-1})	Soluble NH_3-N (mg L^{-1})	pH
Sludge 1	2.7	61	972	254	7.56
Sludge 2	1.6	97	755	573	7.22
Sludge 3	1.9	70	534	361	7.64
Sludge 4	3.2	61	755	447	7.51
Sludge 5	2.1	62	755	567	6.95
Sludge 6	2.1	64	1080	505	7.25
Average	2.3	69	809	451	7.36
Stdev	0.6	14	192	125	0.26

3.2. Amino Acid Degradation

All of the amino acids measured showed a decrease in concentration from Day 0 to Day 14 (Figure 1). Asparagine, aspartate, glutamate, glycine, histidine, lysine, serine, and tyrosine showed almost complete degradation (>90%). All these amino acids have pathways described for their degradation without Stickland reactions. Asparagine can serve as the sole carbon and nitrogen source for multiple organisms via a two-step pathway that produces fumarate [26,27]. Aspartate can be degraded by Campylobacter, which produces succinate and formate [12]. Glutamate fermentation via multiple routes, including the methylaspartate and hydroxyglutarate pathways, have been described and produce acetic and butyric acids [28]. Histidine can be fermented to glutamate, which is fermented via the pathways described previously but also produces formamide with unknown fates in anaerobic environments [29]. Lysine fermentation to acetate and butyrate via multiple metabolic routes has been described, and serine fermentation to acetate and propionate has been shown [12,26,29]. Fermentation of tyrosine has only been shown via Stickland-type reactions previously, but our results suggest that it may be degraded without additional amino acids. Given the high electron density of tyrosine (1.68 g COD g^{-1} tyrosine), it may be using the acetate provided in the media as an electron acceptor.

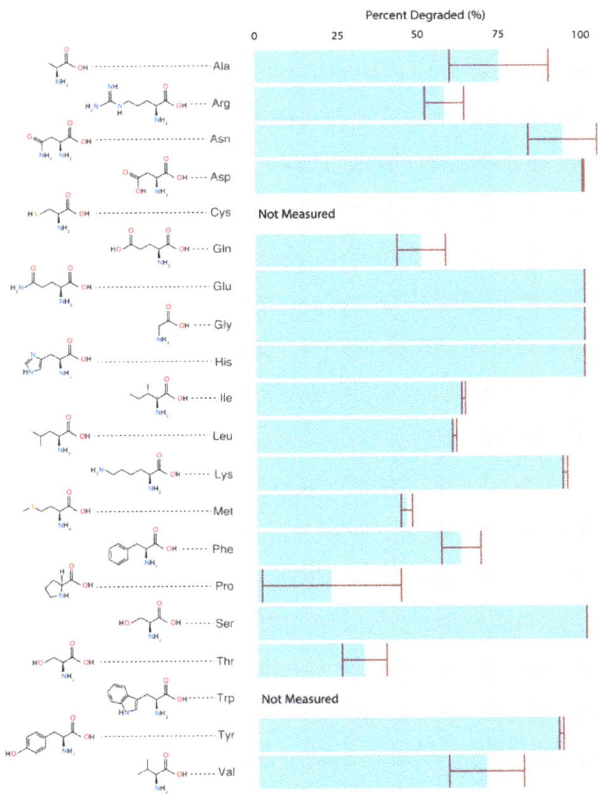

Figure 1. Extent of amino acid degradation during batch experiments with individual amino acids after 14 days of incubation. Error bars represent the standard deviation of biological replicates ($n = 3$). Abbreviations are standard amino acid three-letter abbreviations as follows: alanine (Ala), arginine (Arg), asparagine (Asn), aspartate (Asp), cysteine (Cys), glutamine (Gln), glutamate (Glu), glycine (Gly), histidine (His), isoleucine (Ile), leucine (Leu), lysine (Lys), methionine (Met), phenylalanine (Phe), proline (Pro), serine (Ser), threonine (Thr), tryptophan (Try), and valine (Val).

To assess amino acid consumption byproducts, methane, pH, and ammoniacal nitrogen were analyzed (Figure 2A–C). The methane composition of the headspace after 14 days was always less than 5%, while most incubations resulted in a composition of less than 1% (Figure 2A). This suggests that BRES was dosed at an appropriate level to inhibit methanogenesis. The pH typically dropped during incubations (Figure 2B) and the pH was similar between the biological replicates. The lowest pH was 5.9, suggesting that the pH was maintained at appropriate values for mixed VFA production, which typically occurs above a pH of 5 [30]. Ammoniacal nitrogen concentrations increased during the degradation of all amino acids (Figure 2C). During anaerobic degradation, amino acids undergo deamination, releasing ammonia that may participate in the formation of other compounds, such as urea [31]. Asparagine, aspartate, glycine, histidine, lysine, and serine had the highest increase in ammonia concentrations. Asparagine is utilized in some organisms as the sole carbon and nitrogen source through the action of two enzymes. The first one catalyzes the breakdown of asparagine into aspartate and ammonium. The second enzyme, aspartase ammonia-lyase, facilitates the reversible deamination of aspartate, resulting in the production of fumarate and ammonium [27]. In all incubations, acetic acid was consumed (Figure 2D) and decreased from the initial concentration of 9000 mg COD L^{-1}, but residual acetic acid remained in all incubations.

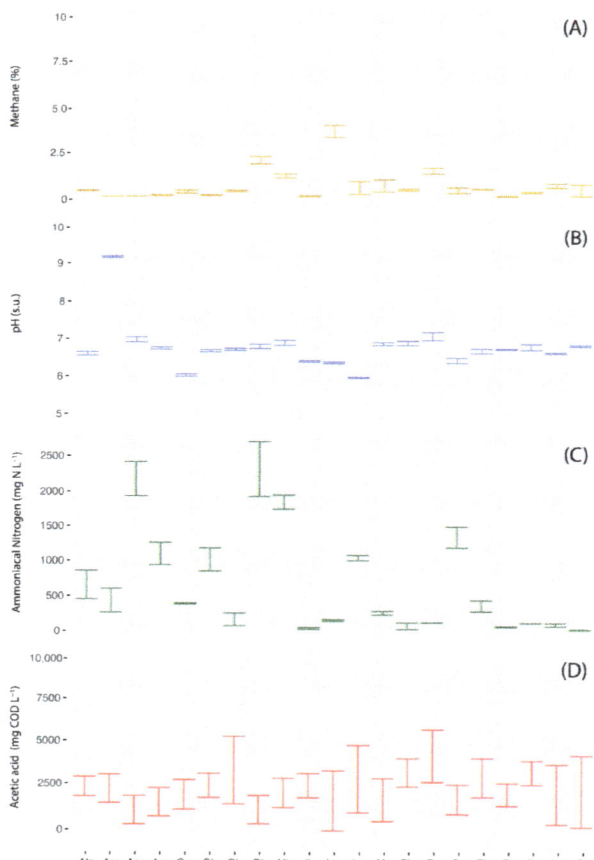

Figure 2. The measured headspace methane composition (**A**), pH (**B**), ammoniacal nitrogen concentration (**C**), and acetic acid concentration (**D**) after 14 days of incubation. Error bars represent the standard

deviation of biological replicates ($n = 3$). Abbreviations are standard amino acid three-letter abbreviations as follows: alanine (Ala), arginine (Arg), asparagine (Asn), aspartate (Asp), cysteine (Cys), glutamine (Gln), glutamate (Glu), glycine (Gly), histidine (His), isoleucine (Ile), leucine (Leu), lysine (Lys), methionine (Met), phenylalanine (Phe), proline (Pro), serine (Ser), threonine (Thr), tryptophan (Try), and valine (Val).

3.3. Production of Volatile Fatty Acids

Of the 20 amino acids tested (Figure 3), 12 produced a significantly higher amount of VFAs ($p < 0.05$) compared to the controls (Figure 4) without amino acids added, but the quantity and type of VFAs produced varied between the amino acids. The overall COD conversion to VFAs was low, with a maximum of 8% of the COD added as acetic acid and amino acids being converted to VFAs in incubations with lysine. Among VFAs other than acetic acid, butyric acid was the most commonly produced VFA, followed by propionic, iso-butyric, and iso-valeric acids (Figure 3). Only trace amounts of other analyzed VFAs (e.g., valeric, hexanoic, and octanoic acids) were found.

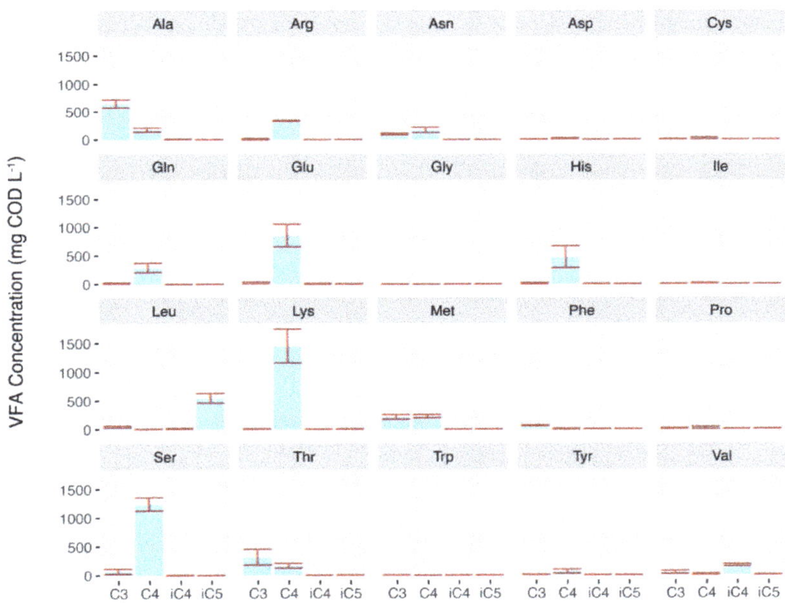

Figure 3. Propionic (C3), butyric, (C4), iso-butyric (iC4), and iso-valeric (iC5) acid concentrations after 14 days of incubation. Error bars represent the standard deviation of biological replicates ($n = 3$). Abbreviations are standard amino acid three-letter abbreviations as follows: alanine (Ala), arginine (Arg), asparagine (Asn), aspartate (Asp), cysteine (Cys), glutamine (Gln), glutamate (Glu), glycine (Gly), histidine (His), isoleucine (Ile), leucine (Leu), lysine (Lys), methionine (Met), phenylalanine (Phe), proline (Pro), serine (Ser), threonine (Thr), tryptophan (Try), and valine (Val).

Lysine resulted in the highest butyrate production of 1455 ± 292 mg COD L^{-1} followed by serine and glutamate. A pathway for lysine conversion to butyrate was described for *Clostridium sticklandii* in 1954 [32] and more recently has been proposed to be fermented to butyric acid by human gut microbes [33]. Serine conversion to butyrate is less well described but has been found to be performed by *Cloacibacillus porcorum* isolated from the swine intestinal tract [34]. Serine is likely to be initially converted to pyruvate as an intermediate via a single enzymatic step with L-serine ammonia lyase [35] prior to being directed to butyric acid production via reverse β-oxidation. Histidine incubations also showed the production of butyric acid. Histidine has several proposed degradation routes with glutamate as an intermediate [36], and glutamate has previously been shown to produce butyric acid as a fermentation end product utilizing electron-bifurcating enzymes to conserve energy in anaerobic bacteria [28,37].

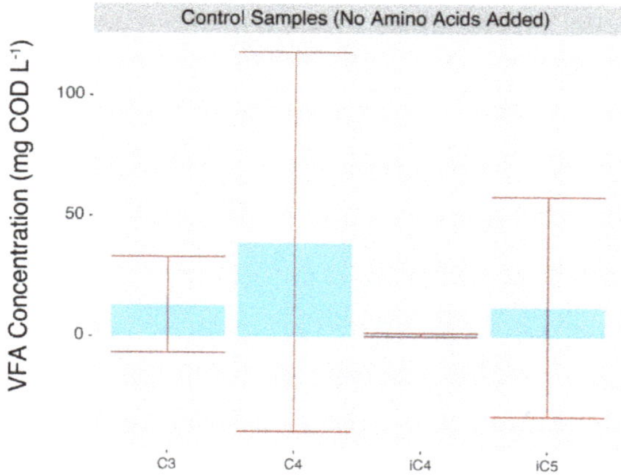

Figure 4. Propionic (C3), butyric, (C4), iso-butyric (iC4), and iso-valeric (iC5) acid concentrations after 14 days of incubation with controls that had sludge and media without amino acids. Error bars represent the standard deviation of biological replicates (n = 6). Abbreviations are standard amino acid three-letter abbreviations as follows: alanine (Ala), arginine (Arg), asparagine (Asn), aspartate (Asp), cysteine (Cys), glutamine (Gln), glutamate (Glu), glycine (Gly), histidine (His), isoleucine (Ile), leucine (Leu), lysine (Lys), methionine (Met), phenylalanine (Phe), proline (Pro), serine (Ser), threonine (Thr), tryptophan (Try), and valine (Val).

While butyric acid was the most commonly produced VFA, propionic, iso-butyric, and iso-valeric acids were also produced in some incubations (Table 3, Figure 3). Alanine had the highest propionate production of 652 $\pm$ 74.5 mg COD L^{-1}. Others have shown that propionate can be produced from alanine [38], but the metabolic routes of alanine conversion to propionate remain unclear. Threonine and methionine incubations also produced propionate, although at lower concentrations. Valine had the highest iso-butyric acid production of 167 $\pm$ 18.6 mg COD L^{-1} and was the only amino acid to produce significantly higher amounts of iso-butyric acid than the controls in experiments with multiple strains of *Clostridium* species [12] and goat ruminal fluid [39]. Likewise, anaerobic degradation of L-leucine produced the highest amount of iso-valeric acid, and this has also been demonstrated with multiple strains of *Clostridium* species [12] and goat ruminal fluid [39].

In total, the batch experiments demonstrated that VFAs are produced to varying extents from multiple amino acids. Alanine and threonine produced mostly propionic acid, while butyric acid was the primary VFA product of arginine, glutamine, glutamate, lysine, and serine incubations. Lastly, when incubated with the branched-chain amino acids valine and leucine, the predominate products were the branched-chain VFAs iso-butyric and iso-valeric acids, respectively. While only a small portion of the COD was converted to VFA end products, these results demonstrate that amino acids can be fermented by anaerobic microbiomes.

Table 3. Electron balance as COD during amino acid degradation.

Amino Acid	Final Amino Acid Concentration (mg COD L^{-1})	Percent Conversion (%) [1]			
		Propionic Acid	Butyric Acid	Iso-Butyric Acid	Iso-Valeric Acid
Alanine	2280 ± 1230	7.2 ± 0.83	1.9 ±0.40	0.08 ± 0.05	0.02 ± 0.01
Arginine	3900 ± 822	0.10 ± 0.13	3.7 ± 0.10	0.01 ± 0.00	0.01 ± 0.01
Asparagine	571 ± 920	1.1 ± 0.17	1.9 ± 0.53	0.00 ± 0.00	0.01 ± 0.00
Aspartate	30.7 ± 19.4	0.06 ± 0.02	0.21 ± 0.11	0.01 ± 0.00	0.02 ± 0.01
Cysteine	Not Measured	0.03 ± 0.03	0.20 ± 0.20	0.00 ± 0.00	0.00 ± 0.00
Glutamine	4660 ± 545	0.23 ± 0.11	3.3 ± 0.94	0.01 ± 0.01	0.02 ± 0.01
Glutamate	1.46 ± 2.53	0.27 ± 0.17	9.5 ± 2.2	0.06 ± 0.07	0.06 ± 0.03
Glycine	3.93 ± 0.83	0.02 ± 0.02	BDL [2]	0.01 ± 0.01	0.03 ± 0.01
Histidine	14.0 ± 1.67	0.13 ± 0.12	5.3 ± 2.1	0.04 ± 0.03	0.03 ± 0.02
Isoleucine	3340 ± 97.5	0.04 ± 0.02	0.11 ± 0.05	BDL [2]	BDL [2]
Leucine	3570 ± 97.2	0.55 ± 0.18	0.09 ± 0.02	0.16 ± 0.08	6.1 ± 0.97
Lysine	559 ± 67.6	0.07 ± 0.04	16 ± 3.2	0.02 ± 0.01	0.05 ± 0.04
Methionine	4870 ± 63.3	2.4 ± 0.45	2.5 ± 0.32	0.00 ± 0.00	0.01 ± 0.01
Phenylalanine	3340 ± 492	0.70 ± 0.10	0.04 ± 0.07	0.01 ± 0.01	0.01 ± 0.01
Proline	6954 ± 1800	0.10 ± 0.07	0.17 ± 0.21	0.01 ± 0.01	0.03 ± 0.02
Serine	5.18 ± 1.53	0.79 ± 0.51	14 ± 1.3	0.02 ± 0.03	0.01 ± 0.01
Threonine	5980 ± 598	3.5 ± 1.6	1.8 ± 0.47	0.01 ± 0.00	0.04 ± 0.02
Tryptophan	Not Measured	0.03 ± 0.00	0.0 ± 0.00	0.01 ± 0.00	0.01 ± 0.00
Tyrosine	744 ± 91.5	0.08 ± 0.03	0.67 ± 0.14	0.01 ± 0.00	0.01 ± 0.00
Valine	2740 ± 2370	0.48 ± 0.31	0.10 ± 0.09	1.9 ± 0.21	0.03 ± 0.02

[1] Percent conversion is based on COD of amino acid added. [2] BDL indicates that the final VFA concentration was below the detection limit.

3.4. Microbial Community

To further understand the conversion of amino acids to VFAs during the incubation experiments, an analysis of the microbial community was performed using 16S rRNA gene amplicon sequencing. This analysis resulted in 49 distinct taxa being identified at the genus level with at least 3% relative abundance in at least one sample (Figure 5). While 49 genera were identified, 10 were unclassified at the genus level but could be classified at a higher level of taxonomy. Organisms related to *Coprothermobacter*, *Fervidobacterium*, Uncultured *Spirochaetaceae*, *Lentimicrobium*, *Acetomicrobium*, *Aminicenantales*, *Comamonadaceae* (Unclassified), and *Bacteroides* (Unclassified) were the predominant bacterial genera during all batch experiments. All these taxa were also abundant in the seed sludge samples.

Coprothermobacter species were abundant in all incubations and are known for their ability to consume a wide variety of proteins and amino acids and produce hydrogen gas [40]. A metaproteomic analysis suggested that *Coprothermobacter* produces a variety of fermentation end products, including formate, pyruvate, acetate, and butanol [41]. *Coprothermobacter* genera were abundant in the following amino acid batch tests: aspartate, phenylalanine, isoleucine, threonine, and valine. Except for threonine, all the other amino acids showed a degradation percentage higher than 60% in the batch tests. *Fervidobacterium* species have been found to be abundant in anaerobic digesters [42]. *Fervidobacterium* abundance has been found to correlate positively with VFA production from protein-rich feedstocks [43]. This genus was abundant during the fermentation of the following amino acids in our batch experiments: phenylalanine, proline, and tryptophan. All these amino did not produce significant amounts of VFAs.

Alanine incubations, which produced an increase in butyrate (Figure 3), experienced an enrichment in *Anaerostignum* and *Desulfovibrio*. *Desulfovibrio* species were also enriched during incubations with leucine, tryptophan, and tyrosine. *Anaerostignum* (aka *Anaerotignum*) species have been shown to produce acetate, propionate, and butyrate using alanine, serine, and threonine, and the type species of this genus, *Anaerostignum aminivornas*, was isolated from an anaerobic digester treating food-processing wastewater [44]. Further, co-cultures of

Anaerotignum neopropionicum and *Clostirum kluyveri* have been used to produce odd-chain carboxylic acids from dilute ethanol [45]. *Desulfovibrio* species are canonical sulfur-reducing bacteria, but *Desulfovibrio* aminophilus, isolated from an anaerobic dairy-waste lagoon, has been shown to degrade alanine, aspartate, leucine, isoleucine, valine, and methionine [46].

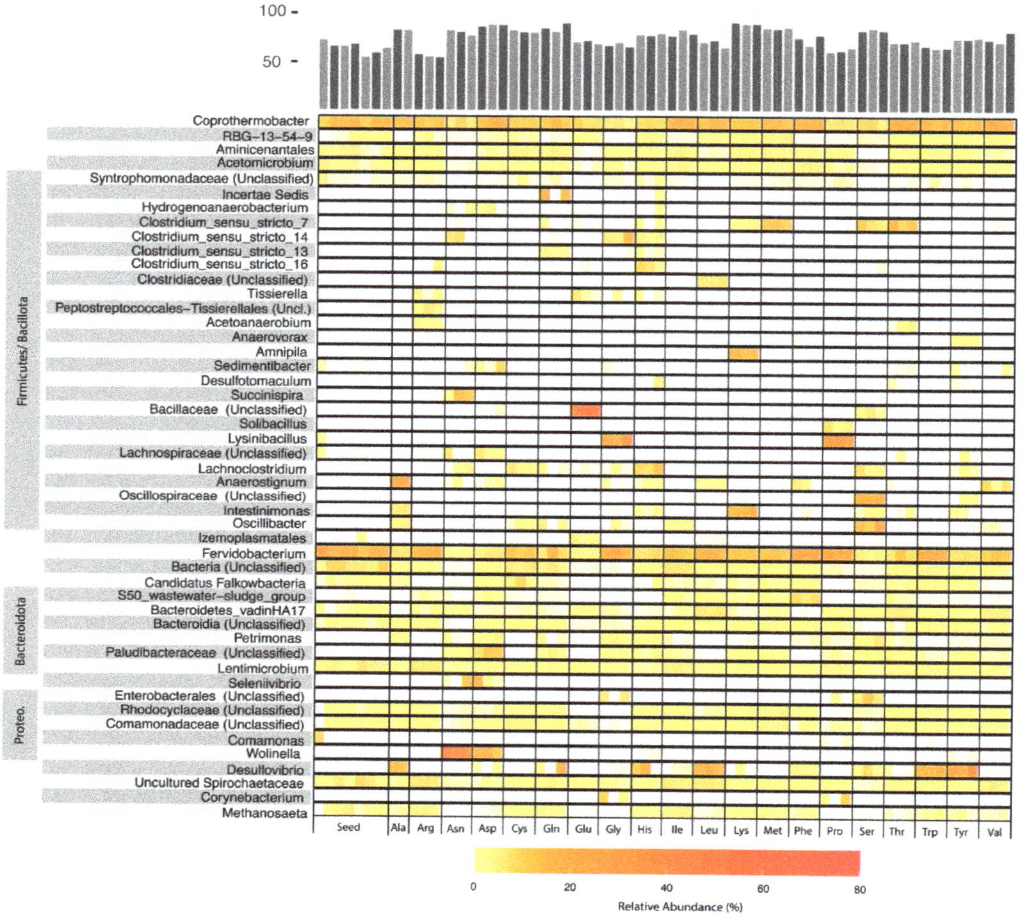

Figure 5. Relative abundance of genera based on 16S rRNA gene amplicon sequencing. Red intensity indicated relative abundance. Triplicates are provided for all twenty amino acids except alanine, which has only two sequencing replicates. The inoculum sludge was sequenced for each round of batch experiments and is shown as "seed". Abbreviations are standard amino acid three-letter abbreviations as follows: alanine (Ala), arginine (Arg), asparagine (Asn), aspartate (Asp), cysteine (Cys), glutamine (Gln), glutamate (Glu), glycine (Gly), histidine (His), isoleucine (Ile), leucine (Leu), lysine (Lys), methionine (Met), phenylalanine (Phe), proline (Pro), serine (Ser), threonine (Thr), tryptophan (Try), and valine (Val).

Lysine incubations resulted in an increased abundance of *Intestinomonas* species and *Amnipila* species. *Intestimonas* species, including *Intestimonas butyricproducens*, are well-studied for their ability to produce butyrate in the human gut and have been shown to produce butyrate from both amino acids and sugars [47]. This genera has also been proposed to consume lysine in infant and adult humans, as it was highly enriched from stool when incubated with lysine [48]. A metabolic pathway for lysine conversion to butyrate has been proposed that relies on the conversion of 5-amino-3-heaxanoate and acetyl-CoA to L-3-amino-butyryl-CoA and then deamination of L-3-amino-butyryl-CoA to crotonyl-CoA, a key intermediate of reverse β-oxidation [33]. *Amnipila* (aka, *Aminipila*) is a genus within the Peptostreptococcales–Tissierellales order, and species such as *Aminipila luticellari* have been shown to produce butyrate and consume a variety of amino acids, including L-lysine [49].

In addition to lysine, both glutamate and serine incubations resulted in the accumulation of butyrate. Glutamate incubations resulted in a sharp increase in the abundance of unclassified Bacillaceae. The Bacillaceae family falls within the Firmicutes phylum and encompasses a wide variety of organisms isolated from vastly different ecosystems (e.g., soils, hypersaline lakes, and hydrothermal vents) [50]. *Lysinibacillus* is a genus within this family that was enriched during incubations with glycine. This genus is known to oxidize several amino acids [51], but there is no known characterization of its fermentation end products. Serine incubations resulted in an enrichment of an Unclassified Oscillospiraceae as well as a genus within this family, *Oscillibacter*. The type species of *Oscillibacter*, *Oscillibacter valericigenes*, is known to produce valerate (C5) as the primary product of glucose fermentation [52]. However, this genus has been identified in other microbial communities producing butyrate enriched from the human colon microbiome [53].

Although asparagine and aspartate incubations did not result in high VFA production, *Wolinella* species were highly enriched during incubations (Figures 3 and 5). This genus has been shown to oxidize formate and reduce fumarate to succinate, and the species, *Wolinella succinogenes*, was isolated from the cow rumen [54]. The genus does not degrade carbohydrates, but there are no known studies on its ability to ferment amino acids. Wolinella species have been identified in anaerobic digesters previously [55,56], but their functional role and degradation capabilities are largely unknown.

4. Conclusions

This work demonstrates that, when methanogenesis is inhibited, amino acids can be converted to VFAs. Although the conversions represented only a small fraction of the total COD (up to 8%), no attempt was made to optimize the production of VFAs other than inhibiting methanogenesis and providing a buffer to keep the pH above 5.0 during the 14-day incubations. Future work with chemostats is expected to further increase the production of VFAs from amino acids. The microbial community analyses revealed several taxa already described to convert amino acids to VFAs including *Anaerostignum*, *Intestimonas*, *Aminipila*, and *Oscillibacter*; all have the potential to play a role in the conversion of amino acids to butyrate, which could subsequently be elongated to even higher value medium-chain carboxylic acids. The specific roles of abundant *Coprothermobacter*, *Fervidobacterium*, *Desulfovibrio*, and *Wolinella* remain unknown, and these genera should be studied more for their potential role in the fermentation of amino acids and proteins to VFAs.

Author Contributions: Conceptualization, L.C. and M.S.; Data curation, L.C.; Formal analysis, L.C.; Funding acquisition, M.S.; Investigation, L.C., J.M., L.R., M.D. and P.S.; Methodology, L.C., P.S. and M.S.; Project administration, M.S.; Resources, M.S.; Software, P.S.; Supervision, M.S.; Validation, M.S.; Visualization, L.C.; Writing—original draft, L.C. and M.S.; Writing—review and editing, L.C., J.M., M.D., P.S. and M.S. All authors have read and agreed to the published version of the manuscript.

Funding: This research was funded by the United States National Science Foundation, grant number 2221828.

Institutional Review Board Statement: Not applicable.

Data Availability Statement: The raw data supporting the conclusions of this article will be made available by the authors upon request.

Acknowledgments: The authors thank Peyton Lienhart and Kennedy Brown for their assistance in sample collection.

Conflicts of Interest: The authors declare no conflicts of interest.

References

1. Scarborough, M.J.; Lawson, C.E.; DeCola, A.C.; Gois, I.M. Microbiomes for sustainable biomanufacturing. *Curr. Opin. Microbiol.* **2022**, *65*, 8–14. [CrossRef]
2. Roy, P.; Mohanty, A.K.; Dick, P.; Misra, M. A Review on the Challenges and Choices for Food Waste Valorization: Environmental and Economic Impacts. *ACS Environ. Au* **2023**, *3*, 58–75. [CrossRef] [PubMed]
3. Chen, S.; Dong, B.; Dai, X.; Wang, H.; Li, N.; Yang, D. Effects of thermal hydrolysis on the metabolism of amino acids in sewage sludge in anaerobic digestion. *Waste Manag.* **2019**, *88*, 309–318. [CrossRef] [PubMed]
4. Liu, H.; Chen, Y.; Ye, J.; Xu, H.; Zhu, Z.; Xu, T. Effects of different amino acids and their configurations on methane yield and biotransformation of intermediate metabolites during anaerobic digestion. *J. Environ. Manag.* **2021**, *296*, 113152. [CrossRef] [PubMed]
5. Miholits, E.M.; Malina, J.F. Effects of Amino Acids on Anaerobic Digestion. *J. (Water Pollut. Control Fed.)* **1968**, *40*, R42–R53.
6. Agler, M.T.; Wrenn, B.A.; Zinder, S.H.; Angenent, L.T. Waste to bioproduct conversion with undefined mixed cultures: The carboxylate platform. *Trends Biotechnol.* **2011**, *29*, 70–78. [CrossRef]
7. Angenent, L.T.; Richter, H.; Buckel, W.; Spirito, C.M.; Steinbusch, K.J.; Plugge, C.M.; Strik, D.P.; Grootscholten, T.I.; Buisman, C.J.; Hamelers, H.V. Chain Elongation with Reactor Microbiomes: Open-Culture Biotechnology to Produce Biochemicals. *Environ. Sci. Technol.* **2016**, *50*, 2796–2810. [CrossRef] [PubMed]
8. Gálvez-Martos, J.-L.; Greses, S.; Magdalena, J.-A.; Iribarren, D.; Tomás-Pejó, E.; González-Fernández, C. Life cycle assessment of volatile fatty acids production from protein-and carbohydrate-rich organic wastes. *Bioresour. Technol.* **2021**, *321*, 124528. [CrossRef]
9. Alibardi, L.; Cossu, R. Effects of carbohydrate, protein and lipid content of organic waste on hydrogen production and fermentation products. *Waste Manag.* **2016**, *47*, 69–77. [CrossRef]
10. Wang, S.; Ping, Q.; Li, Y. Comprehensively understanding metabolic pathways of protein during the anaerobic digestion of waste activated sludge. *Chemosphere* **2022**, *297*, 134117. [CrossRef]
11. Sangavai, C.; Bharathi, M.; Ganesh, S.P.; Chellapandi, P. Kinetic modeling of Stickland reactions-coupled methanogenesis for a methanogenic culture. *AMB Express* **2019**, *9*, 82. [CrossRef]
12. Barker, H. Amino acid degradation by anaerobic bacteria. *Annu. Rev. Biochem.* **1981**, *50*, 23–40. [CrossRef] [PubMed]
13. Nisman, B. The stickland reaction. *Bacteriol. Rev.* **1954**, *18*, 16–42. [CrossRef] [PubMed]
14. Wang, M.; Zhang, X.; Huang, H.; Qin, Z.; Liu, C.; Chen, Y. Amino Acid Configuration Affects Volatile Fatty Acid Production during Proteinaceous Waste Valorization: Chemotaxis, Quorum Sensing, and Metabolism. *Environ. Sci. Technol.* **2022**, *56*, 8702–8711. [CrossRef]
15. Regueira, A.; Lema, J.M.; Carballa, M.; Mauricio-Iglesias, M. Metabolic modeling for predicting VFA production from protein-rich substrates by mixed-culture fermentation. *Biotechnol. Bioeng.* **2020**, *117*, 73–84. [CrossRef]
16. Salvador, A.F.; Cavaleiro, A.J.; Paulo, A.M.S.; Silva, S.A.; Guedes, A.P.; Pereira, M.A.; Stams, A.J.M.; Sousa, D.Z.; Alves, M.M. Inhibition Studies with 2-Bromoethanesulfonate Reveal a Novel Syntrophic Relationship in Anaerobic Oleate Degradation. *Appl. Environ. Microbiol.* **2019**, *85*, e01733-18. [CrossRef]
17. Wu, G.; Bazer, F.W.; Davis, T.A.; Kim, S.W.; Li, P.; Marc Rhoads, J.; Carey Satterfield, M.; Smith, S.B.; Spencer, T.E.; Yin, Y. Arginine metabolism and nutrition in growth, health and disease. *Amino Acids* **2009**, *37*, 153–168. [CrossRef]
18. Rice, E.W.; Bridgewater, L.; Association, A.P.H. *Standard Methods for the Examination of Water and Wastewater*; American Public Health Association: Washington, DC, USA, 2012; Volume 10.
19. Dobrowolski, J.C.; Rode, J.E.; Sadlej, J. Cysteine conformations revisited. *Comput. Theor. Chem.* **2007**, *810*, 129–134. [CrossRef]
20. La Cour, R.; Jørgensen, H.; Schjoerring, J.K. Improvement of tryptophan analysis by liquid chromatography-single quadrupole mass spectrometry through the evaluation of multiple parameters. *Front. Chem.* **2019**, *7*, 797. [CrossRef]
21. Klindworth, A.; Pruesse, E.; Schweer, T.; Peplies, J.; Quast, C.; Horn, M.; Glöckner, F.O. Evaluation of general 16S ribosomal RNA gene PCR primers for classical and next-generation sequencing-based diversity studies. *Nucleic Acids Res.* **2013**, *41*, e1. [CrossRef]
22. Bolyen, E.; Rideout, J.R.; Dillon, M.R.; Bokulich, N.A.; Abnet, C.C.; Al-Ghalith, G.A.; Alexander, H.; Alm, E.J.; Arumugam, M.; Asnicar, F.; et al. Reproducible, interactive, scalable and extensible microbiome data science using QIIME 2. *Nat. Biotechnol.* **2019**, *37*, 852–857. [CrossRef] [PubMed]
23. Callahan, B.J.; McMurdie, P.J.; Rosen, M.J.; Han, A.W.; Johnson, A.J.A.; Holmes, S.P. DADA2: High-resolution sample inference from Illumina amplicon data. *Nat. Methods* **2016**, *13*, 581–583. [CrossRef] [PubMed]
24. Quast, C.; Pruesse, E.; Yilmaz, P.; Gerken, J.; Schweer, T.; Yarza, P.; Peplies, J.; Glöckner, F.O. The SILVA ribosomal RNA gene database project: Improved data processing and web-based tools. *Nucleic Acids Res.* **2012**, *41*, D590–D596. [CrossRef] [PubMed]
25. Barter, R.L.; Yu, B. Superheat: An R package for creating beautiful and extendable heatmaps for visualizing complex data. *J. Comput. Graph. Stat.* **2018**, *27*, 910–922. [CrossRef] [PubMed]
26. Ramsay, I.R.; Pullammanappallil, P.C. Protein degradation during anaerobic wastewater treatment: Derivation of stoichiometry. *Biodegradation* **2001**, *12*, 247–257. [CrossRef] [PubMed]
27. Huerta-Zepeda, A.; Durán, S.; Du Pont, G.; Calderón, J. Asparagine degradation in *Rhizobium etli*. *Microbiology* **1996**, *142*, 1071–1076. [CrossRef] [PubMed]

28. Buckel, W.; Barker, H.A. Two pathways of glutamate fermentation by anaerobic bacteria. *J. Bacteriol.* **1974**, *117*, 1248–1260. [CrossRef] [PubMed]
29. Andreesen, J.R.; Bahl, H.; Gottschalk, G. Introduction to the physiology and biochemistry of the genus *Clostridium*. In *Clostridia*; Springer: Boston, MA, USA, 1989; pp. 27–62.
30. Scarborough, M.J.; Lynch, G.; Dickson, M.; McGee, M.; Donohue, T.J.; Noguera, D.R. Increasing the economic value of lignocellulosic stillage through medium-chain fatty acid production. *Biotechnol. Biofuels* **2018**, *11*, 200. [CrossRef]
31. Mohiuddin, S.S.; Khattar, D. *Biochemistry, Ammonia*; StatPearls Publishing: Treasure Island, FL, USA, 2019.
32. Stadtman, T.C.; White Jr, F. Tracer studies on ornithine, lysine, and formate metabolism in an amino acid fermenting *Clostridium*. *J. Bacteriol.* **1954**, *67*, 651–657. [CrossRef]
33. Bui, T.P.N.; Ritari, J.; Boeren, S.; de Waard, P.; Plugge, C.M.; de Vos, W.M. Production of butyrate from lysine and the Amadori product fructoselysine by a human gut commensal. *Nat. Commun.* **2015**, *6*, 10062. [CrossRef]
34. Levine, U.Y.; Looft, T.; Allen, H.K.; Stanton, T.B. Butyrate-producing bacteria, including mucin degraders, from the swine intestinal tract. *Appl. Environ. Microbiol.* **2013**, *79*, 3879–3881. [CrossRef]
35. Alföldi, L.; Raskó, I.; Kerekes, E. L-serine deaminase of *Escherichia coli*. *J. Bacteriol.* **1968**, *96*, 1512–1518. [CrossRef] [PubMed]
36. Borek, B.A.; Waelsch, H. The enzymatic degradation of histidine. *J. Biol. Chem.* **1953**, *205*, 459–474. [CrossRef] [PubMed]
37. Buckel, W.; Thauer, R.K. Energy conservation via electron bifurcating ferredoxin reduction and proton/Na+ translocating ferredoxin oxidation. *Biochim. Biophys. Acta (BBA)-Bioenerg.* **2013**, *1827*, 94–113. [CrossRef] [PubMed]
38. Brindle, P.A.; Baker, F.C.; Tsai, L.W.; Reuter, C.C.; Schooley, D.A. Sources of Propionate for the Biogenesis of Ethyl-Branched Insect Juvenile Hormones: Role of Isoleucine and Valine. *Proc. Natl. Acad. Sci. USA* **1987**, *84*, 7906–7910. [CrossRef] [PubMed]
39. Menahan, L.; Schultz, L. Metabolism of leucine and valine within the rumen. *J. Dairy. Sci.* **1964**, *47*, 1080–1085. [CrossRef]
40. Gagliano, M.; Braguglia, C.; Petruccioli, M.; Rossetti, S. Ecology and biotechnological potential of the thermophilic fermentative *Coprothermobacter* spp. *FEMS Microbiol. Ecol.* **2015**, *91*, fiv018. [CrossRef] [PubMed]
41. Lü, F.; Bize, A.; Guillot, A.; Monnet, V.; Madigou, C.; Chapleur, O.; Mazéas, L.; He, P.; Bouchez, T. Metaproteomics of cellulose methanisation under thermophilic conditions reveals a surprisingly high proteolytic activity. *ISME J.* **2014**, *8*, 88–102. [CrossRef] [PubMed]
42. Javier-Lopez, R.; Mandolini, E.; Dzhuraeva, M.; Bobodzhanova, K.; Birkeland, N.-K. *Fervidobacterium pennivorans* subsp. *keratinolyticus* subsp. nov., a Novel Feather-Degrading Anaerobic Thermophile. *Microorganisms* **2022**, *11*, 22. [CrossRef]
43. Wu, B.; Wang, X.; Deng, Y.-Y.; He, X.-L.; Li, Z.-W.; Li, Q.; Qin, H.; Chen, J.-T.; He, M.-X.; Zhang, M. Adaption of microbial community during the start-up stage of a thermophilic anaerobic digester treating food waste. *Biosci. Biotechnol. Biochem.* **2016**, *80*, 2025–2032. [CrossRef]
44. Ueki, A.; Goto, K.; Ohtaki, Y.; Kaku, N.; Ueki, K. Description of *Anaerotignum aminivorans* gen. nov., sp. nov., a strictly anaerobic, amino-acid-decomposing bacterium isolated from a methanogenic reactor, and reclassification of *Clostridium propionicum*, *Clostridium neopropionicum* and *Clostridium lactatifermentans* as species of the genus *Anaerotignum*. *Int. J. Syst. Evol. Microbiol.* **2017**, *67*, 4146–4153. [CrossRef] [PubMed]
45. Parera Olm, I.; Sousa, D.Z. Upgrading dilute ethanol to odd-chain carboxylic acids by a synthetic co-culture of *Anaerotignum neopropionicum* and *Clostridium kluyveri*. *Biotechnol. Biofuels Bioprod.* **2023**, *16*, 83. [CrossRef] [PubMed]
46. Baena, S.; Fardeau, M.L.; Labat, M.; Ollivier, B.; Garcia, J.L.; Patel, B.K. *Desulfovibrio aminophilus* sp. nov., a novel amino acid degrading and sulfate reducing bacterium from an anaerobic dairy wastewater lagoon. *Syst. Appl. Microbiol.* **1998**, *21*, 498–504. [CrossRef] [PubMed]
47. Bui, T.P.N.; Shetty, S.A.; Lagkouvardos, I.; Ritari, J.; Chamlagain, B.; Douillard, F.P.; Paulin, L.; Piironen, V.; Clavel, T.; Plugge, C.M.; et al. Comparative genomics and physiology of the butyrate-producing bacterium Intestinimonas butyriciproducens. *Environ. Microbiol. Rep.* **2016**, *8*, 1024–1037. [CrossRef] [PubMed]
48. Bui, T.P.N.; Troise, A.D.; Nijsse, B.; Roviello, G.N.; Fogliano, V.; de Vos, W.M. Intestinimonas-like bacteria are important butyrate producers that utilize Nε-fructosyllysine and lysine in formula-fed infants and adults. *J. Funct. Foods* **2020**, *70*, 103974. [CrossRef]
49. Wei, Z.; Ma, S.; Chen, R.; Wu, W.; Fan, H.; Dai, L.; Deng, Y. *Aminipila luticellarii* sp. nov., an anaerobic bacterium isolated from the pit mud of strong aromatic Chinese liquor, and emended description of the genus Aminipila. *Int. J. Syst. Evol. Microbiol.* **2021**, *71*, 004710. [CrossRef] [PubMed]
50. Mandic-Mulec, I.; Stefanic, P.; van Elsas Jan, D. Ecology of Bacillaceae. *Microbiol. Spectr.* **2015**, *3*, TBS-0017-2013. [CrossRef]
51. Ahmed, I.; Yokota, A.; Yamazoe, A.; Fujiwara, T. Proposal of *Lysinibacillus boronitolerans* gen. nov. sp. nov., and transfer of *Bacillus fusiformis* to *Lysinibacillus fusiformis* comb. nov. and *Bacillus sphaericus* to *Lysinibacillus sphaericus* comb. nov. *Int. J. Syst. Evol. Microbiol.* **2007**, *57*, 1117–1125. [CrossRef]
52. Iino, T.; Mori, K.; Tanaka, K.; Suzuki, K.I.; Harayama, S. *Oscillibacter valericigenes* gen. nov., sp. nov., a valerate-producing anaerobic bacterium isolated from the alimentary canal of a Japanese corbicula clam. *Int. J. Syst. Evol. Microbiol.* **2007**, *57*, 1840–1845. [CrossRef]
53. Vital, M.; Karch, A.; Pieper, D.H. Colonic Butyrate-Producing Communities in Humans: An Overview Using Omics Data. *mSystems* **2017**, *2*. [CrossRef]

54. Tanner, A.C.R.; Badger, S.; Lai, C.-H.; Listgarten, M.A.; Visconti, R.A.; Socransky, S.S. *Wolinella* gen. nov., *Wolinella succinogenes* (Vibrio succinogenes Wolin et al.) comb. nov., and Description of *Bacteroides gracilis* sp. nov., *Wolinella recta* sp. nov., *Campylobacter concisus* sp. nov., and *Eikenella corrodens* from Humans with Periodontal Disease. *Int. J. Syst. Evol. Microbiol.* **1981**, *31*, 432–445. [CrossRef]
55. Shanmugam, S.R.; Adhikari, S.; Nam, H.; Kar Sajib, S. Effect of bio-char on methane generation from glucose and aqueous phase of algae liquefaction using mixed anaerobic cultures. *Biomass Bioenergy* **2018**, *108*, 479–486. [CrossRef]
56. Sinbuathong, N.; Sirirote, P.; Watts, D.; Chulalaksananukul, S. Heavy metal resistant anaerobic bacterial strains from brewery digester sludge. *Int. J. Glob. Warm.* **2013**, *5*, 127–134. [CrossRef]

fermentation

Article

Enhancing Anaerobic Digestion with an UASB Reactor of the Winery Wastewater for Producing Volatile Fatty Acid Effluent Enriched in Caproic Acid

M. Eugenia Ibáñez-López [1], Nicola Frison [2], David Bolzonella [2,*] and José L. García-Morales [1]

[1] Department of Environmental Technologies, IVAGRO-Wine and Agrifood Research Institute, University of Cádiz-Puerto Real Campus, 11510 Cádiz, Spain
[2] Department of Biotechnology, University of Verona, 37134 Verona, Italy
* Correspondence: david.bolzonella@univr.it

Abstract: The production of Volatile Fatty Acids (VFAs) from wastewater holds significant importance in the context of biorefinery concepts due to their potential as valuable precursors for various bio-based processes. Therefore, the primary objective of this research is to investigate the fermentation of Winery Wastewater (WW) in an Upflow Anaerobic Sludge Blanket (UASB) reactor to generate VFAs, with particular emphasis on Caproic Acid (HCa) production and the dynamics of the microbiota, under varying Hydraulic Retention Time (HRT) periods (8, 5, and 2.5 h). The change from an 8 h to a 5 h HRT period resulted in an approximately 20% increase in total VFA production. However, when the HRT was further reduced to 2.5 h, total VFA production decreased by approximately 50%. Concerning the specific production of HCa, expressed in grams of Chemical Oxygen Demand (gCOD), the maximum yield was observed at around 0.9 gCOD/L for a 5-h HRT. Microbial population analysis revealed that Eubacteria outnumbered Archaea across all HRTs. Population dynamics analysis indicated that the Firmicutes Phylum was predominant in all cases. Within this phylum, bacteria such as *Clostridium kluyveri* and *Clostridium* sp., known for their ability to produce HCa, were identified. Based on the results obtained, the application of the UASB reactor for WW treatment, within the biorefinery framework, has the potential to provide a practical alternative for HCa production when operated with a 5 h HRT.

Keywords: Winery Wastewater; anaerobic digestion; Dark Fermentation; hydraulic retention time; caproic acid

Citation: Ibáñez-López, M.E.; Frison, N.; Bolzonella, D.; García-Morales, J.L. Enhancing Anaerobic Digestion with an UASB Reactor of the Winery Wastewater for Producing Volatile Fatty Acid Effluent Enriched in Caproic Acid. *Fermentation* **2023**, *9*, 958. https://doi.org/10.3390/fermentation9110958

Academic Editor: Jishi Zhang

Received: 26 September 2023
Revised: 6 November 2023
Accepted: 6 November 2023
Published: 9 November 2023

1. Introduction

The circular economy represents the emerging socio-economic and environmental model emphasizing a production and consumption approach centred on sharing, lending, reusing, repairing, revitalising, and recycling existing materials and products, rather than disposing of them as waste [1,2]. Indeed, considering that the EU generates over 2 billion tonnes of waste each year, effectively addressing this challenge is critical for environmental security and combating climate change [2].

This substantial volume of materials originates from various sectors of the economy, including agriculture, industry, energy, services, and even households. Cow's milk, grapes, and olives serve as the primary components in the creation of some of the world's most prized food products, including cheese, wine, and olive oil. These foods items, particularly grapes and olives, are mainly found in the Mediterranean region, notably in France, Greece, Italy, Spain, and Tunisia. However, their production generates a significant amount of waste, with wine production in particular resulting in significant production of Winery Wastewater (WW) [3,4].

The application of circular economy principles to this WW can involve its utilization as secondary raw materials [3–6]. Incorporating WW and its derived products into the concept of a biorefinery is one way to achieve this. A biorefinery is a facility that integrates processes and equipment to convert biomass into fuels, energy, and chemicals, as defined by the National Renewable Energy Laboratory (NREL). Within this concept, we can identify the anaerobic biorefinery, which is based on Anaerobic Digestion (AD) of organic matter to produce a range of value-added products, including Volatile Fatty Acids (VFAs) and biogas (bio-hydrogen (H_2) and/or methane (CH_4)) [7]. Biogas, generated through AD, serves as a significant source of renewable energy, finding applications in electricity generation and heat supply. Additionally, VFAs are produced as a product of this process and can be used as a direct feedstock for the production of various bio-based products (e.g., bio-H_2, CH_4, solvents, or biopolymers). Therefore, both biogas (CH_4 and bio-H_2) and VFAs can be integrated into a biorefinery scenario [8]. Within this concept, the production and utilisation of biogas as bioenergy (renewable energy) represent a significant technological advancement, and AD from waste presents an appealing pathway to attain this goal [9].

AD is a multistage process involving hydrolysis, acidogenesis, acetogenesis, and methanogenesis, with biogas as the final product. In contrast, Acidogenic (dark) Fermentation (AF) encompasses only the hydrolysis and acidogenesis stages within AD, resulting in the production of VFAs and bio-H_2 as the final product. The bacteria engaged in these initial stages belong to several phyla, including Firmicutes, Bacteriodetes, Spirochaetes, Actinobacteria, Chloroflexi, and Proteobacteria [10,11].

Microorganisms play a pivotal role in AD, as they are indispensable for the process. Analysing microbial communities can be accomplished through various techniques, including DNA sequencing, metagenomics, metatranscriptomics, and metaproteomics. These methods offer insights into the diversity, dynamics, and functional capabilities of microbial populations, thereby contributing to a comprehensive comprehension of microbial ecology and their interactions within the studied ecosystem, which is vital for achieving optimal performance.

In recent decades, the effectiveness of anaerobic technology in wastewater treatment has become increasingly evident [12]. Various technologies are available for AD of wastes and effluents, with options including suspended and adhered biomass reactors. Adhered biomass reactors (advanced reactors) offer several advantages compared to suspended biomass reactors. These advantages include a higher specific weight, enabling them to stay in the system even with upflow conditions, and the ability to handle higher feed rates without significantly increasing reactor volume [10]. Among these options, the Upflow Anaerobic Sludge Blanket (UASB) reactor emerges as an effective anaerobic process for treating wastewater with high organic loads, as evidenced by its successful applications [6,13]. The start-up of the UASB reactor requires a period of acclimatisation of the anaerobic sludge contained within the reactor (approximately one third of the reactor volume). This start-up phase holds critical importance for all types of anaerobic reactors [14]. In high-rate anaerobic technology reactors, like the UASB, the activation phase of anaerobic sludge must be carefully managed to preserve the existing structure (microbial communities present in different layers) and prevent the detachment of non-attached biomass. The integrity of granular sludge stands as the critical feature in the UASB reactor, ultimately shaping the success of processes within its reaction column [15]. Nevertheless, it is important to note that biomass does not always have to be in granular form or attached biomass. In cases where the UASB reactor is employed to treat domestic wastewaters/blackwaters. biomass can be in suspension form [16].

The Hydraulic Retention Time (HRT) exerts a significant influence on the performance of the AD process. In UASB reactors with supported biomass, it is important to distinguish between the HRT and the Solids Retention Time (SRT) of the biomass. As the HRT decreases, the Organic Loading Rate (OLR) increases [6,17]. It is important to investigate the impact of HRT for each operated AD system, considering the specific substrates or substrate mixtures used as feedstock. Gaining insight into the optimal HRT is vital for comprehending each process, establishing the best operating conditions, and ultimately achieving maximum yields. This knowledge also contributes to a broader perspective that is essential for successful future industrial scale-up [18].

The production of methane, and, thus, the possibility of energy recovery, is one of the advantages of the UASB reactor [19]. Additionally, the production and recovery of VFAs is of great interest due to their high potential as a renewable carbon source. VFAs have a wide range of applications in various industries (pharmaceutical, food, or chemical), and they can serve as valuable feedstock (biogas, biodiesel, or bioplastics). Consequently, VFA recovery through AD is becoming a prominent area of research [20]. While UASB reactors have been explored for VFA production, the emphasis has primarily been on Short Chain Fatty Acids (SCFAs) [19]. There is a growing focus on the production of valuable and versatile SCFAs such as Acetic Acid (HAc), Butyric Acid (HBu), and Propionic Acid (HPr) [21,22].

More recently, there has been a growing interest in the production of Medium-Chain Fatty Acids (MCFA), such as Caproic Acid (HCa), from organic waste and wastewater [23]. This interest arises from the potential to convert soluble SCFAs into less water soluble MCFAs through a process known as Chain Elongation (CE). In VFA extraction, the cost associated with acid solubility accounts for approximately 30% to 40% of the overall process cost, rendering the entire endeavour economically unfeasible. As an alternative, the production of less soluble acids could be a viable solution [8]. MCFAs can be produced through chemical processes that traditionally use fossil sources [11,24,25]. This process involves adding two carbons to the carbon chain length of SCFAs and utilising ethanol (EtOH), which is often present in winery effluent, as an Electron Donor (ED) to facilitate this conversion [22,23,26,27].

The primary advantage of utilizing EtOH to produce HCa lies in the energy gained during the formation of the product. According to stoichiometric reactions, one mole of HCa contains approximately 3452 kJ of energy [24]. Therefore, when aiming to produce HCa through fermentation, it is crucial to consider the requirements of the microbiota involved in the process. Specific anaerobic bacteria, such as *Clostridium kluyveri* and *Clostridium* sp. are capable of producing HCa from various carbon sources. *C. kluyveri* employs ethanol and acetate or succinate, whereas C. sp. utilizes galactitol or glucose for this purpose [25].

The current study aimed to investigate the effects of Hydraulic Retention Times (HRTs) in an Upflow Anaerobic Sludge Blanket (UASB) reactor with Winery Wastewater (WW), to produce a Volatile Fatty Acids (VFAs) effluent enriched in caproic acid (HCa), an important Medium-Chain Fatty Acid (MCFA). Furthermore, the influence of Hydraulic Retention Times (HRTs) changes on microbial communities was investigated.

2. Materials and Method

2.1. Substrate Characterization

The substrate used in this study was the synthetic WW (see Table 1), which was prepared by combining white and red table wines in a 50:50 ratio to achieve a concentration of 17.1 mL/L of wine in water.

Table 1. Initial characterization of synthetic Winery Wastewater (WW).

Parameters	WW
TCOD (g/L)	2.14 ± 0.01
SCOD (g/L)	1.81 ± 0.13
TSS (g/L)	0.24 ± 0.01
VSS (g/L)	0.16 ± 0.01
pH	7.45 ± 0.27
EtOH (gEtOH/L)	0.036 ± 0.002
PO_4-P (mgPO_4-P/L)	5.83 ± 0.26
TAN (gNH_4/L)	0.24 ± 0.02
TVFA (gCOD/L)	0.20 ± 0.01

Additionally, 0.5 g/L ammonium chloride was added to serve as a nitrogen source in the system, which is essential for microbial growth [28]. Given that the optimal Carbon/Nitrogen (C/N) ratio should be around 30:1 and above 8:1 to prevent inhibition of microbial activity due to ammonia accumulation [29,30].

Finally, the pH was adjusted using a 5 M sodium hydroxide solution until neutrality was achieved (approximately 0.25 mL/L).

Winery Wastewater (WW), Total Chemical Oxygen Demand (TCOD), Soluble Chemical Oxygen Demand (SCOD), Total Suspended Solids (TSS), Volatile Suspended Solids (VSS), Ethanol (EtOH), Phosphorus Phosphate (PO_4-P), Total Ammonia Nitrogen (TAN), Total Volatile Fatty Acid (TVFA).

2.2. Operating Conditions

The design of the UASB reactor is relatively simple [31]. The experiments were conducted using a 5-litre capacity UASB digestion reactor, as shown in Figure 1, representing a design previously developed by the research group. In this case, the system had been acclimatised and adapted to the substrate. The initial anaerobic inoculum was sourced from an industrial UASB reactor with a well-developed granular structure. The inoculum consisted of anaerobic granular sludge from an industrial UASB reactor used for treating paper mill wastewater, maintained at a temperature of 30–32 °C. During the start-up phase, one-third of the lab-scale UASB reactor's volume was inoculated with anaerobic granular sludge to achieve an initial concentration in the reactor of around 2%.

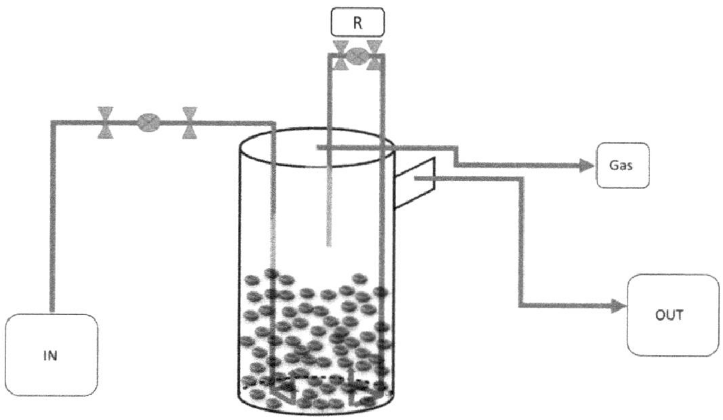

Figure 1. Diagram of the reactor with labelled components: IN: for feeding the reactor from a tank using a pump; OUT: for the treated effluent overflow; Gas: for biogas output and collection in a Tedlar bag; R: for recirculation inside the reactor facilitated by a peristaltic pump.

The top of the reactor is equipped with five different ports, each serving a specific purpose. One port is designated for feeding the feedstock, while two other ports facilitate the recirculation of the effluent. The fourth port is utilised for the release of biogas. The treated effluent is collected from the overflow at the top of the UASB reactor and directed into a separate tank. Most of the solids accumulate at the bottom of the reactor, forming a sludge bed. A triphasic separator is located at the top of the reactor to prevent solids from leaving the system with the effluent.

Both the feed and recirculation processes operated continuously, with a peristaltic pump driving the operations. The pump speed for recirculation was carefully adjusted to maintain a total biomass uptake velocity (Vuptot) of approximately 0.6 m/h as a typical operating condition for this type of reactor [32].

The reactor had previously been operated under different HRTs for screening purposes. Using this as a reference, the experimental design for this study was proposed with three distinct HRTs; one optimal, one critical, and one in between. Additionally, it was considered that macroscopic changes in the control parameters could be observed between one HRT and the other.

A high OLR and a low HRT promote the growth of hydrolytic and acidogenic bacteria, facilitating the accumulation of VFAs [11]. Therefore, a descending HRT sequence of 8, 5, and 2.5 h was employed. These values corresponded to OLRs, expressed in grams of Chemical Oxygen Demand (gCOD), of 6.4, 10.3, and 20.6 gCOD/(L·d), respectively, as shown in Table 2. Each HRT was maintained for a minimum of one month to ensure stable performance during each operating period.

Table 2. Variation of Organic Loading Rate (OLR) and feed flow at different Hydraulic Retention Times (HRTs).

HRT (h)	**8**	**5**	**2.5**
OLR (gCOD/(L·d))	6.4 ± 0.01	10.3 ± 0.01	20.6 ± 0.01
FV (L/d)	15	24	48
FR (mL/min)	10.41	16.67	33.33

Grams of Chemical Oxygen Demand (gCOD), Feed Volume (FV), and Flow Rate (FR). Organic Loading Rate (OLR) in gCOD/(L·d) and Hydraulic Retention Times (HRT) in h.

2.3. Analytical Methods

To start-up and monitor the reactor, an initial assessment of the substrate and routine evaluations of critical parameters throughout the process were performed. The following steps were carried out: pH was measured using a benchtop pH meter (Mettler Toledo, USA), and the Total Ammonia Nitrogen (TAN) was determined using the distillation method in adherence to standard guidelines [33].

Total Suspended Solids (TSS), Volatile Suspended Solids (VSS), Total Chemical Oxygen Demand (TCOD), and Soluble Chemical Oxygen Demand (SCOD) were assessed using the Standard Methods [33].

Ethanol (EtOH) and Phosphorus Phosphate (PO_4-P) were analysed using a commercial kit. EtOH analysis was conducted with a Megazyme kit (AOAC Method 2019.08). For PO_4-P quantification, the HACH TNT+ (TNT844-LM) reagent kit and a HACH® spectrophotometer were employed.

To determine the concentration of individual Acetic Acid (HAc), Propionic Acid (HPr), Butyric Acid (HBu), Valeric Acid (HVa), and Caproic Acid (HCa), and cumulative Volatile Fatty Acids (VFAs), the samples were diluted with distilled water at a ratio of 1:50 or 1:100. Subsequently, they were filtered through a 0.20 μm filter and 0.1 M hydrochloric acid was added. VFA concentrations were then determined using ionic chromatography (Dionex ICS-1100 with AS23 column, Thermo Fisher Scientific, Waltham, MA, USA).

To monitor the biogas produced in the Tedlar bags, the gas composition was measured daily using a methanimeter (Biogas 5000 gas analyser Landtec), and the daily gas volume was determined using an acid water displacement volumetric gasometer.

2.4. Microbial Community Analysis

Microbial community analysis involves the study and characterisation of the various microorganisms present in a specific environment or sample. These analyses typically encompass the identification of microorganism types, their relative abundance. and their potential roles within the ecosystem.

The microbial community analysis involved DNA sequencing of the biomass granules. A total of seven samples were collected for microbiological analysis: the first for pre-start-up characterization, followed by subsequent samples in the middle of each period and on the last day of each experimental period before the new HRT was set. DNA extraction was performed using a solid sample PowerSoil DNA Isolation extraction kit (Qiagen DNeasy

PowerSoil Pro 2495), following the manufacturer's instructions. The extracted samples were then sent to an external laboratory for sequencing. The samples were stored in a cold and aqueous solution until DNA extraction. To ensure the accuracy of the DNA extraction, a confirmation gel test was performed. The region of interest was the V3-V4 region of the 16S rRNA gene, and the extracted DNA was stored at −80 °C until being sent to BMR Genomics S.R.L. for sequencing.

Data analysis was carried out using BMR Genomics software (Qiime2 16S-V3V4 analysis Silva based).

3. Results and Discussion

Due to the strong influence of HRT and the OLR on process performance and acid profile, researchers are exploring alternatives to Continuous Stirred Tank Reactors (CSTR). Some of these being investigated are: the Sequence Batch Reactor (SBR), the Anaerobic Membrane Bioreactor (AnMBR), and the UASB. These reactors offer the advantage of separating SRT from HRT. This means that high feed rates can be achieved without significantly increasing the reactor volume, thereby increasing the cost-effectiveness of the process [11]. In this study, the UASB reactor was chosen, as it is well-suited for treating wastewater with low solids concentrations [4,15,34]. A wide range of industrial effluents have been successfully treated using the UASB reactor [31].

3.1. *Influence of HRT on Effluent Characteristics*

The stability threshold for a high-rate reactor, such as the UASB, in AD is usually identified as 8–6 h. Operating the reactor below this time interval can lead to destabilisation of the microbiota [14]. For this reason, the first HRT was chosen at the upper boundary of the specified range, set at 8 h. The second HRT was set just around the recommended optimum value, at 5 h. Finally, an HRT of 2.5 h, below the theoretical optimal interval, was chosen.

The analysis was carried out on the effluent of each HRT tested. Daily measurements of pH and VFA were performed. TCOD, SCOD, TSS, VSS, and TAN were measured twice a week. PO_4-P and EtOH measurements were conducted once a week. The results presented below in Table 3 for each HRT represent the averages of the values obtained during each period after stabilization.

Table 3. Medium values of the characterization of the feed (influent) and the effluent for each HRT.

	Influent	**Effluent**		
		HRT (h)		
Parameters		**8**	**5**	**2.5**
pH	7.45 ± 0.27	5.05 ± 0.24	4.81 ± 0.10	4.89 ± 0.10
TCOD (g/L)	2.14 ± 0.01	2.48 ± 0.21	2.47 ± 0.20	2.47 ± 0.05
SCOD (g/L)	1.81 ± 0.13	2.34 ± 0.34	2.34 ± 0.16	2.28 ± 0.18
TSS (g/L)	0.24 ± 0.01	0.06 ± 0.01	0.17 ± 0.02	0.39 ± 0.07
VSS (g/L)	0.16 ± 0.01	0.05 ± 0.01	0.06 ± 0.03	0.12 ± 0.02
TAN (gNH_4/L)	0.24 ± 0.02	0.25 ± 0.02	0.12 ± 0.00	0.12 ± 0.00
PO_4-P ($mgPO_4$-P/L)	5.83 ± 0.26	2.36 ± 0.12	2.32 ± 0.07	3.51 ± 2.18
EtOH (gEtOH/L)	0.036 ± 0.002	0.012 ± 0.001	0.005 ± 0.000	0.035 ± 0.002
TVFA (gCOD/L)	0.20 ± 0.01	1.37 ± 0.16	1.72 ± 0.15	0.93 ± 0.33
HAc (gCOD/L)	0.04 ± 0.00	0.29 ± 0.07	0.17 ± 0.06	0.25 ± 0.05
HPr (gCOD/L)	0.00 ± 0.00	0.05 ± 0.01	0.08 ± 0.03	0.03 ± 0.00
HBu (gCOD/L)	0.16 ± 0.10	0.27 ± 0.05	0.63 ± 0.12	0.40 ± 0.22
HVa (gCOD/L)	0.00 ± 0.00	0.07 ± 0.04	0.06 ± 0.03	0.00 ± 0.01
HCa (gCOD/L)	0.00 ± 0.00	0.69 ± 0.13	0.78 ± 0.08	0.25 ± 0.17

TCOD: Total Chemical Oxygen Demand, SCOD: Soluble Chemical Oxygen Demand, TSS: Total Suspended Solids, VSS: Volatile Suspended Solids, TAN: Total Ammonia Nitrogen, PO_4-P: Phosphorus Phosphate, EtOH: Ethanol, TVFA: Total Volatile Fatty Acid, HAc: Acetic Acid, HPr: Propionic Acid, HBu: Butyric Acid, HVa: Valeric Acid and HCa: Caproic Acid.

Process pH affects not only raw material degradability, but also the acid profile [11]. The pH in the reactor is a critical factor for VFA production because most acidogenic microorganisms cannot survive in extreme environments (pH 3 or pH 12) [35]. Prior research has indicated that a slightly acidic to neutral pH range (5.5–7.0) favours higher VFA production during AF [36–38]. In order to produce VFAs from wastewater, the pH should be kept between 5 and 6 [39]. Throughout our experimental period, the pH remained close to 5 (5.29–4.71), which falls within the optimal range for this stage.

Most of the organic matter present in wastewater is soluble and easily biodegradable, but it is deficient in nitrogen and phosphorus (P) [4,40]. For this reason, ammonium chloride was added to compensate this nitrogen deficiency [40]. In particular, TAN can cause toxicity and inhibition of anaerobic systems at concentrations above 4000–5000 mg/L [17,41,42]. In our reactor, since the maximum concentration reached is 250 mg/L, this phenomenon was not observed. TAN concentration remained constant during the 8 h HRT but decreased by almost half for the other two HRTs due to net metabolic incorporation into the new biomass within the granules [15].

As discussed in several studies, the high concentration of COD in comparison to nutrients levels, underscores the significance of the COD:N:P ratio for biomass growth in AD [4,43,44]. Consequently, N and P concentrations are limiting compared to COD [4,40,43]. PO_4-P concentration in WW fluctuates between 40 to 5 mg/L during the various annual stages of wine production, so, phosphorus is not a limiting factor for WW [4,40,45].

In the first two HRT, there is a net incorporation of PO_4-P into the biomass, with an effluent concentration of approximately 2.3 mg/L compared to a feed concentration of about 6 mg/L [45]. This indicates net consumption of 60%. However, in the last HRT (2.5 h), the mean value of PO_4-P in the effluent increases to 3.5 mg/L, with a variability percentage exceeding 60%. This suggests an unstable reactor performance and destabilisation, likely attributed to the degradation of the granules inside the digester. The same trend can be seen in the evolution of the solids (as discussed in Section 3.1.1), where they initially decrease and then increase with shorter HRT. This behaviour is associated with a certain degradation of the granule structure and, consequently, a destabilisation of the system.

The TCOD in WW is notably higher than that found in municipal wastewater, with peaks above 35 g/L during the harvesting period [4]. The usual concentrations are 2–10 g/L, during the harvest period the range is 6–8 g/L, while for the rest of the year the concentrations are 1–3 g/L [4,40]. In all tested HRTs, both TCOD and SCOD remained stable, with mean values of 2.47 g/L for TCOD and 2.34 g/L for SCOD. In general, waste materials commonly used to produce VFAs are rich in organic matter, with a TCOD above 4 g/L [17]. In our case, WW exhibited a TCOD concentration of around 2.14 g/L, a value that falls within the typical range for non-seasonal winery effluents [4].

Some studies suggest that EtOH can contribute up to 30% of the soluble COD in WW [4]. In this study, a net consumption of EtOH is observed in the first two HRTs, with efficiencies reaching up to 75%. These efficiencies are directly linked to the production of HCa, accounting for approximately 45–50% of the total VFA production (see Section 3.1.2). On the other hand, there is no EtOH consumption in the last HRT, which is another indicator of system destabilisation.

3.1.1. Removal Efficiencies of TSS and VSS with HRT

As the HRT decreased, there was an increase in the concentration of TSS and VSS, as indicated in Table 3. In general, it was found that the reduction of VSS in the system exceeded 60% during HRT of 8 h and 5 h for different OLR. The efficiency of AD treatment can be evaluated by examining the reduction of solids, as shown in Figure 2. According to the literature, the removal efficiency of TSS is reported to be around 70% for HRT of 8–6 h, which is consistent with the values observed in Figure 2 [34]. The removal efficiency of TSS decreased from 74.3% to 29.9% as the OLR increased from 6.4 to 10.3 gCOD/(L·d), and it even reached negative values when the OLR reached 20.5 gCOD/(L·d). The increase in the OLR, and consequently the higher upflow velocity, may be associated with a rise in the particles being dragged into the effluent due to the entrainment of the accumulated particles in the sludge bed. Therefore, an increase in OLR resulted in a decrease in TSS and VSS reduction, as reported in other studies [17], suggesting system instability, as previously discussed for other parameters. Furthermore, it can be observed that VSS/TSS in the effluent exhibits a linear correlation, increasing as HRT decreases, as per Equation (1) with a $R^2 = 0.97$.

$$VSS/TSS = -0.39 \cdot HRT + 4.41 \quad (1)$$

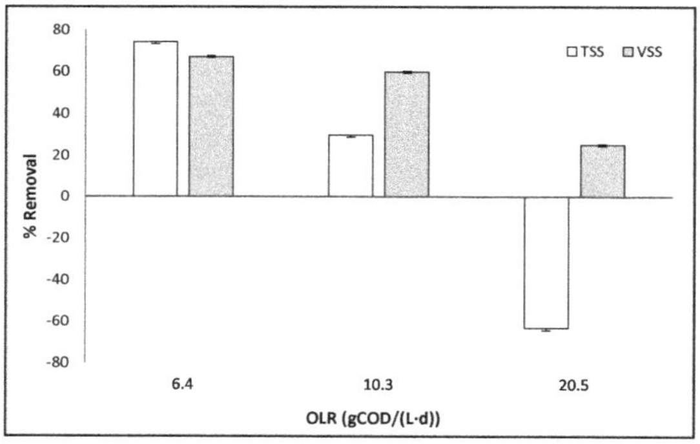

Figure 2. Evolution of suspended solid removal, total and volatile, with OLR (gCOD/(L·d)).

This parameter and its evolution indicate a net release of active biomass as HRT decreases, which means that these microorganisms leave the system [15]. This phenomenon is directly related to the destabilisation observed in the system, as it indicates a significant loss of active microorganisms, which can negatively affect the efficiency and capacity of the AD process [16].

3.1.2. VFAs Production

In AF for VFA production, retention time and mixed microbial cultures in the anaerobic reactor are critical operational parameters [28]. The development of different VFAs shows variability based on these operational parameters. Previous research studies have highlighted that acidic pH conditions and higher OLR promote VFA production [46,47].

Finding the optimum HRT is essential for achieving the highest possible production yield. Working at higher HRTs could be advantageous for VFA production as microorganisms have more time to react and hydrolysis is favoured. However, prolonged HRT could lead to disruption of VFA production and growth of undesirable microorganisms in the AF, such as archaea, which degrade VFAs. Therefore, acetogenesis and methanogenesis must be prevented in order for VFAs to accumulate [11,17]. As can be seen in Figure 3, when the HRT is augmented from 2.5 h to 5 h, there is a 46% increase in TVFA production, but when HRT is extended from 5 h to 8 h, there is only a 26% increase. Regarding the amount of VFA produced relative to feed (OLR), the %COD to VFA increased with increasing HRT and was 5%, 17%, and 21% for 2.5 h, 5 h, and 8 h HRT, respectively.

HAc showed fluctuations throughout the process, initially increasing to values close to 300 mgCOD/L, decreasing during the medium HRT and then increasing again for the shorter HRT. HPr and HVa remained at low levels in all effluents. Finally, both HBu and HCa followed the same trend as TVFA, reaching maximum values of 630 mgCOD/L for HBu and 780 mgCOD/L for HCa with a 5 h HRT and decreasing by 36.5% and 68%, respectively, when the HRT was reduced to 2.5 h.

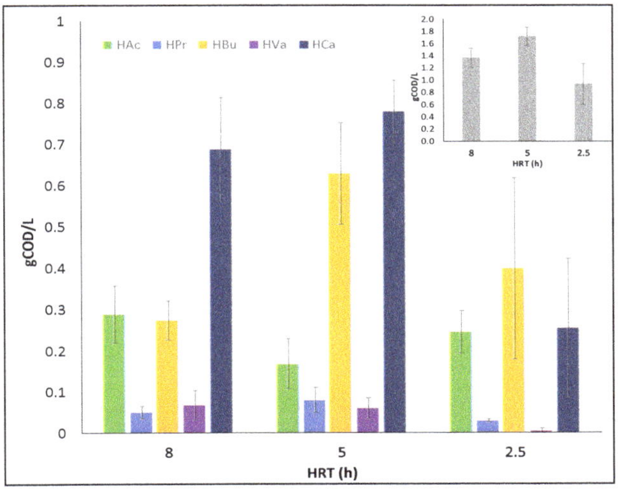

Figure 3. Evolution of the concentration of VFAs produced in each HRT: Acetic Acid (HAc), Propionic Acid (HPr), Butyric Acid (HBu), Valeric Acid (HVa), and Caproic Acid (HCa). In the upper right corner of the graph, Total Volatile Fatty Acids (TVFAs) are represented.

The three most common mixed VFAs produced by AD of waste streams are HAc, HPr, and HBu [17,36]. However, in our study, the predominant VFA obtained is HCa, as can be seen in Figure 3. This prevalence of HCa is attributed to the presence of EtOH in the WW.

VFAs containing more than four carbons (such as butyric, isobutyric, valeric, and isovaleric acids) are not directly utilised by methanogenic microorganisms. They require preliminary conversion into HAc, which serves as a precursor through direct metabolic conversion into CH_4 and carbon dioxide (CO_2) [48,49].

One of the ways in which HBu and HAc are produced is through the metabolic β-oxidation process of HCa and other VFAs [50]. This suggests that HCa is converted to HAc [51]. For an HRT of 8, the concentration of HCa was 50%, while those of HAc and HBu were 21% and 20%, respectively. In contrast, for HRTs of 5 and 2.5, the production of HCa was 45% and 27%, with an increase in HBu to 37% and 43%, respectively.

HAc is the most widely used organic acid and one of the most commercially important VFAs [36,51]. However, the global HCa market is undergoing rapid change and is projected to reach 358.8 million by 2030, growing at an average annual rate of 8.1% [36,52]. Therefore, achieving maximum efficiencies of this acid under the conditions described has a significant positive impact.

There are several studies suggesting that EtOH contributes as an ED, thereby promoting the production of MCFAs, including HCa. EtOH in the influent can promote hydrolysis, with small, degraded molecules undergoing further fermentation during acidogenesis, ultimately producing VFAs [21,22,26,27]. These studies suggest that EtOH can be used as an ED to facilitate CE [22,24,50]. In Figure 4, it can be observed that a higher presence of HCa in the medium corresponds to a decrease of EtOH in the effluents. According to the stoichiometry, if there is not enough EtOH to react with all the HAc (2EtOH:1HAc), the production of HBu and HCa is interrupted [24].

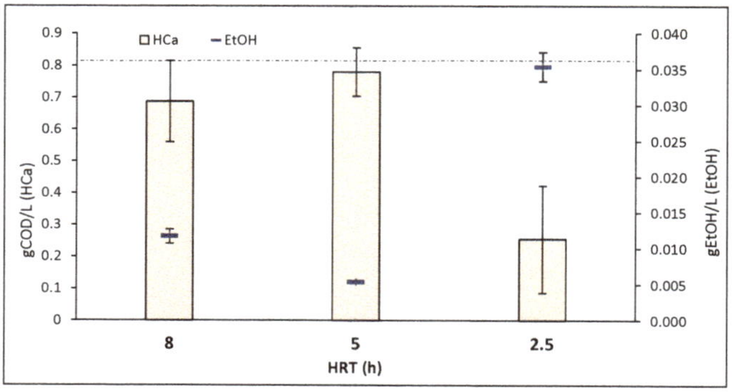

Figure 4. Average concentrations of caproic acid and ethanol at each HRT. The dotted line indicates the medium ethanol concentration in the influent (0.036 gEtOH/L).

This suggests that the use of EtOH for CE, as proposed in previous studies, aligns with our results. For the maximum concentration of HCa at 5 h HRT, the presence of EtOH in the medium is negligible (Figure 4). It can be observed that the initial EtOH concentration has been reduced by approximately 68% for the 8 h HRT and 85% for the 5 h HRT. However, for the 2.5 h HRT, its reduction is negligible and the production of HCa for this time also decreases by 67% compared to the maximum obtained at 5 h HRT. EtOH acts as an ED and is assimilated by microbial consortia, which use it to produce VFAs, particularly favouring the production of MCFAs. Among the well-studied genera in this context, *C. kluyveri* stands out, as it can use EtOH as an ED to produce long-chain VFAs. Similarly, the genus *Megasphaera elsdeni* demonstrates the ability to employ lactate for the same purpose [20].

3.2. Biogas Production

Biogas production occurs in anaerobic environments through the action of microorganisms, using VFAs as precursors [17]. It is, therefore, challenging to establish optimal conditions for all groups of microorganisms in a digester [17,53].

AF is used to convert biodegradable biomass into VFAs, CO_2, and H_2 through the action of anaerobic acidogenic microorganisms. However, the production of H_2 can be limited by the accumulation of VFAs in the system (end product inhibition) [54,55].

Regarding the biogas production (Table 4), daily analyses were carried out to control the stability of the reactor and to establish a relationship between the biogas and the microbial community inside the reactor.

Table 4. Medium daily values of the percentages and volume of biogas for each HRT.

HRT (h)	Total Volume Biogas (mL/d)	CH_4/CO_2	CO_2 (mL)	CH_4 (mL)	% CO_2	% CH_4	% COD_{CH4} (*)
8	600 ± 35.69	12.81 ± 1.09	43.39 ± 5.98	556.18 ± 29.70	7.22 ± 0.57	92.78 ± 0.57	24.83 ± 1.35
5	897 ± 31.25	16.11 ± 1.71	52.41 ± 7.02	844.69 ± 24.23	5.83 ± 0.58	94.17 ± 0.58	23.78 ± 0.64
2.5	51 ± 61.70	6.96 ± 8.77	6.41 ± 8.54	44.57 ± 53.15	9.08 ± 5.77	90.92 ± 5.77	0.63 ± 0.74

Carbon dioxide (CO_2), methane (CH_4), and Chemical Oxygen Demand (COD). (*) CH_4 generation expressed as % of COD converted from OLR in each HRT.

The results show that there was a 33% increase in total biogas production when transitioning from an 8 h HRT to a 5 h HRT. However, a significant decline of approximately 94% in biogas production was observed at the 2.5 h HRT, signifying a substantial change in reactor stability. It can be observed that for the first two HRTs, the CH_4 generation expressed as % of the converted COD is around 24%, whereas for the 2.5 h HRT, this % decreases to around 1%.

The data indicate that although there were variations in the total volume of biogas produced, the proportions of gas composition remained relatively stable. The % CO_2 content remained consistent, ranging from 6% to 9% across all HRTs. Volatile Fatty Acids (VFAs) are formed during AD, which also produces biogas as a product. Biogas is rich in CH_4, with a minimum content of 65–70%. The CH_4 levels obtained in this study were quite high, exceeding the upper boundary of 85% [31,36]. Values were around 91–94% in all HRTs studied. Such high CH_4 levels, over 90%, have also been reported in other studies [56,57]. However, the variability of the 2.5 h HRT caused this value to fluctuate and to decrease to 85%. This fluctuation confirmed the destabilisation of the system.

The significant variability observed in all parameters related to biogas production at the lowest HRT was also indicative of system destabilisation.

3.3. *Microbial Community Dynamics*

AD is a multi-step process involving a wide variety of microorganisms, including fermentative bacteria (acidogens), hydrogen-producing, acetate-forming bacteria (acetogens), and archaea that convert acetate or hydrogen to methane (methanogens). The imbalance in any one step can lead to the collapse of the entire system [58,59].

Operating parameters such as pH, HRT, and OLR, among others, have a synergistic effect on the microbial communities involved in the fermentation processes, influencing cellular metabolism [60].

In UASB reactors, it has been observed that bacteria can naturally form aggregates in the form of flocs and granules [34]. By extracting DNA from the granules, it was possible to identify and analyse the microbial populations. It has also allowed the study and identification of the dominant groups of microorganisms, both their diversity and abundance present when environmental changes occur, as well as due to the variation of HRTs.

The microbiota was categorized into archaea and bacteria populations, resulting in a total of 240,413 high-quality archaeal sequences and 77,632 bacterial sequences. The relative abundance of bacterial and archaeal communities during fermentation is shown in Figure 5a.

No significant differences were observed in the distribution of archaeal phyla among the various HRTs. Euryarchaeota was the predominant phylum, comprising over 96% of the archaeal population (Figure 5b). This dominance aligns with findings from other studies. The remaining two phyla each represented less than 3% of the total archaeal population across all HRTs [61]. The predominant genera with an abundance >15% detected in the Pre-Start-Up (PSU) were Methanosaeta (16%), Methanobacterium (18.5%), and Methanobrevibacter (63%). Methanosaeta is one of the major genera of acetoclastic methanogens when the microbial community is acclimated to HBu [62]. This genus belongs to the phylum Halobacterota, as shown in Figure 5b, where its abundance is approximately 17% in the PSU. However, in the different HRTs studied, its presence does not exceed 1%, and in some cases, it is absent. On the other hand, both Methanobacterium and Methanobrevibacter belong to the family Methanobacteriaceae (hydrogenotrophic methanogens), which reduce CO_2 with H_2 [63]. For the 8 h HRT, the most abundant genus is Methanobrevibacter, accounting for over 90%, followed by Methanobacterium at 4–8%. The remaining genera (Methanosaeta and Methanomethylophilus) have negligible abundances, sometimes even disappearing. In the case of 5 h HRT, we find that, as in the 8 h HRT, the two dominant genera are Methanobrevibacter (93–79%) and Methanobacterium (7–19%) for the intermediate and final samples of this HRT. Although these genera remain dominant, there is a decrease in the abundance Methanobrevibacter accompanied by an increase in Methanobacterium. The other two genera, as observed in other HRTs, have abundances of <2%. Finally, for the 2.5 h HRT, Methanobrevibacter initially decreases to 52% and then increases to 92%. In contrast, Methanobacterium increases to 45% and then decreases to 8%. This shift in the microbial community, as observed in previous sections, confirms the destabilization of the system at this HRT.

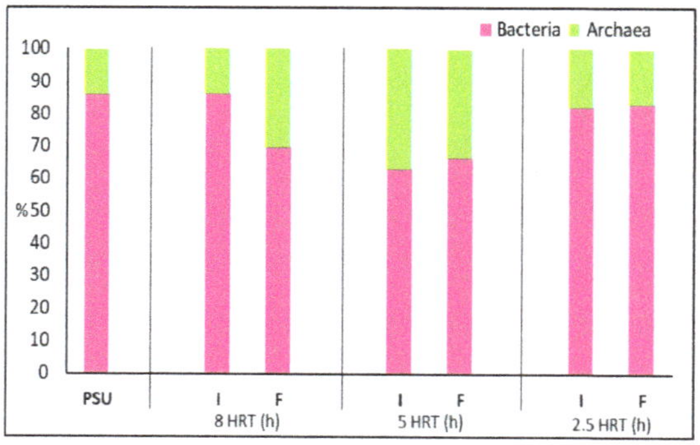

(**a**)

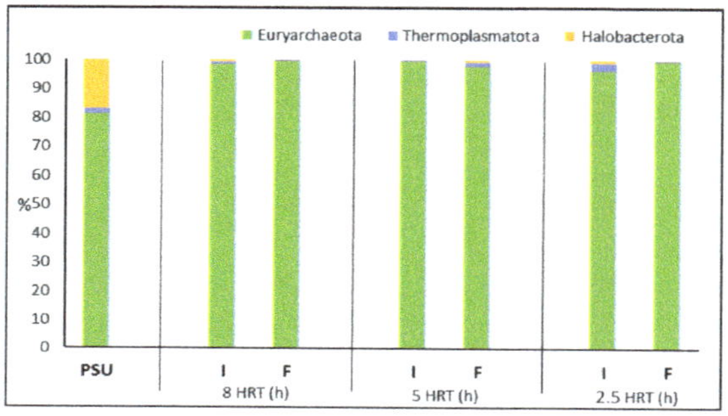

(**b**)

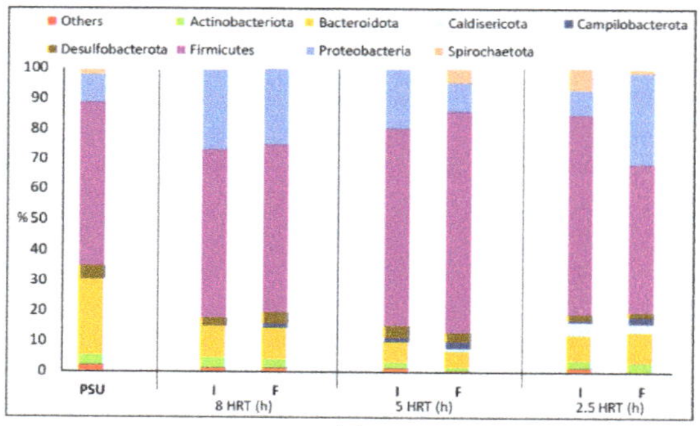

(**c**)

Figure 5. (**a**) Evolution of the percentage abundance of Archaea versus Bacteria at each HRT; (**b**) Evolution of the percentage of abundance of the Phylum Bacterium in each HRT; (**c**) Evolution of the percentage of abundance of the Phylum Archaea in each HRT. (PSU (Pre-Start-Up), I (Intermediate), and F (Final)).

Figure 5c shows the main bacterial phyla found in the different HRTs studied. Over 11 bacterial phyla were detected, and only those with a significant abundance of more than 1% are shown, while the rest were grouped together. The most abundant bacterial phyla in all fermentation samples were Firmicutes (>48%), Proteobacteria (>8%), and Bacteroidota (<5%). Dominance of Firmicutes and Proteobacteria as dominant phyla has also been reported in other studies, and in particular bacteria belonging to phylum Firmicutes have been identified as important microorganisms for maximising VFA formation [11,61]. Bacteroidota is a phylum involved in the degradation of complex polymers and the hydrolysis of proteins into

VFAs (HAc) and NH_3 [61,64,65]. This phylum is present with an abundance of approximately 10% for the 8 h HRT, decreases to 6% for the 5 h HRT and then increases again to 9% when the HRT is reduced to 2.5 h. It is worth noting that in the sample taken for PSU, this phylum had a predominance of 25%, resulting in a reduction of about 33% compared to the initial relative abundance.

Firmicutes is a Gram-positive bacterium widely reported in AD and is known to be important in acid hydrolysis [61,64]. The high abundance of Firmicutes, approximately 50%, may be related to its adaptation to high VFAs and low pH conditions [66]. This phylum was found in all HRTs with the highest relative abundances, which were 55%, 69%, and 57% for 8 h, 5 h, and 2.5 h HRTs, respectively. This phylum usually has the highest relative abundance (65–83%) in AF [11]. Within this phylum, we can find some of the genera known to be VFAs producers, as shown in Table 5.

The Proteobacteria phylum consists of acidogenic bacteria involved in the degradation of organic matter and the consumption of VFAs. It was one of the dominant phyla in the AD of sewage sludge [65,67]. We found this phylum with an abundance of 9% in the PSU, which increased by over 50% for the 8 h and 5 h HRTs.

Spirochaetota is a phylum of syntrophic acetate-oxidising bacteria that are often present in bacterial communities in reactors. Treponema belongs to this phylum, possesses the capability to hydrolyse cellulose and hemicellulose to produce HAc from H_2 and CO_2 [64]. In the HRTs tested, the abundance of this phylum is low, never exceeding 8% and even disappearing in the 8 h HRT. In the specific case of the genus Treponema, it is below 2% in all cases.

Within the Desulfobacterota phylum, with an average abundance of about 3%, the metal-reducing genus can be found, such as the species *Desulfovibrio*, with almost complete dominance within its phylum. This species can grow syntrophically, producing H_2, or act as an electron carrier between species [62,68].

The first reported case of carbon CE was in 1942, in an experiment using a pure culture of *C. kluyveri*, where EtOH was converted to HCa. EtOH was oxidised to HAc, then to HBu and finally to HCa [59,69]. *C. kluyveri*, belonging to the order Clostridiales and family Clostridiaceae. It was found in the reactor with a relative abundance of 0.1% at the beginning of the reactor, 1% for the intermediate 8 h HRT and 2.7% for the final of 8 h HRT, 1.2% for the intermediate 5 h HRT, 0.5% for the final 5 h HRT, 0.8% for the intermediate 2.5 h HRT, and 0.7% for the final 2.5 h HRT. However, if we calculate the specific percentage of this bacterium within the relative abundance of its genus (Table 5), we see a predominance of 36.89% for the 5 h HRT, 24.57% for the 8 h HRT, and 7.25% for the 2.5 h HRT, for the intermediate states of each HRT. An HRT of 5 h shows the highest predominance of these bacteria, which is consistent with the results obtained for HCa concentrations. The detection of *Clostridium butyricum*, known for its ability to produce HBu and CO_2, was observed. However, its presence was consistently at or below 1% in all cases. In the class Clostridia, but belonging to another order, Oscillospirales, and family Ruminococcaceae, the genus Caproiciproducens was found (Table 5). It has been described that the strain *Caproiciproducens galactitolivorans BS-1T* produces H_2, CO_2, EtOH, HAc, HBu, and HCa as the final metabolic products of anaerobic fermentation, which corresponds to the production of HCa [50,59]. HCa producers, such as C. sp. (genus Caproiciproducens) [59] among others, were found in the following proportions: 3.5% for 8 h HRT, 2.3% for 5 h HRT, and 1.7% for 2.5 h HRT.

Table 5 shows the frequency of certain genera known to produce certain VFAs. In general, we can observe variations in abundance between the intermediate and final periods of the 2.5 h HRT, as was observed for the Archaea.

Table 5. Percentage of abundance of bacteria producing the different VFAs for each HRT and authors with similar genus identification.

VFAs	Genus	% Abundance							Ref.
			8 HRT		5 HRT		2.5 HRT		
		PSU	I	F	I	F	I	F	
HAc	Acetobacter	0.21	4.08	5.77	2.02	3.36	2.93	2.24	[36,50,51,53]
	Clostridium	2.11	4.07	9.13	4.88	3.65	11.03	1.01	[36,50,51,53]
HPr	Propionibacterium		0.38	0.54	0.29	0.21	1.18	0.11	[36,50]
	Acidaminococcus	0.86	0.72	0.56	0.33	0.17	0.23	0.22	[70]
HBu	Butyricicoccus	0	0.02	0.04	0.1	0.03	0.05	0.1	[71]
	Butyrivibrio	0.02	0.02	0.01	0	0	0.01	0.05	[36,50]
	Acidaminococcus	0.86	0.72	0.56	0.33	0.17	0.23	0.22	[70]
	Clostridium	2.11	4.07	9.13	4.88	3.65	11.03	1.01	[36,50]
	Eubacterium	0.5	0.22	0.14	0.06	0.06	0	0.08	[36,50]
HCa	Caproiciproducens	7.27	4.72	4.11	3.83	3.19	3.14	0.97	[59]
	Clostridium	2.11	4.07	9.13	4.88	3.65	11.03	1.01	[36,50,59]

4. Conclusions

The effect of Hydraulic Retention Time (HRT) on the acidogenic fermentation stage of a UASB technology for the treatment of synthetic Winery Wastewater was investigated, assessing various system-specific parameters and microbiota evolution.

A net generation of caproic acid was observed in all the studied HRTs. There is a direct correlation between the final concentration of this acid and the removal of ethanol in the effluent.

In all sceneries, the population of Eubacteria was higher than that of Archaea. The dominant phyla within their respective domains were Euryarchaeota and Firmicutes.

The optimal operational results were obtained at 5 h HRT, where the following observations were made.

- The maximum total production of Volatile Fatty Acids in the effluent was 1.7 gCOD/L, with a 45% of HCa (0.9 gCOD/L).
- Approximately 21% of feed COD is converted to VFA.
- The Archaeal population remained stable at over 35%, most of them (>96%) belonged to the phylum Euryarchaeota, within which the family Methanobacteriaceae (hydrogenotrophic methanogens) dominated with more than 98%.
- Within the Eubacteria population, we identified genera known for their HCa production capabilities, including *Clostridium kluyveri* and *Clostridium* sp. Both genera belong to the dominant phylum Firmicutes, which constituted 69% of the population.
- The biogas has a high methane content (>94%), with 24% of the fed COD converted into CH_4.

When operating at the lowest HRT (2.5 h), system destabilization is observed, as evidenced by various parameters, such as the removal efficiencies of suspended solids and ethanol, a decrease in VFAs, and reduced biogas production.

Based on the results obtained, the application of a UASB reactor for the treatment of WW could potentially serve as a practical alternative within the context of bio-refinery and the circular economy for the production of caproic acid.

Author Contributions: M.E.I.-L.: software, formal analysis, investigation, visualization, and writing—original draft preparation. N.F.: conceptualization, methodology, resources, supervision, writing—review and editing. D.B.: conceptualization, methodology, resources, writing—review and editing, supervision, project administration and funding acquisition. J.L.G.-M.: supervision, writing—review and editing, funding acquisition. All authors have read and agreed to the published version of the manuscript.

Funding: The stage of M. Eugenia Ibañez-López has been co-financed by the European Union within the framework of the ERDF Operational Program 2014–2020 and by the Ministry of Economic Transformation, Industry, Knowledge and Universities of the Junta de Andalucía. (Project reference: FEDER-UCA18-107460), Verinsur, S.A., ZonoSistem. Ingeniería del Ozono S.L. and the Own Research Plan—UCA 2022-2023.

Institutional Review Board Statement: Not applicable.

Informed Consent Statement: Not applicable.

Data Availability Statement: The authors do not have permission to share data.

Acknowledgments: The stay of M. Eugenia Ibañez-López has been co-financed by the European Union within the framework of the ERDF Operational Program 2014–2020 and by the Ministry of Economic Trans-formation, Industry, Knowledge and Universities of the Junta de Andalucía. (Project reference: FEDER-UCA18-107460), Verinsur, S.A., ZonoSistem. Ingeniería del ozono S.L. and the Own Plan—UCA 2022–2023. Thanks to the laboratory of the University of Verona 'LabICAB' (Laboratory of Chemical Plants for the Environment and Bioprocesses) for allowing me to carry out my stay with them. To Davide Bertasini and Riccardo Lo Coco for showing me how to work in that laboratory.

Conflicts of Interest: The authors declare that they have no known competing financial interests or personal relationships that could have appeared to influence the work reported in this paper.

References

1. Kuci, A.; Fogarassy, C. European Green Deal Policy for the Circular Economy: Opportunities and Challenges. *Hung. Agric. Eng.* **2021**, *39*, 65–73. [CrossRef]
2. European Parliament. Available online: https://www.europarl.europa.eu/news/en/headlines/economy/20151201STO05603/circular-economy-definition-importance-and-benefits?&at_campaign=20234-Economy&at_medium=Google_Ads&at_platform=Search&at_creation=RSA&at_goal=TR_G&at_audience=circular%20economy&at_topic=Circular_Economy&at_location=ES&gclid=CjwKCAjwxOymBhAFEiwAnodBLO_wVOkTIxnucxu-_fh1c-y_r_XMJibQvl1RBl8h7h7fK-GVP2W8iRoCMFkQAvD_BwE (accessed on 21 September 2023).
3. Gottardo, M.; Bolzonella, D.; Tuci, G.A.; Valentino, F.; Majone, M.; Pavan, P.; Battista, F. Producing volatile fatty acids and polyhydroxyalkanoates from foods by-products and waste: A review. *Bioresour. Technol.* **2022**, *361*, 127716. [CrossRef]
4. Bolzonella, D.; Papa, M.; Da Ros, C.; Muthukumar, L.A.; Rosso, D. Winery wastewater treatment: A critical overview of advanced biological processes. *Crit. Rev. Biotechnol.* **2019**, *39*, 489–507. [CrossRef]
5. Food and Agriculture Organization of the United Nations. General Introduction. Gender and Land Rights Database, WWW Document. Available online: https://www.fao.org/home/en/ (accessed on 21 September 2023).
6. Sivaprakasam, S.; Balaji, K. Effect of HRT on biogas yield in treating dairy industry wastewater using Upflow Anaerobic Sludge Fixed Film reactor. *Mater. Today Proc.* **2020**, *43*, 1443–1448. [CrossRef]
7. Surendra, K.; Sawatdeenarunat, C.; Shrestha, S.; Sung, S.; Khanal, S.K. Anaerobic Digestion-Based Biorefinery for Bioenergy and Biobased Products. *Ind. Biotechnol.* **2015**, *11*, 103–112. [CrossRef]
8. Patel, A.; Sarkar, O.; Rova, U.; Christakopoulos, P.; Matsakas, L. Valorization of volatile fatty acids derived from low-cost organic waste for lipogenesis in oleaginous microorganisms—A review. *Bioresour. Technol.* **2021**, *321*, 124457. [CrossRef] [PubMed]
9. Del Nery, V.; Alves, I.; Damianovic, M.H.R.Z.; Pires, E.C. Hydraulic and organic rates applied to pilot scale UASB reactor for sugar cane vinasse degradation and biogas generation. *Biomass-Bioenergy* **2018**, *119*, 411–417. [CrossRef]
10. Morales-Palomo, S.; González-Fernández, C.; Tomás-Pejó, E. Prevailing acid determines the efficiency of oleaginous fermentation from volatile fatty acids. *J. Environ. Chem. Eng.* **2022**, *10*, 107354. [CrossRef]
11. Tomás-Pejó, E.; González-Fernández, C.; Greses, S.; Kennes, C.; Otero-Logilde, N.; Veiga, M.C.; Bolzonella, D.; Müller, B.; Passoth, V. Production of short-chain fatty acids (SCFAs) as chemicals or substrates for microbes to obtain biochemicals. *Biotechnol. Biofuels Bioprod.* **2023**, *16*, 1–17. [CrossRef]
12. Ganesh, P.S.; Ramasamy, E.V.; Gajalakshmi, S.; Sanjeevi, R.; Abbasi, S.A. Studies on treatment of low-strength effluents by UASB reactor and its application to dairy industry wash waters. *Indian J. Biotechnol.* **2007**, *6*, 234–238.
13. Gao, Y.; Cai, T.; Yin, J.; Li, H.; Liu, X.; Lu, X.; Tang, H.; Hu, W.; Zhen, G. Insights into biodegradation behaviors of methanolic wastewater in up-flow anaerobic sludge bed (UASB) reactor coupled with in-situ bioelectrocatalysis. *Bioresour. Technol.* **2023**, *376*, 128835. [CrossRef] [PubMed]
14. Mainardis, M.; Buttazzoni, M.; Goi, D. Up-Flow Anaerobic Sludge Blanket (UASB) Technology for Energy Recovery: A Review on State-of-the-Art and Recent Technological Advances. *Bioengineering* **2020**, *7*, 43. [CrossRef]
15. Tiwari, M.K.; Guha, S.; Harendranath, C.S.; Tripathi, S. Influence of extrinsic factors on granulation in UASB reactor. *Appl. Microbiol. Biotechnol.* **2006**, *71*, 145–154. [CrossRef]
16. Boiocchi, R.; Zhang, Q.; Gao, M.; Liu, Y. Modeling and optimization of an upflow anaerobic sludge blanket (UASB) system treating blackwaters. *J. Environ. Chem. Eng.* **2022**, *10*, 107614. [CrossRef]

17. Chuenchart, W.; Logan, M.; Leelayouthayotin, C.; Visvanathan, C. Enhancement of food waste thermophilic anaerobic digestion through synergistic effect with chicken manure. *Biomass-Bioenergy* **2020**, *136*, 105541. [CrossRef]
18. Sillero, L.; Solera, R.; Pérez, M. Thermophilic-mesophilic temperature phase anaerobic co-digestion of sewage sludge, wine vinasse and poultry manure: Effect of hydraulic retention time on mesophilic-methanogenic stage. *Chem. Eng. J.* **2023**, *451*, 138478. [CrossRef]
19. Collivignarelli, M.C.; Abbà, A.; Caccamo, F.M.; Calatroni, S.; Torretta, V.; Katsoyiannis, I.A.; Miino, M.C.; Rada, E.C. Applications of Up-Flow Anaerobic Sludge Blanket (UASB) and Characteristics of Its Microbial Community: A Review of Bibliometric Trend and Recent Findings. *Int. J. Environ. Res. Public Health* **2021**, *18*, 10326. [CrossRef]
20. Atasoy, M.; Owusu-Agyeman, I.; Plaza, E.; Cetecioglu, Z. Bio-based volatile fatty acid production and recovery from waste streams: Current status and future challenges. *Bioresour. Technol.* **2018**, *268*, 773–786. [CrossRef]
21. Agler, M.T.; Wrenn, B.A.; Zinder, S.H.; Angenent, L.T. Waste to bioproduct conversion with undefined mixed cultures: The carboxylate platform. *Trends Biotechnol.* **2011**, *29*, 70–78. [CrossRef]
22. He, J.; Shi, Z.; Luo, T.; Zhang, S.; Liu, Y.; Luo, G. Phenol promoted caproate production via two-stage batch anaerobic fermentation of organic substance with ethanol as electron donor for chain elongation. *Water Res.* **2021**, *204*, 117601. [CrossRef]
23. Chen, W.-S.; Strik, D.P.; Buisman, C.J.; Kroeze, C. Production of Caproic Acid from Mixed Organic Waste: An Environmental Life Cycle Perspective. *Environ. Sci. Technol.* **2017**, *51*, 7159–7168. [CrossRef] [PubMed]
24. Cavalcante, W.d.A.; Leitão, R.C.; Gehring, T.A.; Angenent, L.T.; Santaella, S.T. Anaerobic fermentation for n-caproic acid production: A review. *Process. Biochem.* **2017**, *54*, 106–119. [CrossRef]
25. Choi, K.; Jeon, B.S.; Kim, B.-C.; Oh, M.-K.; Um, Y.; Sang, B.-I. In Situ Biphasic Extractive Fermentation for Hexanoic Acid Production from Sucrose by Megasphaera elsdenii NCIMB 702410. *Appl. Biochem. Biotechnol.* **2013**, *171*, 1094–1107. [CrossRef] [PubMed]
26. Wu, Q.; Bao, X.; Guo, W.; Wang, B.; Li, Y.; Luo, H.; Wang, H.; Ren, N. Medium chain carboxylic acids production from waste biomass: Current advances and perspectives. *Biotechnol. Adv.* **2019**, *37*, 599–615. [CrossRef] [PubMed]
27. Wu, S.-L.; Wei, W.; Sun, J.; Xu, Q.; Dai, X.; Ni, B.-J. Medium-Chain fatty acids and long-chain alcohols production from waste activated sludge via two-stage anaerobic fermentation. *Water Res.* **2020**, *186*, 116381. [CrossRef] [PubMed]
28. Lee, W.S.; Chua, A.S.M.; Yeoh, H.K.; Ngoh, G.C. A review of the production and applications of waste-derived volatile fatty acids. *Chem. Eng. J.* **2014**, *235*, 83–99. [CrossRef]
29. Guarino, G.; Carotenuto, C.; Di Cristofaro, F.; Papa, S.; Morrone, B.; Minale, M. Does the C/N ratio really affect the bio-methane yield? a three years investigation of buffalo manure digestion. *Chem. Eng. Trans.* **2016**, *49*, 463–468. [CrossRef]
30. Melis, E.; Asquer, C.; Carboni, G.; Scano, E.A. Role of *Cannabis sativa* L. in energy production: Residues as a potential lignocellulosic biomass in anaerobic digestion plants. In *Current Applications, Approaches, and Potential Perspectives for Hemp*; Academic Press: Cambridge, MA, USA, 2023; pp. 111–199. [CrossRef]
31. Poh, P.E.; Chong, M.F. Development of anaerobic digestion methods for palm oil mill effluent (POME) treatment. *Bioresour. Technol.* **2009**, *100*, 1–9. [CrossRef]
32. Lim, S.J.; Kim, T.H. Applicability and trends of anaerobic granular sludge treatment processes. *Biomass Bioenergy* **2014**, *60*, 189–202. [CrossRef]
33. Eaton, A.D. *Standard Methods for the Examination of Water and Wastewater*; American Public Health Association: Washington, DC, USA, 1999.
34. Seghezzo, L.; Zeeman, G.; van Lier, J.B.; Hamelers, H.V.M.; Lettinga, G. A Review: The anaerobic treatment of sewage in uasb and EGSB reactors. *Bioresour. Technol.* **1998**, *65*, 175–190. [CrossRef]
35. Liu, H.; Wang, J.; Liu, X.; Fu, B.; Chen, J.; Yu, H.Q. Acidogenic fermentation of proteinaceous sewage sludge: Effect of Ph. *Water Res.* **2012**, *46*, 799–807. [CrossRef] [PubMed]
36. Yin, S.A.D.M.; Mahboubi, A.; Sapmaz, T.; Varjani, S.; Qiao, W.; Koseoglu-Imer, D.Y.; Taherzadeh, M.J. A Glimpse of the World of Volatile Fatty Acids Production and Application: A review. *Bioengineered* **2022**, *13*, 1249–1275. [CrossRef]
37. Sukphun, P.; Sittijunda, S.; Reungsang, A. Volatile fatty acid production from organic waste with the emphasis on membrane-based recovery. *Fermentation* **2021**, *7*, 159. [CrossRef]
38. Cavinato, C.; Da Ros, C.; Pavan, P.; Bolzonella, D. Influence of temperature and hydraulic retention on the production of volatile fatty acids during anaerobic fermentation of cow manure and maize silage. *Bioresour. Technol.* **2017**, *223*, 59–64. [CrossRef]
39. Bengtsson, S.; Hallquist, J.; Werker, A.; Welander, T. Acidogenic fermentation of industrial wastewaters: Effects of chemostat retention time and pH on volatile fatty acids production. *Biochem. Eng. J.* **2008**, *40*, 492–499. [CrossRef]
40. Latessa, S.H.; Hanley, L.; Tao, W. Characteristics and practical treatment technologies of winery wastewater: A review for wastewater management at small wineries. *J. Environ. Manag.* **2023**, *342*, 118343. [CrossRef]
41. Yu, H.Q.; Fang, H.H.P. Acidification of mid-and high-strength dairy wastewaters. *Water Res.* **2001**, *35*, 3697–3705. [CrossRef]
42. Sillero, L.; Solera, R.; Perez, M. Improvement of the anaerobic digestion of sewage sludge by co-digestion with wine vinasse and poultry manure: Effect of different hydraulic retention times. *Fuel* **2022**, *321*, 124104. [CrossRef]
43. Fernández, B.; Seijo, I.; Ruiz-Filippi, G.; Roca, E.; Tarenzi, I.; Lema, J.M. Characterization, management and treatment of wastewater from white wine production. *Water Sci. Technol.* **2007**, *56*, 121–128. [CrossRef]
44. Moletta, R. Winery and distillery wastewater treatment by anaerobic digestion. *Water Sci. Technol.* **2005**, *51*, 137–144. [CrossRef]

45. Valderrama, C.; Ribera, G.; Bahí, N.; Rovira, M.; Giménez, T.; Nomen, R.; Lluch, S.; Yuste, M.; Martinez-Lladó, X. Winery wastewater treatment for water reuse purpose: Conventional activated sludge versus membrane bioreactor (MBR). A comparative case study. *Desalination* **2012**, *306*, 1–7. [CrossRef]
46. Cheng, F.; Brewer, C.E. Conversion of protein-rich lignocellulosic wastes to bio-energy: Review and recommendations for hydrolysis + fermentation and anaerobic digestion. *Renew. Sustain. Energy Rev.* **2021**, *146*, 111167. [CrossRef]
47. Tena, M.; Luque, B.; Perez, M.; Solera, R. Enhanced hydrogen production from sewage sludge by cofermentation with wine vinasse. *Int. J. Hydrogen Energy* **2020**, *45*, 15977–15984. [CrossRef]
48. Wang, Q.; Kuninobu, M.; Ogawa, H.I.; Kato, Y. Degradation of Volatile Fatty Acids in Highly Efficient Anaerobic Digestion. Available online: www.elsevier.com/locate/biombioe (accessed on 21 September 2023).
49. Malinowsky, C.; Nadaleti, W.; Debiasi, L.R.; Moreira, A.J.G.; Bayard, R.; Borges de Castilhos, A., Jr. Start-up phase optimization of two-phase anaerobic digestion of food waste: Effects of organic loading rate and hydraulic retention time. *J. Environ. Manag.* **2021**, *296*, 113064. [CrossRef]
50. Kim, B.C.; Jeon, B.S.; Kim, S.; Kim, H.; Um, Y.; Sang, B.I. *Caproiciproducens galactitolivorans* gen. Nov., sp. nov., a bacterium capable of producing caproic acid from galactitol, isolated from a wastewater treatment plant. *Int. J. Syst. Evol. Microbiol.* **2015**, *65*, 4902–4908. [CrossRef]
51. Mostafa, N.A. Production and recovery of volatile fatty acids from fermentation broth. *Energy Convers. Manag.* **1999**, *40*, 1543–1553. [CrossRef]
52. Global Industry Analysts, Inc. *Caproic Acid—Global Strategic Business Report*; Global Industry Analysts, Inc.: San Jose, CA, USA, 2023.
53. Lv, W.; Schanbacher, F.L.; Yu, Z. Putting microbes to work in sequence: Recent advances in temperature-phased anaerobic digestion processes. *Bioresour. Technol.* **2010**, *101*, 9409–9414. [CrossRef]
54. Wang, D.; Duan, Y.; Yang, Q.; Liu, Y.; Ni, B.J.; Wang, Q.; Zeng, G.; Li, X.; Yuan, Z. Free ammonia enhances dark fermentative hydrogen production from waste activated sludge. *Water Res.* **2018**, *133*, 272–281. [CrossRef]
55. Rawoof, S.A.A.; Kumar, P.S.; Vo, D.V.N.; Subramanian, S. Sequential production of hydrogen and methane by anaerobic digestion of organic wastes: A review. *Environ. Chem. Lett.* **2021**, *19*, 1043–1063. [CrossRef]
56. Solera, R.; Romero, L.I.; Sales, D. Analysis of the methane production in thermophilic anaerobic reactors: Use of autofluorescence microscopy. *Biotechnol. Lett.* **2001**, *23*, 1889–1892. [CrossRef]
57. García Morales, J.L. Dinámica de Colonización de la *Biopelícula bacteriana* en Reactores Anaerobios Termofílicos. Ph.D. Thesis, Universidad de Cádiz, Cádiz, Spain, 1998.
58. Ramirez, I.; Volcke, E.I.P.; Rajinikanth, R.; Steyer, J.P. Modeling microbial diversity in anaerobic digestion through an extended ADM1 model. *Water Res.* **2009**, *43*, 2787–2800. [CrossRef]
59. Dong, W.; Yang, Y.; Liu, C.; Zhang, J.; Pan, J.; Luo, L.; Wu, G.; Awasthi, M.K.; Yan, B. Caproic acid production from anaerobic fermentation of organic waste—Pathways and microbial perspective. *Renew. Sustain. Energy Rev.* **2023**, *175*, 113181. [CrossRef]
60. Strazzera, G.; Battista, F.; Garcia, N.H.; Frison, N.; Bolzonella, D. Volatile fatty acids production from food wastes for biorefinery platforms: A review. *J. Environ. Manag.* **2018**, *226*, 278–288. [CrossRef]
61. Zhou, H.; Brown, R.C.; Wen, Z. Anaerobic digestion of aqueous phase from pyrolysis of biomass: Reducing toxicity and improving microbial tolerance. *Bioresour. Technol.* **2019**, *292*, 121976. [CrossRef] [PubMed]
62. Zhao, Z.; Zhang, Y.; Yu, Q.; Dang, Y.; Li, Y.; Quan, X. Communities stimulated with ethanol to perform direct interspecies electron transfer for syntrophic metabolism of propionate and butyrate. *Water Res.* **2016**, *102*, 475–484. [CrossRef] [PubMed]
63. Owusu-Agyeman, I.; Eyice, Ö.; Cetecioglu, Z.; Plaza, E. The study of structure of anaerobic granules and methane producing pathways of pilot-scale UASB reactors treating municipal wastewater under sub-mesophilic conditions. *Bioresour. Technol.* **2019**, *290*, 121733. [CrossRef]
64. Zhao, X.; Liu, J.; Liu, J.; Yang, F.; Zhu, W.; Yuan, X.; Hu, Y.; Ciu, Z.; Wang, X. Effect of ensiling and silage additives on biogas production and microbial community dynamics during anaerobic digestion of switchgrass. *Bioresour. Technol.* **2017**, *241*, 349–359. [CrossRef]
65. Liu, C.; Li, H.; Zhang, Y.; Si, D.; Chen, Q. Evolution of microbial community along with increasing solid concentration during high-solids anaerobic digestion of sewage sludge. *Bioresour. Technol.* **2016**, *216*, 87–94. [CrossRef]
66. Lin, L.; Yu, Z.; Li, Y. Impact of different ratios of feedstock to liquid anaerobic digestion effluent on the performance and microbiome of solid-state anaerobic digesters digesting corn stover. *Bioresour. Technol.* **2016**, *200*, 744–752. [CrossRef]
67. Lin, L.; Yu, Z.; Li, Y. Sequential batch thermophilic solid-state anaerobic digestion of lignocellulosic biomass via recirculating digestate as inoculum—Part II: Microbial diversity and succession. *Bioresour. Technol.* **2017**, *241*, 1027–1035. [CrossRef]
68. Walker, C.B.; He, Z.; Yang, Z.K.; He, Q.; Zhou, J.; Voordouw, G.; Wall, J.D.; Arkin, A.P.; Hazen, T.C.; Stolyar, S.; et al. The electron transfer system of syntrophically grown *Desulfovibrio vulgaris*. *J. Bacteriol.* **2009**, *191*, 5793–5801. [CrossRef] [PubMed]
69. Barker, H.A.; Taha, S.M. *Clostridium kluyverii*, an Organism Concerned in the Formation of Caproic Acid from Ethyl Alcohol. *J. Bacteriol.* **1942**, *43*, 347–363. [CrossRef] [PubMed]

70. Böck, A. Fermentation. In *Encyclopedia of Microbiology*, 3rd ed.; Academic Press: Cambridge, MA, USA, 2009; pp. 132–144. [CrossRef]
71. Chang, S.C.; Shen, M.H.; Liu, C.Y.; Pu, C.M.; Hu, J.M.; Huang, C.J. A gut butyrate-producing bacterium *Butyricicoccus pullicaecorum* regulates short-chain fatty acid transporter and receptor to reduce the progression of 1,2-dimethylhydrazine-associated colorectal cancer. *Oncol. Lett.* **2020**, *20*, 327. [CrossRef]

fermentation

MDPI

Article

Exploitation of Cocoa Pod Residues for the Production of Antioxidants, Polyhydroxyalkanoates, and Ethanol

Licelander Hennessey Ramos [1,2], Miluska Cisneros-Yupanqui [3], Diana Vanessa Santisteban Soto [3], Anna Lante [3], Lorenzo Favaro [3], Sergio Casella [3] and Marina Basaglia [3,*]

1 GIPRONUT, Departamento de Química, Facultad de Ciencias, Universidad del Tolima, Ibagué 730006, Colombia; licelander@sena.edu.co
2 Área de Agroindustria, Servicio Nacional de Aprendizaje—SENA, km 5, vía El Espinal—Ibagué, Dindalito 733527, Colombia
3 DAFNAE—Department of Agronomy Food Natural Resources Animals and Environment, Viale dell'Università 16, 35020 Padova, Italy; dianavanessa@santistebansoto.univr.it (D.V.S.S.); anna.lante@unipd.it (A.L.); lorenzo.favaro@unipd.it (L.F.); sergio.casella@unipd.it (S.C.)
* Correspondence: marina.basaglia@unipd.it

Abstract: Cocoa pod husks (CPH) and cocoa bean shells (CBS) are the main by-products of the cocoa industry and a source of bioactive compounds. These residues are not completely used and thrown in the fields without any treatment, causing environmental problems. Looking for a holistic valorization, the aim of this work was first to deeply characterize CPH and CBS in their chemical composition, amino acid, and fatty acid profiles, as well as their application as antioxidants. CBS had a high level of protein (17.98% DM) and lipids (16.24% DM) compared with CPH (4.79 and 0.35% DM respectively). Glutamic acid and aspartic acid were the predominant amino acids. The total phenolic compounds (TPC) detected in the ethanolic extracts of CPH and CBS were similar to pyrogallol as the main detected polyphenol (72.57 mg/L). CBS ethanolic extract showed a higher antioxidant activity than CPH. Both extracts increased the oxidation stability of soybean oil by 48% (CPH) and 32% (CBS). In addition, alkaline pretreatment of CPH was found suitable for the release of 15.52 ± 0.78 g glucose/L after subsequent saccharification with the commercial enzyme Cellic®. CTec2. Alkaline hydrolyzed and saccharified CPH (Ahs-CPH) was assessed for the first time to obtain polyhydroxy alkanoate (PHAs) and bioethanol. Ahs-CPH allowed the growth of both *Cupriavidus necator* DSM 545 and *Saccharomyces cerevisiae* Fm17, well-known as PHA- and bioethanol-producing microbes, respectively. The obtained results suggest that such agricultural wastes have interesting characteristics with new potential industrial uses that could be a better alternative for the utilization of biomass generated as million tons of waste annually.

Keywords: *Cupriavidus necator* DSM 545; *Saccharomyces cerevisiae* Fm17; polyhydroxyalkanoate; bioethanol; cocoa by-products; oxidative stability

Citation: Ramos, L.H.; Cisneros-Yupanqui, M.; Santisteban Soto, D.V.; Lante, A.; Favaro, L.; Casella, S.; Basaglia, M. Exploitation of Cocoa Pod Residues for the Production of Antioxidants, Polyhydroxyalkanoates, and Ethanol. *Fermentation* **2023**, *9*, 843. https://doi.org/10.3390/fermentation9090843

Academic Editors: Jose Luis García-Morales and Francisco Jesús Fernández Morales

Received: 13 August 2023
Revised: 11 September 2023
Accepted: 12 September 2023
Published: 14 September 2023

1. Introduction

Theobroma cacao is a perennial tropical tree, belonging to the Sterculiaceae family, native to the tropical forests of the upper Amazon region [1]. The cultivation of cocoa is of high economic importance. Indeed, according to the International Cocoa Organization (ICCO), over fifty million people depend on cocoa for their livelihood with a global production capacity of 68% in Africa, 17% in Asia, and 15% in the Americas [2]. In 2021, the world's annual cocoa bean production was approximately 4.2 million metric tons with the Ivory Coast, Ghana, Indonesia, Brazil, Nigeria, Cameroon, Ecuador, and Colombia as major producers [3,4]. After two years, the cacao tree produces large pod-shaped fruits, with cocoa beans contained in the cocoa pod, consisting of the shell, kernel (or cotyledon), and germ. Around 75% of the total weight of the fruit is due to pods [5].

The cocoa shell is removed along with the germ before or after roasting and the broken cotyledon fragments, called nibs, free from the shell are used in the production of chocolate [6,7]. Cocoa pod husks (CPH) and cocoa beans shells (CBS) are the main by-products of the cocoa industry. After cocoa beans are extracted from the fruit, CPH are generated. It has been estimated that for each ton of dry beans produced, 10 tons of wet CPH are spawned. CBS represent up to 20% of cocoa beans and are generally underutilized. As a result, the growing market has brought the cocoa industry to massive production levels, causing the excessive generation of waste [8–10].

Due to the large quantities, managing cocoa waste is challenging. Large amounts of the generated biomass are burned by farmers, simply chopped and incorporated into the soil as a fertilizer, or left directly into the ground until their decomposition, generating smelly odors, causing soil contamination and the emission of greenhouse gasses [11]. Additionally, under certain climatic conditions, this biomass can cause an excessive proliferation of fungi, including potential pathogens [1]. For this crop, Good Agricultural Practices (GAP) recommend at least shredding and composting [12]. Therefore, it is mandatory to explore alternative utilizations of CPH and CBS. These materials are composed mainly of fiber, carbohydrates, lignin, proteins, and minerals [13–15]. Furthermore, although considered as wastes, these residues are extremely rich in biologically active molecules, often with nutraceutical properties such as phenolic compounds.

The Sustainable Development Goals Report 2023 by the United Nations emphasizes the immense potential for the utilization of lignocellulosic biomass "contributing to create a brighter future for all" [16]. Indeed, after the development of effective pretreatment methods and biocatalytic systems, renewable lignocellulosic biomass could be abundant feedstocks for the sustainable production of a wide range of molecules in an integrated biorefinery platform, as well reviewed by Yadav [17] and Mujtaba et al. [18].

Thus, in the framework of the circular economy, CPH and CBS are attracting increasing attention as possible starting materials to obtain added value products in the food sector, as well as in other contexts. Thanks to their chemical characteristics, several studies proposed strategies for the exploitation of cocoa wastes, pods, and husks as soil fertilizers, sources of pectin and polyphenols, animal feed, and in the production of soap or activated carbon [10]. The nutritional and biotechnological applications of cocoa wastes, together with health benefits and possible therapeutic roles in cancer, have been considerably reviewed by Sanchez et al. [19] and Cinar et al. [20].

The search for new applications of cocoa wastes integrates well into the current concepts of bioeconomy as postulated also by the European Union "Green Deal", aimed to restructure the industrial sector, promoting the circular economy to minimize negative environmental impacts, drastically reducing plastic pollution and greenhouse gas emissions [21].

As also determined in the present work, the chemical composition of CPH and CBS is mainly cellulose, hemicellulose, fibers, amino acid, and fatty acids; therefore, these cocoa residues grant large volumes of lignocellulosic biomass, a green and cheap resource material to develop a wide portfolio of bioproducts. Thus, due to the growing interest in replacing synthetic food antioxidants with natural ones [22], dry samples of CPH and CBS were extracted by ultrasound-assisted extraction (UAE) to recover their total phenolic compounds (TPC), successively tested in soybean oil as natural antioxidants to delay lipid oxidation. Besides, this residual material could be a promising feedstock to obtain medium to high value-added molecules [13–15] such as bioethanol [13], other biofuels, and bioplastics of medical, pharmaceutical, agricultural, or food interest [5].

Specifically, bio-based bioplastics are numbered among the most promising products obtainable from renewable sources. Indeed, the indiscriminate use of conventional fossil-derived plastics generates significant environmental pollution, especially due to their extremely difficult degradation; for this reason, these commodities are considered as a potential alternative [23]. Polyhydroxyalkanoate (PHAs) can be included among the most promising bioplastics due to their high biodegradability, biocompatibility, and versatility [24]. PHAs are natural, biodegradable, and compostable polyesters accumulated by numerous microorganisms in the form of intracellular granules and are formed from 600 to 35,000 monomer units of hydroxyalkanoic acids. PHAs have functional characteristics similar to those of many of the most common fossil-based plastics; at the same time, they are completely biodegradable in soil, fresh water, and marine environments and are both industrially and domestically compostable. LCA (Life Cycle Assessment) studies estimate that the replacement of 1 kg of fossil plastic with PHA could reduce the amount of CO_2 emitted by 2 kg [21].

In 2022, PHAs consisted of only 3.9% of global bioplastic production [25] and the commercialization of PHAs has been slowed down by their high production costs, largely due to carbon substrates which represent 30–40% of the total [26]. Therefore, to reduce the price and make PHAs more economically sustainable, it is crucial to search for novel and low-cost carbon-rich substrates [27,28]. Thus, cocoa wastes might represent an attractive alternative to the pure sugars generally used to obtain PHAs.

Among the biofuels, bioethanol is obtained from renewable sources and can be used as a fuel, chemical, or solvent. To date, first-generation bioethanol is mainly obtained from feedstocks that contain simple sugars or starch, raising concerns related to the use of soil, the consumption of water resources, and the subtraction of the grain to the production of food or feed. The International Energy Agency (2010) [29] reports that in the coming years, it will be mandatory to produce ethanol from the waste of the food industry and agroforestry, such as those from the cocoa manufacturing.

In order to attain the above bio-products, both waste streams were chemically and enzymatically treated to obtain substrates suitable for the growth of suitable microbial strains. Particularly, *Cupriavidus necator* DSM 545 for the production of PHAs and *Saccharomyces cerevisiae* Fm17 for the production of bioethanol.

Although CBS has been has been successfully evaluated as a substrate for PHAs synthesis by *Bacillus thermus* after sulfuric acid thermal treatment [30] and CPH for ethanol production [13,31], to the best of our knowledge, this is the first account reporting the exploitation of CPH and CBS through the recovery of phenolic compounds and the sustainable microbial production of PHAs and bioethanol after efficient mild pre-treatments and the enzymatic hydrolysis of cocoa waste.

2. Materials and Methods

2.1. Feedstocks and Chemicals

CPH and CBS were supplied by the Servicio Nacional de Aprendizaje, Espinal, Colombia. After harvesting, CPH and CBS were washed with distilled water, cut into small pieces, dried at 48 °C for 24 h, ground by a professional mill (MF 10 basic Microfine grinder IKA-Werke, Staufen, Germany), finally sieved through a 500 μm sieve and stored at 4 °C. The material was derived from *T. cacao* clone IMC-67 (Iquitos Maranon Collection), cropped in Espinal, Tolima, Colombia, in 2010. The site is located at a latitude of 4°10′10″ N, a longitude of 74°55′52″ W, and an elevation of 348 m. Soybean commercial oil was purchased in a local market (Padova, Italy). All chemicals were of analytical grade and were obtained from Sigma-Aldrich, unless stated otherwise.

2.2. Chemical Analyses of CPH and CBS

CPH and CBS were analyzed in terms of ash, starch, hemicellulose, cellulose, lignin, and protein content, according to international standard methods, as described in [19]. In short, total ash was determined by calcinating the residues at 550 °C as described in methods 942.05 and 934.01 by Association of Official Analytical Chemists (AOAC). Cellulose, hemicellulose, and lignin were measured according to Van Soest et al. methodology [32]. Starch contents were calculated according to AOAC method 920.40. Total nitrogen was determined by the Kjeldahl method, followed by the protein calculation using the general factor of 6.25 (AOAC, Method 981.10). The dry matter (DM) content was obtained by drying triplicate samples for 48 h at 100 °C in an oven. The amino acid and fatty acid profiles were determined as previously reported [20,21]. The dry matter (DM) content was obtained by drying triplicate samples for 48 h at 100 °C in an oven. The amino acid and fatty acid profiles were determined as previously reported [33,34].

2.3. Phenolic Extraction of Cocoa Pods and Shells

Two grams of dry CPH and CBS were solubilized in 10 mL of 70% v/v ethanol, mixed for 20 min using an orbital shaker at room temperature, and subjected to ultrasound for 2 min, at intervals of 30 s with SONOPULS ultrasonic homogenizer at 20 kHz ± 500 Hz frequency. The KE76 tip was used for the sonication. The samples were then centrifuged for 5 min at 9500× g at 4 °C (Hettich Zentrifugen, MOD: Universal 320R, Tuttlingen, Germany). The supernatants (ethanolic extracts) were recovered, filtered using Whatman paper No. 1, and stored at −18 °C until used.

2.4. Quantification of the Total Phenolic Compounds (TPC)

The TPC were quantified in the phenolic extracts using the FolinCiocalteu method [35]. Briefly, 1 mL of diluted ethanolic extract was mixed with 5 mL of $NaCO_3$ 10%, containing NaOH 1 M and 500 µL of Folin–Ciocalteau reagent previously diluted twice in distilled water. A blank solution with the dilution solvent was also set up. After 30 min under darkness, the samples were filtered using 0.45-µm Millipore acetate cellulose filters (Merk Life Science S.r.l., Milano, Italy). Hence, the absorbance was measured at 650 nm using a Varian Carry 50 Bio UV/Vis spectrophotometer. The results were expressed as mg of gallic acid equivalent per g of sample (mg GA/g).

2.5. Phenolic Compounds Identification

The phenolic profile of CPH and CBS was determined by HPLC analysis [35] using a Thermo Finnigan SpectraSystem UV6000LP (Thermo Finnigan, San Jose, CA, USA) HPLC system with diode array detector. Before injection in the column, the samples were filtered with a 0.22-µm cellulose acetate filter (Merk Life Science S.r.l., Milano, Italy). The phenols present in the sample were identified based on the retention time of the corresponding commercial standards (pyrogallol, hydroxybenzoic acid, protocatechuic acid, caffeic acid, syringic acid, and ferulic acid), previously solubilized in absolute methanol using the SupelcosilTM LC-18 column at the following operating conditions: mobile phase, 18 mL n-butanol (solvent A)/1.5 mL 50% v/v acetic acid (solvent B); flow rate, 0.6 mL/min; isocratic flow; wavelength, 214 nm, 275 nm, and 310 nm; temperature, 25 °C; and running time, 60 min.

2.6. Antioxidant Activity

The assay was performed by using the ferric reducing antioxidant potential (FRAP) method in agreement with Stratil et al. [36]. The FRAP reagent was prepared by mixing 2.5 mL of 0.01 M of TPTZ in HCl 40 mM, 2.5 mL of aqueous solution of 0.02 M $FeCl_3$, and 25 mL of 0.3 M of sodium acetate buffer (0.2 M sodium acetate/0.2 M acetic acid). A FRAP volume of 900 µL was mixed with 100 µL of sample and incubated at 37 °C for 30 min. Also, a blank solution was prepared with the dilution solvent. The absorbance was measured at 593 nm using a Varian Carry 50 Bio UV/Vis spectrophotometer. The results were expressed as mg of 6-hydroxy-2,5,7,8-tetramethylchroman-2-carboxylic acid (Trolox) equivalent per g of sample (mg TE/g).

2.7. Determination of the Oxidative Stability

Ethanol extracts from CPH and CBS at 40% were added to soybean oils and mixed for 20 min using an orbital shaker. After a sonication treatment for 2 min (4 intervals of 30 s), oxidative stability was evaluated in both oil mixtures using the official Rancimat method (AOCS, 2012), according to the procedure previously described by Tinello et al. [37]. As controls, soybean oil without supplementation was introduced in the Rancimat assay. A quantity of 3 g of samples (control or supplemented oil) was weighed in the Rancimat apparatus (Metrohm, model 743, Herisau, Switzerland) and subjected to a stream of air at the rate of 20 L/h kept at a constant temperature of 110 °C, causing an accelerated oxidation process. The oxidative stability was expressed as the induction time (IT) corresponding to the time (h) at which the water conductivity (µS/min) starts increasing because of the production of 11 compounds involved in the lipid oxidation. The antioxidant activity index (AAI) was calculated by the following equation:

$$AAI = \frac{\text{IT of oil with GPP}}{\text{IT of oil without GPP}}$$

2.8. Pretreatments and Enzymatic Hydrolysis

Pretreatments of lignocellulosic substrates are needed to break the structure of lignin and remove the hemicellulose structure to increase exposure to cellulolytic enzymes in the subsequent hydrolysis phases. Therefore, before enzymatic hydrolysis, pretreatments with hydrogen peroxide and alkaline were evaluated:

To determine the best conditions for hydrogen peroxide pretreatment, 2.5 g of CPH in 100 mL flasks were treated with increasing doses of H_2O_2 (0, 2.5, 5, and 7.5% v/w). Water was added to a total volume of 25 mL. The flasks were incubated in an agitated bath (Mod. SW22 Julabo, Seelbach, Germany) at 55° C. After 4 h, the pH of the pretreated CPH were adjusted to 5.0 using concentrated HCl before the following saccharification phase.

For the alkaline pretreatment, 2.5 g CPH were incubated in a 100 mL flask with 25 mL of 4% NaOH. After boiling for 30 min, the pretreated CPHs were centrifuged at 5000 rpm for 15 min and the supernatant discarded. The pellet was washed twice with distilled water, the pH was adjusted with 0.5 M HCl at 7.0, washed again, dried at 100 for 16 h, and finally resuspended in 25 mL of 50 mM citrate buffer [38].

For the following enzymatic saccharification, Cellic® CTec2 (Novozymes, Bagsvaerd, Denmark) 12% w/w (g/g cellulose) was added to hydrogen peroxide or alkaline pretreated CPH, according to the supplier's instructions. The suspensions were stirred (100 rpm) at 50 °C. Samples (2 mL) were withdrawn after 0, 24, 48, and 72 h and boiled for 10 min to inactivate the enzymes. A blank without the enzyme addition was used to assess autohydrolysis. The suspensions were then centrifuged and the glucose content in the supernatants (pretreated CPH) was measured using the Megazyme glucose assay kit in a UV-Vis spectrophotometer (Megazyme International Ireland Ltd., Wicklow, Ireland) following instructions of the manufacturer. Optical densities were converted in g glucose/L.

On the basis of the glucose released from cellulose after treatment with Cellic® CTec2, the saccharification degree after alkaline or hydrogen peroxide pretreatment was calculated according to the following equation.

$$Ds_{glucan} = \frac{[\text{glucose g/L}] \times 0.9}{[\text{cellulose g/L}]} \times 100\%$$

In addition, reference experiments were performed using Cellic® CTec2 with non-pretreated CPH.

2.9. Microbial Strains

C. necator DSM 545, one of the most efficient PHAs producers and *S. cerevisiae* Fm17, outperforming bioethanol yeast, was obtained from DSMZ (Deutsche Sammlung von Mikroorganismen und Zellkulturen, Braunschweig, Germany) and the collection of DAFNAE (University of Padova, Italy), respectively [39,40].

2.10. PHAs Production by C. necator DSM 545 from Alkaline Pretreated CPH Hydrolyzate (Ahs-CPH)

Inocula of *C. necator* DSM545 were obtained in MM medium [41] amended with 30 g/L glucose in aerobic conditions at 30 °C under shaking (145 rpm) for 24 h. Bacteria were then centrifuged (5500 rpm for 15 min), washed twice with 0.9% NaCl to remove any carbon sources, and re-suspended in sterile 0.9% NaCl. The experiments were conducted in 125 mL flasks containing 30 mL of Ahs-CPH, sterilized by 0.22 µm Whatman filters. In some flasks, the salts contained in MM medium were added to the supernatants. After inoculation, the flasks were incubated at 30 °C and agitated at 150 rpm. For an assessment of the cell growth, samples were withdrawn and OD_{600nm} was monitored by a spectrophotometer (Spectronic® Genesys™ 2PC, Vimercate, Italy). After 72 h, cultures were centrifuged, the pellets were frozen at −80 °C, and lyophilized for cell dry mass (CDM) determination and analysis of PHAs.

2.11. Cell Dry Matter and PHAs Analysis

To determine the cell dry matter (CDM), the freeze-dried bacterial pellets were weighed. PHAs were analyzed by Gas chromatography according to the protocol described by Braunegg et al. [42]. In brief, 10 mg of freeze-dried cells were treated for 4 h at 100 °C in a mixture of 2 mL of methanol containing 3% H_2SO_4 and 2 mL of chloroform. The resulting methyl esters of hydroxy alkanoic acids were analyzed by gas chromatography as previously described [43]. A Thermo Finnigan Trace GC gas chromatograph (Mundelein, Illinois, USA) was used with a AT-WAX fused silica capillary column (Alltech Italia s.r.l., Milan, Italy) and a flame ionization. The carrier gas was helium (He) and the operating temperatures during the analysis were: 250 °C for the injection chamber, 270 °C for the detector, and 150 °C for the oven. The internal standard was benzoic acid (2.5 g/L), while the external standard was 3-hydroxybutyric acid [43]. The results obtained were expressed as a percentage of PHAs of CDM.

2.12. Bioethanol Production by *S. cerevisiae* Fm17 from Ahs-CPH

S. cerevisiae Fm17 pre-cultures were obtained in 500 mL flasks containing 250 mL YPD medium incubated at 30 °C on a rotary shaker set at 130 rpm for 24 h. The pre-culture was centrifuged at 5500 rpm for 10 min and the pellet was washed twice in sterile demineralized water, re-suspended in 5 mL, and used as inoculum. Then, 50 mL of Ahs-CPH was transferred in 60 mL serum bottles and inoculated with yeast at a concentration of around 5×10^8 CFU/mL. Ampicillin and streptomycin (each 100 µg/mL) were added to avoid possible bacterial contamination. Rubber stoppers were used to set up oxygen-limited conditions and a needle was inserted for CO_2 removal. Bottles were then incubated at 150 rpm and 30 °C; 2 mL of samples was withdrawn at 0, 4, 24, 48, and 72 h for subsequent chemical analysis. Control fermentations were performed in YPD with 15 g/L glucose as a carbon source.

Samples from bioethanol production were analyzed for ethanol and residual glucose through liquid chromatography using a Shimadzu Nexera HPLC system equipped with a RID-10A refractive index detector (Shimadzu, Kyoto, Japan), and a Phenomenex Rezex ROA-Organic Acid H+ (8%) column (300 mm × 7.8 mm) was used as described in Cagnin et al. [44]. The column temperature was set at 65 °C and the flow rate was 0.6 mL/min using isocratic elution, with 0.005 M H_2SO_4 as a mobile phase. The ethanol yield (g of ethanol/g of used glucose equivalent) was calculated based on the amount of glucose utilized during the fermentation and compared to the maximum theoretical yield of 0.51 g of ethanol/g of utilized glucose.

2.13. Statistical Analysis

The results are presented as mean ± standard deviation of three independent experiments (n = 3) by using an analysis of variance (ANOVA) one way, followed by the post-hoc Duncan test ($p < 0.05$) using Statgraphics Centurion XIX (StatPoint Inc., Rockville, MD, USA).

3. Results and Discussion

3.1. Chemical Composition of Cocoa Pods Husk

Table 1 shows the proximate composition of the CBS and CPH. This step was crucial to trigger a possible valorization project of these by-products as potential sources of ingredients which can be used in food, as well as animal feed, and as cheap substrates for microbial fermentations. The dry matter content of the CPH and CBS was found to be 91.84 and 94.88% respectively. Both by-products are rich in non-starchy polysaccharides with cellulose contents of 22.32% and 12.30% of DM, respectively. Hemicellulose was around 10% of dry matter for both residues and lignin 21.15 and 14.80% for the CPH and CBS, respectively. All values are in agreement with those previously described by de Souza [15] and Mendoza- Meneses [45].

Table 1. Chemical composition (% of dry matter) of the CPH and CBS. Results of chemical analyses are the means of three replicates with standard deviation below 5%.

	CPH	CBS
Cellulose	22.32	12.30
Hemicellulose	10.10	10.07
Acid Detergent Lignin (ADL)	21.15	14.80
Neutral Detergent Fiber (NDF)	53.85	37.91
Acid Detergent Fiber (ADF)	43.75	27.85
Acid Insoluble Ash (AIA)	0.28	0.66
Protein	4.79	17.98
Ashes	10.41	7.28
Lipids	0.35	16.24

The CPH here studied contained low amounts of lipids (0.35%), while other authors reported contents reaching values of 1.5–2.34%.

A high ash content (10.41% and 7.28%) was detected in both residues. Other authors reported that the ash content could be referred as mineral content in some foods [46], suggesting an additional possible utilization of these residues as a mineral boost. Regarding other components, the moisture (8.16% for the CPH) and protein (4.79% for the CPH) contents are in agreement with previous reports [46,47]. The total fiber, measured by the NDF method, showed values of 53.85 and 37.91%, for the CPH and CBS, respectively. Martinez et al. [47] account total dietary fiber (TDF) values for cocoa pods in congruence with this study (55.99–56.10%). Vrismann et al. [8] reported a lower TDF for cocoa pods (36.6%), and lignin values similar to those found in this study (21.4%). Regarding the CBS, the NDF, ADF, and ADL values are similar to those found by Campione et al. [48].

In conclusion, although the chemical composition of the cocoa wastes can vary according to the plant varieties, growing conditions, soil, and type of material, the results here obtained were comparable to those shown in previous studies [7,38,49].

Regarding protein, the values here found are in the range presented by other authors (15.59–20.9%) [46,50,51] with the highest content in the CBS (18%), which could therefore represent a potential cheap source of sustainable vegetal protein.

Results concerning amino acids and fatty acid profiles are reported in Tables S1 and S2, respectively. Table S1 indicates that, compared to the CBS, CPH has a lower content of acidic (aspartic and glutamic acid) and basic (arginine and lysine) amino acids. Proline and valine are more present in the CBS as well as other aromatic amino acids (phenylalanine, tyrosine, histidine, and tryptophan). These results are consistent with those of other authors that found aspartic and glutamic acids as predominant in the CPH [52,53]. The CBS contains higher quantities of essential amino acids than the CPH with leucine as the most concentrated, reaching values of 243.17 and 870.03 mg/100 g for the CPH and CBS, respectively.

Regarding the lipid component (Table S2), the predominant fatty acids are palmitic, stearic, oleic, and linoleic acids in both residues. The saturated fatty acid (myristic, palmitic, and stearic) content is similar in both the CPH and CBS. Regarding unsaturated fatty acids, the CBS has a higher content of oleic, while the CPH contains more linoleic acid. These results are similar to those of the literature [50,54,55], which reported that the main fatty acids found in cocoa bean shells were oleic, stearic, and palmitic acids. Further studies showed that linoleic acid was the predominant component of the oil from cocoa pods [56] or in the CPH powder extract [57].

3.2. Quantification and Identification of Total Phenolic Compounds (TPC) and Antioxidant Activity

The amounts of TPC in the ethanolic extracts of the CPH and CBS are shown in Table 2. According to various authors, the TPC extracted from the CPH with different solvents and conditions range between 2.07 and 107.3 mg GAE/g [58–62]. The values found in the present study are within the mentioned ranges and are not significantly different ($p > 0.05$) between the CPH and CBS (10.08 ± 1.40 mg GA/g and 13.04 ± 0.10 mg GA/g, respectively).

Table 2. Total phenolic compounds (mg GA/g) and antioxidant activity (mg TE/g) from the CPH and CBS. Mean values (n = 3) and standard deviations are presented. Different letters within the same column indicate significant differences, according to ANOVA (one-way) and the Duncan test ($p < 0.05$).

Residue	TPC (mg GA/g)	Antioxidant Activity (mg TE/g)
CPH	10.08 ± 1.40^{a}	9.93 ± 0.38^{b}
CBS	13.04 ± 0.10^{a}	16.24 ± 0.61^{a}

TPC: total phenolic compounds, GA: gallic acid, TE: Trolox.

However, higher or lower values were recorded by several previous studies, clearly affected by cultivation area, cocoa variety, postharvest processes, extraction methods, used solvent, etc., [46,58,60,63,64]. For example, Sotelo et al. [61] extracted significantly ($p < 0.05$) more phenols using ultrasound (23.0 ± 0.9 mg GAE/g) in comparison to a conventional method (16.4 ± 0.41 mg GAE/g).

Table 3 shows the phenolic compounds identified in the CPH and CBS by HPLC. In this study, the main detected polyphenols in both substrates were pyrogallol (72.567 mg/L) and vanillic acid (7.207 mg/L). p-hydroxybenzoic acid (2.467 mg/L) and syringic acid (0.765 mg/L) were found only in the CPH. The Antioxidant activity (AOA) was 9.93 ± 0.38 and 16.24 ± 0.61 mg TE/g for the CPH and CBS, respectively (Table 4). However, the concentration of the solvent may deeply affect the results [59,65]. In this work, 70% ethanol was used as a solvent and high concentrations of pyrogallol were found in the CBS (164.474 mg/L) and in CPH (72.567 mg/L). The higher antioxidant capacity of the CBS could be therefore attributed to this molecule.

Table 3. Phenolic compounds characterization of the CBS and CPH by HPLC. The data are expressed in mg/L.

Compound	CBS	CPH
Pyrogallol	164.47	72.57
Syringic acid	8.59	0.77
p-hydroxybenzoic acid	-	2.47
Vanillic acid	-	7.21

Table 4. Induction time (IT) and Antioxidant Activity index (AAI) of the CPH and CBS. Mean values (n = 3) and standard deviations are presented. Different letters within the same column indicate significant differences. according to ANOVA (one-way) and the Duncan test ($p < 0.05$). IT refers to the time (h) at the break point of the two extrapolated straight parts of the curve obtained by Rancimat apparatus. AAI = IT of corn oil with antioxidant/IT of soybean oil (control).

Sample	IT (hours)	AAI (mg TE/g)
CPH	9.46 ± 0.04^{a}	1.48
CPS	8.45 ± 0.10^{b}	1.32
Control (soybean oil)	6.40 ± 0.06^{c}	1.00

3.3. Oxidative Stability

As reported in Table 4, the oxidative stability of soybean oil supplemented with the ethanolic extracts of the CPH and CBS was evaluated by measuring the induction time (IT) and the antioxidant activity index (AAI). The IT of the cocoa pod's extract (9.46 ± 0.04 h) was significantly longer than the cocoa shell's extract (8.45 ± 0.10 h), both being significantly longer than the control, consisting of soybean oil without supplementation (6.40 ± 0.06 h).

Indeed, according to the AIA values, the CPH achieved an increase of 48% and the cocoa shell of 32%, as compared to the control ($p < 0.05$).

Similarly, Boungo Teboukeu et al. [65] found that the CPH phenolic extract was effective in delaying the oxidation of palm oil during heating at 180 °C and reported a maximum value of AAI of 1.20 when the extract was added at 200 ppm. Other similar results were obtained with ethanolic extracts from by-products such as rapeseed [35] and red chicory powder [66] on soybean oil or grape pomace powder on corn oil [67].

These findings further support that the exploitation of raw materials of residual origin as sources of cheap and renewable proteins and antioxidants plays an essential role in the emerging eco-products and healthy ingredients market. Although additional research is necessary to determine the optimal conditions for an efficient extraction of antioxidant compounds from these by-products, the ultrasound application resulted to be a green solution for the environmentally friendly recovery of these molecules.

3.4. Pretreatment of Cocoa Residues

Based on the analyses reported in Table 1, the amount of glucose that could be released from the hydrolysis of cellulose in the CBS would be too low to support microbial growth. Therefore, pretreatment trials were conducted exclusively on the CPH.

To better expose the feedstock cellulose to the successive enzymatic hydrolysis, two pretreatments of the CPH were attempted: increasing amounts of hydrogen peroxide or 4% *w*/*v* NaOH. After pretreatments, samples were enzymatically saccharified with Cellic® CTec2 and the released glucose was then determined. This hexose is one of the preferred carbon sources for both the microorganisms used in this study (*C. necator* and *S. cerevisiae*) and both microbes cannot use pentoses.

3.4.1. Hydrogen Peroxide (H_2O_2) Pretreatment of CPH

For the reason discussed above, increasing amounts of H_2O_2 were applied to maximize the release of glucose. The largest amount of this sugar (6.5 g/L) was obtained with 7.5% hydrogen peroxide and 24 h of incubation, with a percentage of saccharification of $21.27 \pm 0.29\%$ (Figure 1). A lower concentration of H_2O_2 resulted in quantities of released sugar only slightly above those obtained from the non-treated samples. A kinetic model was developed to describe the dynamics of glucose release after treatment with H_2O_2 and the maximum yield for glucose was 43.49% [68]. However, the H_2O_2 pretreatment conditions were very different; these authors operated in a reactor at 150 °C for 6 h while, in this work, H_2O_2 incubations were conducted at 55 °C for 4 h in a thermal bath. High temperatures seem to favor the release of sugars from the CPH and additional studies would be necessary to determine whether in harsh conditions inhibiting substances that may affect subsequent fermentation processes are generated. Indeed, recent studies reported that high temperature pretreatments of lignocellulose can generate furfurals and 5-hydroxymethylfurfural, which are inhibitory compounds for a variety of microorganisms [69]. In addition, high temperatures for extended periods could be considered uneconomical and not environmentally friendly. Thus, although raising H_2O_2 resulted in increasing, but still low, amounts of enzymatically released glucose, the H_2O_2 pretreatment was not considered suitable for a subsequent sustainable biotechnological use of CPHs as feedstocks.

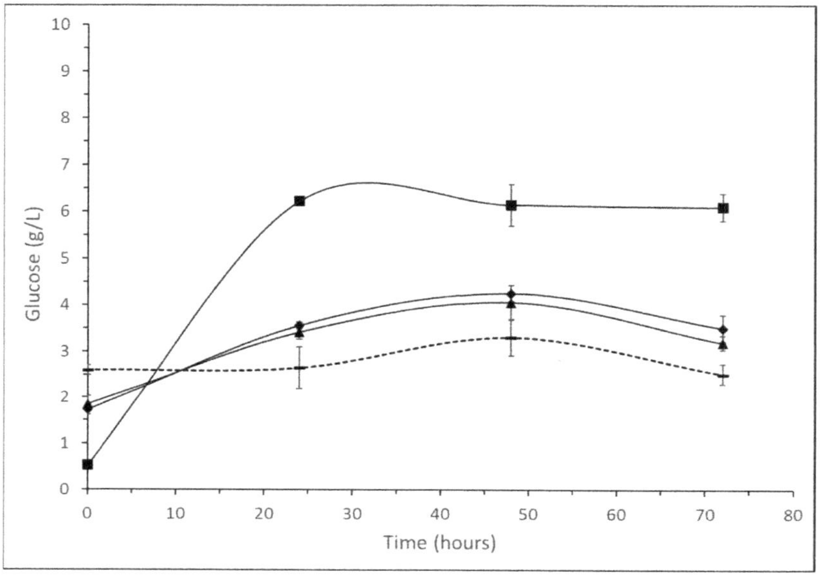

Figure 1. Effect of H_2O_2 pretreatments of CPH on the release of glucose after enzymatic hydrolysis with Cellic® CTec2. H_2O_2 2.5% *v*/*w* (▲). H_2O_2 5.0% *v*/*w* (♦). H_2O_2 7.5% *v*/*w* (■). No H_2O_2 added (-). Mean values (n = 3) and standard deviations are presented.

3.4.2. Alkaline (NaOH) Pretreatment of CPH

Alkaline pretreatment has been demonstrated as suitable to solubilize lignin and, partially, the hemicellulose, and it is traditionally used in pulp processing. This application increases the internal surface of cellulose, contributes to reduce its crystallinity, and, thus, makes the polysaccharide more accessible to further enzymatic attack by cellulases [70]. When the CPH was pretreated with NaOH and the subsequent enzymatic hydrolysis was applied, a higher amount of glucose (15.52 ± 0.78 g/L) was found after 72 h of saccharification; this value corresponds to a degree of saccharification of 62.26%. Non-pretreated samples did not show a spontaneous glucose release (Figure 2). Moreover, comparing the yields with those obtained on H_2O_2 pretreated substrates, alkaline pretreatment revealed to be more effective in making cellulose more accessible to further enzymatic attack by cellulases.

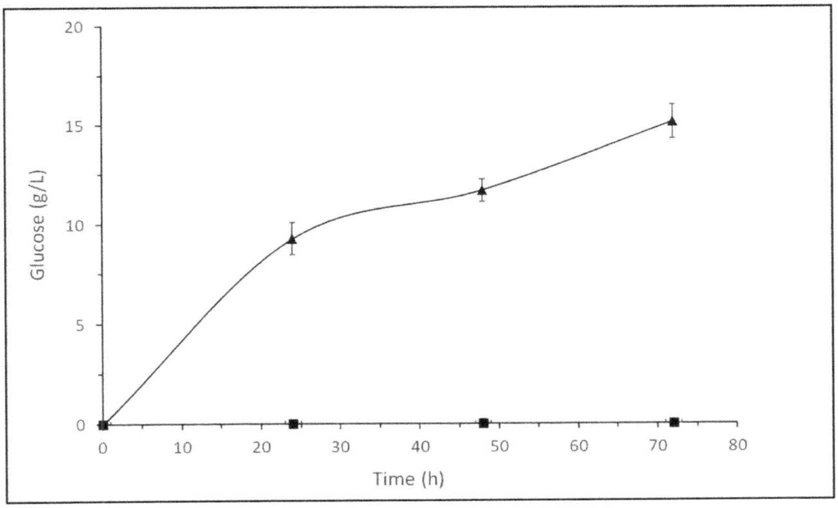

Figure 2. Effect of NaOH pretreatments of the CPH on the release of glucose during enzymatic hydrolysis with Cellic® CTec2. NaOH 4% *v/w* pretreated (▲). No NaOH added (■). Mean values (n = 3) and standard deviations are presented.

These results are partially in agreement with those of other authors that used the harsher NaOH autoclave-assisted hydrolysis. For example, with this pretreatment, Sarmiento-Vasquez et al [71] found a higher maximum concentration of 60.5 g/L of glucose and a yield of 275 mg glucose/g of CPH. Hernández-Mendoza et al. [38] obtained a syrup with 66.80 g/L of the reducing sugars.

Other authors used acidic hydrolysis on cocoa residual biomass but with lower results. For example, Shet et al. [72] carried out an acidic optimized pretreatment with HCl 3.6 M), reporting an initial sugar concentration of only 4.09 g/L.

Overall, in this work, around 160 g of glucose was recovered per g of CPH. Non-pretreated samples did not show spontaneous glucose release (Figure 2), thus confirming that a mild alkaline pretreatment of the CPH is effective in making cellulose more accessible to further enzymatic attack by cellulases.

3.5. PHAs and Bioethanol Production from Alkaline Pretreated Saccharified CPH (Ahs-CPH)

3.5.1. PHAs

C. necator DSM 545 growth and PHAs accumulation were assessed on media containing Ahs-CPH with or without MM salts (Figure 3). Reference growths were performed with amounts of glucose comparable with those contained in the Ahs-CPH. Although pure CPH saccharified hydrolysate contains around 15 g/L glucose (Figure 2), it poorly sustains the development of *C. necator*. On the other hand, the amendment of the hydrolyzate with MM salts resulted in bacterial growth similar to that obtained with MM + 15 g/L glucose, indicating that the hydrolyzate is lacking the essential nutrients contained in MM.

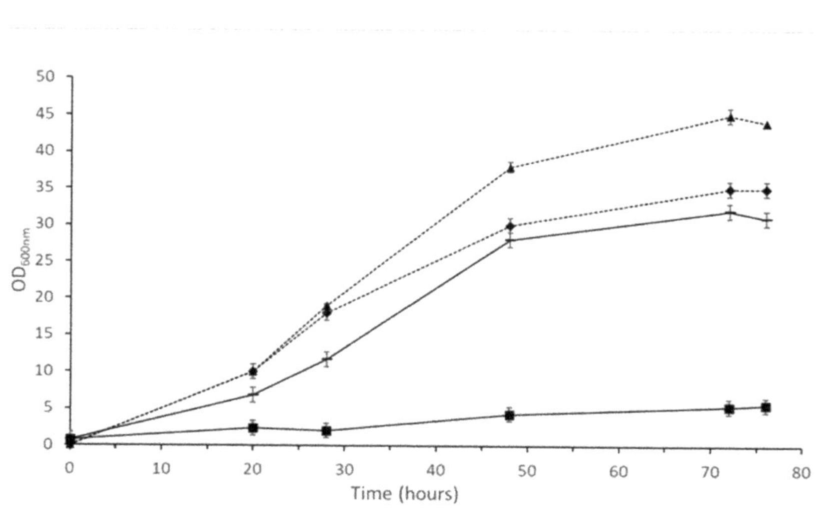

Figure 3. Growth of *C. necator* DSM 545 on media containing Ahs-CPH (solid line) or glucose (dashed line) as a carbon source. MM + 20 g/L glucose (▲). MM + 15 g/L glucose (♦). Ahs-CPH (■). Ahs-CPH + MM salts (-). Mean values (n = 3) and standard deviations are presented.

After 76 h, the culture broths were centrifuged, the pellets collected, and the PHAs content determined (Table 5). As previously reported [73], the PHAs accumulated by *C. necator* DSM 545 on glucose ranged from 72.00 to 74.60% of CDM, confirming the ability of this strain to efficiently accumulate PHAs from this monosaccharide. On Ahs-CPH, the PHAs reached 51.30% of CDM. When the MM salts were supplemented to Ahs-CPH, the percentage of the PHAs increased to 58.60% showing once again that essential nutrients were not present in the substrate.

Table 5. PHAs accumulation of *C. necator* DSM 545 on Ahs-CPH, MM + Ahs-CPH, MM + 15 g/L glucose, and MM + 20 g/L glucose after 76 h incubation. Mean values (n = 3) and standard deviations are presented.

Medium	PHAs (% CDM)
Ahs-CPH	51.30 ± 2.83
MM+ Ahs-CPH	58.60 ± 4.95
MM+ glucose 15 g/L	72.00 ± 1.00
MM + glucose 20 g/L	74.60 ± 0.28

CDM: cell dry matter.

Though studies specifically focused on PHAs production from the CPH are not available in the literature, thus limiting the discussion, the PHAs accumulated by *C. necator* on the Ahs-CPH are comparable with those obtained from other agro-industrial substrates and microorganisms.

For example, Brojanigo et al. [39] reported PHB values up to 44% of CDM using *C. necator* DSM 545 and enzyme-treated broken rice and, for the first time, reported on the Consolidated Bioprocessing of PHB from broken rice (43% CDM) and purple sweet potato waste (36% CDM) [74] by using a specifically engineered *C. necator* DSM 545 strain. *C. necator* H16 was screened for PHAs production from bagasse hydrolyzate and wheat bran hydrolyzate by Brodin et al. and Annamalai & Sivakumar, who found PHAs contents of 54 and 66% of CDM, respectively [75,76]. De Souza et al. [77] found 57.8% PHAs accumulated by *Bacillus megaterium* using pretreated corn husk hydrolyzate as a carbon source. Up to

62% PHAs of CDM produced by *Burkholderia cepacia* from sugarcane bagasse hydrolyzate were also reported [78].

Although further studies are required to optimize the processes, CPH could be considered a promising substrate for the growth of and PHAs synthesis by *C. necator*.

3.5.2. Bioethanol

To test the possible exploitation of CPHs for bioethanol production, the growth and ethanol performances by *S. cerevisiae* Fm17 were assessed for 72 h in a medium with Ahs-CPH (Figure 4A). A glucose (15 g/L) broth was included as benchmark.

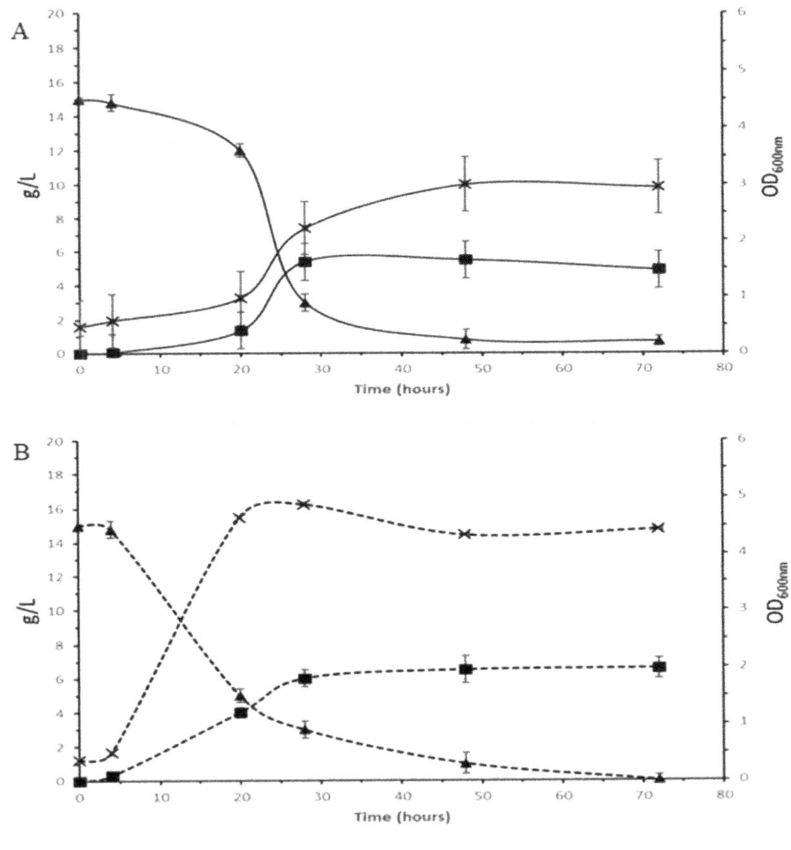

Figure 4. Growth (X), glucose consumption (▲) and ethanol production (■) by *S. cerevisiae* FM17 in Ahs-CPH ((**A**) solid line) and glucose benchmark broth ((**B**) dashed line). Results are the means of three replicates and the standard deviations are reported.

Ethanol production started within the first twenty hours of fermentation in both media. With Ahs-CPH, *S. cerevisiae* Fm17 metabolized 95.7% of released glucose (Figure 4A, Table 6 within the first 50 h of incubation; the ethanol concentration increased, reaching a steady state approximately 30–40 h after inoculation, with a maximum value of 5.50 g/L at 48 h.

Table 6. Consumed sugars. Ethanol and ethanol yields by *S. cerevisiae* Fm17 in glucose and Ahs-CPH. $Y_{EtOH/S}$ is the ethanol yield per gram of consumed substrate calculated on the highest ethanol production. Results are the means of three replicates and, when relevant, the standard deviations are reported.

Parameter	Growth on Glucose	Growth on Ahs-CPH
Sugars concentration (g/L)	15.00 ± 0.08	15.15 ± 0.85
Consumed sugars (%)	100.0	95.3
Highest Ethanol (g/L)	6.60 ± 0.60	5.50 ± 0.30
$Y_{EtOH/S}$	0.44	0.38
% theoretical yield	86	74

In pure glucose with a concentration similar to the amount used in fermentation with Ahs-CPH, the ethanol level (6.00 g/L) was slightly higher than that of Ahs-CPH (5.50 g/L) (Figure 4B and Table 6), probably due to the presence of inhibitors or to the lack of some nutrients in the hydrolyzate. While with pure glucose 100% of sugars were consumed by yeast, with the hydrolyzate, a small amount of sugar remained in the exhausted broth. The maximum ethanol yield (% of the theoretical) and the ethanol yield/consumed carbon (g ethanol/g glucose) obtained with the hydrolyzate were 74% and 0.38 g/g respectively, which is slightly lower if compared with those obtained with pure glucose (86% and 0.44 g/g) (Table 6).

Other authors found a bioethanol yield of 13.66 g/L by *S. cerevisiae* but using a CPH hydrolysate by 1 M HCl, representing a bioconversion efficiency of 87% at 26 h of fermentation [31].

In this work, the pretreatment was performed with 100 g of CPHs per liter and, after saccharification and fermentation, 5.5 g/L of ethanol was obtained. As such, the yield of bioethanol per unit of feedstock should be 0.055 g of ethanol per g of CPH.

Similar results were obtained by Valladares-Diestra et al. [13] using a CPH's hydrothermal pretreatment assisted with citric acid. This author reports an overall yield of 0.07 g of ethanol per g of CPH, but only 0.042 g of ethanol was raised by *S. cerevisiae*.

In optimized conditions, Hernandez-Mendoza et al. [38] obtained from CPHs 18.06 g/L of ethanol, but after a harsher treatment with 5% NaOH and 30 min at 120 °C in autoclave.

Cocoa by-products are generated in large amounts and theoretically could be used as a potential substrate for bioethanol production. The synthesis of bioethanol from CPHs via direct fermentation using *Zymomonas mobilis* was successfully obtained by Yogaswara et al. [79], but with a low value of conversion and a small maximum reaction velocity. The microbial production of bioethanol from lignocellulosic cocoa residues is in fact hardened by the accessibility of fermentable sugars contained in the recalcitrant structure of cellulose and hemicellulose polymers. Thus, pretreatments aimed to facilitate the enzymatic saccharification of biomass are necessary [5]. The acid pretreatment of CPH is easy, cheap, and efficient, but generates fermentation inhibitors such as furfural and hydroxymethylfurfural, which are toxic to the metabolism of fermentative microorganisms [73]. On the contrary, the use of alkali on CPHs in mild conditions, like those used in this work, seems to generate glucose, minimally affecting yeast fermentation. Nevertheless, the glucose concentrations obtained in this study should be considered as a proof-of-concept and need specific investigations dealing with substrate loadings, NaOH concentrations, mixing, as well as enzyme dosage optimization. As such, ethanol levels would become more and more profitable with concentrations higher than 4% (*v*/*v*) [80].

4. Conclusions

This paper demonstrated for the first time that both the CPH and CBS can be efficiently converted into a cluster of valuable products. This biorefinery approach, indeed, resulted in the valorization of cocoa by-products into bioactive compounds, such as phenolic compounds with promising industrial traits, as well as PHAs and bioethanol. Ultrasound-assisted extracts from the CPH and CBS showed a relevant antioxidant ability on the soybean oil here used as a model substrate; this activity is probably due to pyrogallol revealed by the chemical analyses. Furthermore, a mild alkaline pretreatment with 4% NaOH and the subsequent enzymatic saccharification of the CPH and CBS resulted in the release of glucose to support the synthesis of ethanol and PHAs by *S. cerevisiae* Fm17 and *C. necator* DSM 545, respectively.

Further insights are needed to fully exploit both tested waste streams and will deal with up-scaling, the optimization of process parameters, as well as techno-economical analyses. This perspective will be of great impact to boost bioeconomy applications also in developing countries.

Supplementary Materials: The following supporting information can be downloaded at: https://www.mdpi.com/article/10.3390/fermentation9090843/s1, Table S1. Total amino acid profile of CBS and CPH (mg/100 g DM). Results of chemical analyses are the means of three replicates with standard deviation below 5%. Table S2. Fatty acid profile of cocoa pod and cocoa shell (mg/100 g DM) Results are the means of three replicates with standard deviation below 5%.

Author Contributions: Conceptualization, methodology, and resources, L.F., M.B. and A.L.; formal analysis, L.H.R. and M.C.-Y.; investigation, L.H.R., M.C.-Y. and D.V.S.S.; data curation, M.B., A.L. and L.F.; writing—original draft preparation, M.B., A.L., L.H.R. and M.C.-Y.; writing—review and editing, M.B., A.L., L.F. and S.C.; funding acquisition, M.B. and A.L. All authors have read and agreed to the published version of the manuscript.

Funding: This work was partially funded by the University of Padova with the following research projects BIRD234877/23. DOR2352129. DOR2251254/22; DOR2107797/21. DOR2084579/20. DOR1928058/19. BIRD210708.

Institutional Review Board Statement: Not applicable.

Informed Consent Statement: Not applicable.

Data Availability Statement: Data will be made available upon request.

Acknowledgments: The authors wish to thank Organizzazione internazionale Italo-Latino Americana (IILA) for the IILA-MAECI/DGCS 2020–2021 scholarship program and Ambasciata d'Italia in Perù for Scholarship MAECI 2022 (Prot. N. 4831), supporting LHR and DVSS, respectively. Ameya Gupte, PhD (Padova University) is acknowledged for the HPLC analysis.

Conflicts of Interest: The authors declare no conflict of interest.

References

1. Soares, T.; Oliveira, M.B. Cocoa By-Products: Characterization of Bioactive Compounds and Beneficial Health Effects. *Molecules* **2022**, *27*, 1625. [CrossRef] [PubMed]
2. ICCO. ICCO Cocoa Year 2021/22. *Q. Bull. Cocoa Stat.* **2022**, *XLVIII*, 1.
3. Daymond, A.; Bekele, F. Cacao. In *Cash Crops*; Springer: Cham, Switzerland, 2022; pp. 23–53.
4. Afoakwa, E.O.; Paterson, A.; Fowler, M.; Ryan, A. Flavor Formation and Character in Cocoa and Chocolate: A Critical Review. *Crit. Rev. Food Sci. Nutr.* **2008**, *48*, 840–857. [CrossRef] [PubMed]
5. Lu, F.; Rodriguez-Garcia, J.; Van Damme, I.; Westwood, N.J.; Shaw, L.; Robinson, J.S.; Warren, G.; Chatzifragkou, A.; McQueen Mason, S.; Gomez, L.; et al. Valorisation Strategies for Cocoa Pod Husk and Its Fractions. *Curr. Opin. Green Sustain. Chem.* **2018**, *14*, 80–88. [CrossRef]
6. Beckett, S. *The Science of Chocolate*, 3rd ed.; RSC Publishing: Cambridge, UK, 2001; ISBN 9780854049707.
7. Vásquez, Z.S.; de Carvalho Neto, D.P.; Pereira, G.V.M.; Vandenberghe, L.P.S.; de Oliveira, P.Z.; Tiburcio, P.B.; Rogez, H.L.G.; Góes Neto, A.; Soccol, C.R. Biotechnological Approaches for Cocoa Waste Management: A Review. *Waste Manag.* **2019**, *90*, 72–83. [CrossRef] [PubMed]
8. Vriesmann, L.C.; de Mello Castanho Amboni, R.D.; de Oliveira Petkowicz, C.L. Cacao Pod Husks (*Theobroma cacao* L.): Composition and Hot-Water-Soluble Pectins. *Ind. Crops Prod.* **2011**, *34*, 1173–1181. [CrossRef]
9. Oddoye, E.O.K.; Agyente-Badu, C.K.; Gyedu-Akoto, E. Cocoa and Its By-Products: Identification and Utilization. In *Chocolate in Health and Nutrition. Nutrition and Health*; Humana Press: Totowa, NJ, USA, 2013; Volume 7, pp. 23–37. [CrossRef]
10. Acosta, N.; De Vrieze, J.; Sandoval, V.; Sinche, D.; Wierinck, I.; Rabaey, K. Cocoa Residues as Viable Biomass for Renewable Energy Production through Anaerobic Digestion. *Bioresour. Technol.* **2018**, *265*, 568–572. [CrossRef] [PubMed]
11. Antwi, E.; Engler, N.; Narra, S.; Schüch, A.; Nelles, M. Environmental Effect of Cocoa Pods Disposal in 3 West African Countries. In *13th Rostock Bioenergy Forum Proceedings*; Rostock University: Rostock, Germany, 2019; 463p.

12. Hougni, D.-G.J.M.; Schut, A.G.T.; Woittiez, L.S.; Vanlauwe, B.; Giller, K.E. How Nutrient Rich Are Decaying Cocoa Pod Husks? The Kinetics of Nutrient Leaching. *Plant Soil* **2021**, *463*, 155–170. [CrossRef]
13. Kley Valladares-Diestra, K.; Porto de Souza Vandenberghe, L.; Ricardo Soccol, C. A Biorefinery Approach for Pectin Extraction and Second-Generation Bioethanol Production from Cocoa Pod Husk. *Bioresour. Technol.* **2022**, *346*, 126635. [CrossRef]
14. Muñoz-Almagro, N.; Valadez-Carmona, L.; Mendiola, J.A.; Ibáñez, E.; Villamiel, M. Structural Characterisation of Pectin Obtained from Cacao Pod Husk. Comparison of Conventional and Subcritical Water Extraction. *Carbohydr. Polym.* **2019**, *217*, 69–78. [CrossRef]
15. Porto de Souza Vandenberghe, L.; Kley Valladares-Diestra, K.; Amaro Bittencourt, G.; Fátima Murawski de Mello, A.; Sarmiento Vásquez, Z.; Zwiercheczewski de Oliveira, P.; Vinícius de Melo Pereira, G.; Ricardo Soccol, C. Added-Value Biomolecules' Production from Cocoa Pod Husks: A Review. *Bioresour. Technol.* **2022**, *344*, 126252. [CrossRef] [PubMed]
16. United Nations The 17 Goals. Available online: https://sdgs.un.org/goals (accessed on 1 August 2023).
17. Yadav, A.; Sharma, V.; Tsai, M.L.; Chen, C.W.; Sun, P.P.; Nargotra, P.; Wang, J.X.; Dong, C. Di Development of Lignocellulosic Biorefineries for the Sustainable Production of Biofuels: Towards Circular Bioeconomy. *Bioresour. Technol.* **2023**, *381*, 129145. [CrossRef] [PubMed]
18. Mujtaba, M.; Fernandes Fraceto, L.; Fazeli, M.; Mukherjee, S.; Savassa, S.M.; Araujo de Medeiros, G.; do Espírito Santo Pereira, A.; Mancini, S.D.; Lipponen, J.; Vilaplana, F. Lignocellulosic Biomass from Agricultural Waste to the Circular Economy: A Review with Focus on Biofuels, Biocomposites and Bioplastics. *J. Clean. Prod.* **2023**, *402*, 136815. [CrossRef]
19. Sánchez, M.; Laca, A.; Laca, A.; Díaz, M. Cocoa Bean Shell: A By-Product with High Potential for Nutritional and Biotechnological Applications. *Antioxidants* **2023**, *12*, 1028. [CrossRef] [PubMed]
20. Cinar, Z.Ö.; Atanassova, M.; Tumer, T.B.; Caruso, G.; Antika, G.; Sharma, S.; Sharifi-Rad, J.; Pezzani, R. Cocoa and Cocoa Bean Shells Role in Human Health: An Updated Review. *J. Food Compos. Anal.* **2021**, *103*, 104115. [CrossRef]
21. Koller, M.; Mukherjee, A. A New Wave of Industrialization of PHA Biopolyesters. *Bioengineering* **2022**, *9*, 74. [CrossRef] [PubMed]
22. Cadariu, A.I.; Cocan, I.; Negrea, M.; Alexa, E.; Obistioiu, D.; Hotea, I.; Radulov, I.; Poiana, M.A. Exploring the Potential of Tomato Processing Byproduct as a Natural Antioxidant in Reformulated Nitrite-Free Sausages. *Sustainability* **2022**, *14*, 11802. [CrossRef]
23. Atiwesh, G.; Mikhael, A.; Parrish, C.C.; Banoub, J.; Le, T.A.T. Environmental Impact of Bioplastic Use: A Review. *Heliyon* **2021**, *7*, e07918. [CrossRef]
24. Berezina, N. Novel Approach for Productivity Enhancement of Polyhydroxyalkanoates (PHA) Production by *Cupriavidus necator* DSM 545. *New Biotechnol.* **2013**, *30*, 192–195. [CrossRef]
25. Available online: https://www.european-bioplastics.org/market/ (accessed on 1 August 2023).
26. Favaro, L.; Basaglia, M.; Casella, S. Improving Polyhydroxyalkanoate Production from Inexpensive Carbon Sources by Genetic Approaches: A Review. *Biofuels Bioprod. Biorefining* **2019**, *13*, 208–227. [CrossRef]
27. Favaro, L.; Basaglia, M.; Rodriguez, J.E.G.; Morelli, A.; Ibraheem, O.; Pizzocchero, V.; Casella, S. Bacterial Production of PHAs from Lipid-Rich by-Products. *Appl. Food Biotechnol.* **2019**, *6*, 45–52. [CrossRef]
28. Yukesh Kannah, R.; Dinesh Kumar, M.; Kavitha, S.; Rajesh Banu, J.; Kumar Tyagi, V.; Rajaguru, P.; Kumar, G. Production and Recovery of Polyhydroxyalkanoates (PHA) from Waste Streams—A Review. *Bioresour. Technol.* **2022**, *366*, 128203. [CrossRef] [PubMed]
29. International Energy Agency. *World Energy Outlook 2010*; Birol, F., Ed.; International Energy Agency: Paris, France, 2010. Available online: https://iea.blob.core.windows.net/assets/1b090169-1c58-4f5d-9451-ee838f6f00e5/weo2010.pdf (accessed on 1 August 2023).
30. Sánchez, M.; Laca, A.; Laca, A.; Díaz, M. Cocoa Bean Shell as Promising Feedstock for the Production of Poly(3-Hydroxybutyrate) (PHB). *Appl. Sci.* **2023**, *13*, 975. [CrossRef]
31. Samah, O.A.; Sias, S.; Hua, Y.G.; Hussin, N.N. Production of Ethanol from Cocoa Pod Hydrolysate. *J. Math. Fundam. Sci.* **2011**, *43*, 87–94. [CrossRef]
32. Van Soest, P.J.; Robertson, J.B.; Lewis, B.A. Methods for Dietary Fiber, Neutral Detergent Fiber, and Nonstarch Polysaccharides in Relation to Animal Nutrition. *J. Dairy Sci.* **1991**, *74*, 3583–3597. [CrossRef] [PubMed]
33. Ebrahimi, P.; Mihaylova, D.; Marangon, C.M.; Grigoletto, L.; Lante, A. Impact of Sample Pretreatment and Extraction Methods on the Bioactive Compounds of Sugar Beet (*Beta vulgaris* L.) Leaves. *Molecules* **2022**, *27*, 8110. [CrossRef] [PubMed]
34. Amare, E.; Grigoletto, L.; Corich, V.; Giacomini, A.; Lante, A. Fatty Acid Profile, Lipid Quality and Squalene Content of Teff (*Eragrostis teff* (Zucc.) Trotter) and Amaranth (*Amaranthus caudatus* L.) Varieties from Ethiopia. *Appl. Sci.* **2021**, *11*, 3590. [CrossRef]
35. Cisneros-Yupanqui, M.; Chalova, V.I.; Kalaydzhiev, H.R.; Mihaylova, D.; Krastanov, A.I.; Lante, A. Preliminary Characterisation of Wastes Generated from the Rapeseed and Sunflower Protein Isolation Process and Their Valorisation in Delaying Oil Oxidation. *Food Bioprocess Technol.* **2021**, *1*, 1962–1971. [CrossRef]
36. Stratil, P.; Kubáň, V.; Fojtová, J. Comparison of the Phenolic Content and Total Antioxidant Activity in Wines as Determined by Spectrophotometric Methods. *Czech J. Food Sci.* **2008**, *26*, 242–253. [CrossRef]
37. Tinello, F.; Lante, A.; Bernardi, M.; Cappiello, F.; Galgano, F.; Caruso, M.C.; Favati, F. Comparison of OXITEST and RANCIMAT Methods to Evaluate the Oxidative Stability in Frying Oils. *Eur. Food Res. Technol.* **2018**, *244*, 747–755. [CrossRef]
38. Hernández-Mendoza, A.G.; Saldaña-Trinidad, S.; Martínez-Hernández, S.; Pérez-Sariñana, B.Y.; Láinez, M. Optimization of Alkaline Pretreatment and Enzymatic Hydrolysis of Cocoa Pod Husk (*Theobroma cacao* L.) for Ethanol Production. *Biomass Bioenergy* **2021**, *154*, 106268. [CrossRef]

39. Brojanigo, S.; Gronchi, N.; Cazzorla, T.; Wong, T.S.; Basaglia, M.; Favaro, L.; Casella, S. Engineering Cupriavidus Necator DSM 545 for the One-Step Conversion of Starchy Waste into Polyhydroxyalkanoates. *Bioresour. Technol.* **2022**, *347*, 126383. [CrossRef] [PubMed]
40. Cagnin, L.; Gronchi, N.; Basaglia, M.; Favaro, L.; Casella, S. Selection of Superior Yeast Strains for the Fermentation of Lignocellulosic Steam-Exploded Residues. *Front. Microbiol.* **2021**, *12*, 756032. [CrossRef] [PubMed]
41. Ramsay, B.A.; Lomaliza, K.; Chavarie, C.; Dube, B.; Bataille, P.; Ramsay, J.A. Production of Poly-(Beta-Hydroxybutyric-Co-Beta-Hydroxyvaleric) Acids. *Appl. Environ. Microbiol.* **1990**, *56*, 2093. [CrossRef] [PubMed]
42. Braunegg, G.; Sonnleitner, B.; Lafferty, R.M. A Rapid Gas Chromatographic Method for the Determination of Poly-b-Hydroxybutyric Acid in Microbial Biomass. *Eur. J. Appl. Microbiol. Biotechnol.* **1978**, *6*, 29–37. [CrossRef]
43. Rodríguez Gamero, J.E.; Favaro, L.; Pizzocchero, V.; Lomolino, G.; Basaglia, M.; Casella, S. Nuclease Expression in Efficient Polyhydroxyalkanoates-Producing Bacteria Could Yield Cost Reduction during Downstream Processing. *Bioresour. Technol.* **2018**, *261*, 176–181. [CrossRef]
44. Cagnin, L.; Favaro, L.; Gronchi, N.; Rose, S.H.; Basaglia, M.; van Zyl, W.H.; Casella, S. Comparing Laboratory and Industrial Yeast Platforms for the Direct Conversion of Cellobiose into Ethanol under Simulated Industrial Conditions. *FEMS Yeast Res.* **2019**, *19*, foz018. [CrossRef]
45. Mendoza-Meneses, C.J.; Feregrino-Pérez, A.A.; Gutiérrez-Antonio, C. Potential Use of Industrial Cocoa Waste in Biofuel Production. *J. Chem.* **2021**, *2021*, 3388067. [CrossRef]
46. Coimbra, M.C.; Jorge, N. Proximate Composition of Guariroba (*Syagrus oleracea*), Jerivá (*Syagrus romanzoffiana*) and Macaúba (*Acrocomia aculeata*) Palm Fruits. *Food Res. Int.* **2011**, *44*, 2139–2142. [CrossRef]
47. Martínez, R.; Torres, P.; Meneses, M.A.; Figueroa, J.G.; Pérez-Álvarez, J.A.; Viuda-Martos, M. Chemical, Technological and in Vitro Antioxidant Properties of Cocoa (*Theobroma cacao* L.) Co-Products. *Food Res. Int.* **2012**, *49*, 39–45. [CrossRef]
48. Campione, A.; Pauselli, M.; Natalello, A.; Valenti, B.; Pomente, C.; Avondo, M.; Luciano, G.; Caccamo, M.; Morbidini, L. Inclusion of Cocoa By-Product in the Diet of Dairy Sheep: Effect on the Fatty Acid Profile of Ruminal Content and on the Composition of Milk and Cheese. *Animal* **2021**, *15*, 100243. [CrossRef] [PubMed]
49. Alvarez-Barreto, J.F.; Larrea, F.; Pinos, M.C.; Benalcázar, J.; Oña, D.; Andino, C.; Viteri, D.A.; Leon, M.; Almeida-Streitwieser, D. Chemical Pretreatments on Residual Cocoa Pod Shell Biomass for Bioethanol Production. *Bionatura* **2021**, *6*, 1490–1500. [CrossRef]
50. Botella-Martínez, C.; Lucas-Gonzalez, R.; Ballester-Costa, C.; Pérez-álvarez, J.Á.; Fernández-López, J.; Delgado-Ospina, J.; Chaves-López, C.; Viuda-Martos, M. Ghanaian Cocoa (*Theobroma cacao* L.) Bean Shells Coproducts: Effect of Particle Size on Chemical Composition, Bioactive Compound Content and Antioxidant Activity. *Agronomy* **2021**, *11*, 401. [CrossRef]
51. Rojo-Poveda, O.; Barbosa-Pereira, L.; Mateus-Reguengo, L.; Bertolino, M.; Stévigny, C.; Zeppa, G. Effects of Particle Size and Extraction Methods on Cocoa Bean Shell Functional Beverage. *Nutrients* **2019**, *11*, 867. [CrossRef] [PubMed]
52. Donkoh, A.; Atuahene, C.C.; Wilson, B.N.; Adomako, D. Chemical Composition of Cocoa Pod Husk and Its Effect on Growth and Food Efficiency in Broiler Chicks. *Anim. Feed Sci. Technol.* **1991**, *35*, 161–169. [CrossRef]
53. Serra Bonvehí, J.; Ventura Coll, F. Protein Quality Assessment in Cocoa Husk. *Food Res. Int.* **1999**, *32*, 201–208. [CrossRef]
54. Lessa, O.A.; dos Santos Reis, N.; Leite, S.G.F.; Gutarra, M.L.E.; Souza, A.O.; Gualberto, S.A.; de Oliveira, J.R.; Aguiar-Oliveira, E.; Franco, M. Effect of the Solid State Fermentation of Cocoa Shell on the Secondary Metabolites, Antioxidant Activity, and Fatty Acids. *Food Sci. Biotechnol.* **2018**, *27*, 107–113. [CrossRef]
55. Soares, I.D.; Okiyama, D.C.G.; da Costa Rodrigues, C.E. Simultaneous Green Extraction of Fat and Bioactive Compounds of Cocoa Shell and Protein Fraction Functionalities Evaluation. *Food Res. Int.* **2020**, *137*, 109622. [CrossRef]
56. Adjin-Tetteh, M.; Asiedu, N.; Dodoo-Arhin, D.; Karam, A.; Amaniampong, P.N. Thermochemical Conversion and Characterization of Cocoa Pod Husks a Potential Agricultural Waste from Ghana. *Ind. Crops Prod.* **2018**, *119*, 304–312. [CrossRef]
57. Adewole, E.; Ajiboye, B.; Ojo, B.; Ogunmodede, O.; Oso, O. Characterization of Cocoa (*Theobroma cocoa*) Pod. *J. Chem. Inf. Model.* **2013**, *4*.
58. Abdul Karim, A.; Azlan, A.; Ismail, A.; Hashim, P.; Abd Gani, S.S.; Zainudin, B.H.; Abdullah, N.A. Phenolic Composition, Antioxidant, Anti-Wrinkles and Tyrosinase Inhibitory Activities of Cocoa Pod Extract. *BMC Complement. Altern. Med.* **2014**, *14*, 381. [CrossRef] [PubMed]
59. Dewi, S.R.; Stevens, L.A.; Pearson, A.E.; Ferrari, R.; Irvine, D.J.; Binner, E.R. Investigating the Role of Solvent Type and Microwave Selective Heating on the Extraction of Phenolic Compounds from Cacao (*Theobroma cacao* L.) Pod Husk. *Food Bioprod. Process.* **2022**, *134*, 210–222. [CrossRef]
60. Rachmawaty; Mu'Nisa, A.; Hasri; Pagarra, H.; Hartati; Maulana, Z. Active Compounds Extraction of Cocoa Pod Husk (*Thebroma cacao* L.) and Potential as Fungicides. *J. Phys. Conf. Ser.* **2018**, *1028*, 012013. [CrossRef]
61. Sotelo, L.; Alvis, A.; Arrázola, G. Evaluation of Epicatechin, Theobromine and Caffeine in Cacao Husks (*Theobroma cacao* L.), Determination of the Antioxidant Capacity. *Rev. Colomb. Cienc. Hortícolas* **2015**, *9*, 124–134. [CrossRef]
62. Valadez-Carmona, L.; Plazola-Jacinto, C.P.; Hernández-Ortega, M.; Hernández-Navarro, M.D.; Villarreal, F.; Necoechea-Mondragón, H.; Ortiz-Moreno, A.; Ceballos-Reyes, G. Effects of Microwaves, Hot Air and Freeze-Drying on the Phenolic Compounds, Antioxidant Capacity, Enzyme Activity and Microstructure of Cacao Pod Husks (*Theobroma cacao* L.). *Innov. Food Sci. Emerg. Technol.* **2017**, *41*, 378–386. [CrossRef]

63. Bortolini, C.; Patrone, V.; Puglisi, E.; Morelli, L. Detailed Analyses of the Bacterial Populations in Processed Cocoa Beans of Different Geographic Origin, Subject to Varied Fermentation Conditions. *Int. J. Food Microbiol.* **2016**, *236*, 98–106. [CrossRef] [PubMed]
64. Campos-Vega, R.; Nieto-Figueroa, K.H.; Oomah, B.D. Cocoa (*Theobroma cacao* L.) Pod Husk: Renewable Source of Bioactive Compounds. *Trends Food Sci. Technol.* **2018**, *81*, 172–184. [CrossRef]
65. Boungo Teboukeu, G.; Tonfack Djikeng, F.; Klang, M.J.; Houketchang Ndomou, S.; Karuna, M.S.L.; Womeni, H.M. Polyphenol Antioxidants from Cocoa Pods: Extraction Optimization, Effect of the Optimized Extract, and Storage Time on the Stability of Palm Olein during Thermoxidation. *J. Food Process. Preserv.* **2018**, *42*, e13592. [CrossRef]
66. Lante, A.; Nardi, T.; Zocca, F.; Giacomini, A.; Corich, V. Evaluation of Red Chicory Extract as a Natural Antioxidant by Pure Lipid Oxidation and Yeast Oxidative Stress Response as Model Systems. *J. Agric. Food Chem.* **2011**, *59*, 5318–5324. [CrossRef]
67. Cisneros-Yupanqui, M.; Zagotto, A.; Alberton, A.; Lante, A.; Zagotto, G.; Ribaudo, G.; Rizzi, C. Study of the Phenolic Profile of a Grape Pomace Powder and Its Impact on Delaying Corn Oil Oxidation. *Nat. Prod. Res.* **2020**, *36*, 455–459. [CrossRef]
68. Mensah, M.; Yaw Asiedu, N.; Abunde Neba, F.; Nana Amaniampong, P.; Boakye, P.; Addo, A. Modeling, Optimization and Kinetic Analysis of the Hydrolysis Process of Waste Cocoa Pod Husk to Reducing Sugars. *SN Appl. Sci.* **2020**, *2*, 1160. [CrossRef]
69. García, Y.C.; Morales, J.J.; Montalvo, P.A.J.; Figueroa, L.V. Pretratamiento e Hidrólisis Enzimática de La Cascarilla de Arroz. *Rev. Soc. Quím. Perú* **2019**, *85*, 476–488. [CrossRef]
70. Benalcazar Bassante, J.C. *Evaluación de Diferentes Pretratamientos Químicos a la Biomasa de la Cáscara de Cacao Para Procesos de Fermentación Alcohólica*; Universidad San Francisco de Quito: Quito, Ecuador, 2018.
71. Sarmiento-Vásquez, Z.; Vandenberghe, L.; Rodrigues, C.; Tanobe, V.O.A.; Marín, O.; de Melo Pereira, G.V.; Ghislain Rogez, H.L.; Góes-Neto, A.; Soccol, C.R. Cocoa Pod Husk Valorization: Alkaline-Enzymatic Pre-Treatment for Propionic Acid Production. *Cellulose* **2021**, *28*, 4009–4024. [CrossRef]
72. Shet, V.B.; Sanil, N.; Bhat, M.; Naik, M.; Mascarenhas, L.N.; Goveas, L.C.; Rao, C.V.; Ujwal, P.; Sandesh, K.; Aparna, A. Acid Hydrolysis Optimization of Cocoa Pod Shell Using Response Surface Methodology Approach toward Ethanol Production. *Agric. Nat. Resour.* **2018**, *52*, 581–587. [CrossRef]
73. Gutiérrez-Macías, P.; Montañez-Barragán, B.; Barragán-Huerta, B.E. A Review of Agro-Food Waste Transformation into Feedstock for Use in Fermentation. *Fresenius Environ. Bull.* **2015**, *24*, 3703–3716.
74. Brojanigo, S.; Parro, E.; Cazzorla, T.; Favaro, L.; Basaglia, M.; Casella, S. Conversion of Starchy Waste Streams into Polyhydroxyalkanoates Using Cupriavidus Necator DSM 545. *Polymers* **2020**, *12*, 1496. [CrossRef] [PubMed]
75. Brodin, M.; Vallejos, M.; Opedal, M.T.; Area, M.C.; Chinga-Carrasco, G. Lignocellulosics as Sustainable Resources for Production of Bioplastics—A Review. *J. Clean. Prod.* **2017**, *162*, 646–664. [CrossRef]
76. Annamalai, N.; Sivakumar, N. Production of Polyhydroxybutyrate from Wheat Bran Hydrolysate Using Ralstonia Eutropha through Microbial Fermentation. *J. Biotechnol.* **2016**, *237*, 13–17. [CrossRef]
77. de Souza, L.; Manasa, Y.; Shivakumar, S. Bioconversion of Lignocellulosic Substrates for the Production of Polyhydroxyalkanoates. *Biocatal. Agric. Biotechnol.* **2020**, *28*, 101754. [CrossRef]
78. Silva, L.F.; Taciro, M.K.; Michelin Ramos, M.E.; Carter, J.M.; Pradella, J.G.C.; Gomez, J.G.C. Poly-3-Hydroxybutyrate (P3HB) Production by Bacteria from Xylose, Glucose and Sugarcane Bagasse Hydrolysate. *J. Ind. Microbiol. Biotechnol.* **2004**, *31*, 245–254. [CrossRef]
79. Yogaswara, R.R.; Billah, M.; Saputro, E.A.; Erliyanti, N.K. A Kinetic Study in Fermentation of Cocoa Pod Husk Using Zymomonas mobilis. In *Proceedings of the IOP Conference Series: Materials Science and Engineering*; IOP Publishing: Bristol, UK, 2021.
80. Favaro, L.; Jansen, T.; van Zyl, W.H. Exploring Industrial and Natural *Saccharomyces cerevisiae* Strains for the Bio-Based Economy from Biomass: The Case of Bioethanol. *Crit. Rev. Biotechnol.* **2019**, *39*, 800–816. [CrossRef]

fermentation

Brief Report

Modelisation of the Biomethane Accumulation in Anaerobic Co-Digestion of Whey and Sugarcane Molasse Mixtures

Huaita Pacari Arotingo Guandinango [1], Rosario del Carmen Espín Valladares [2], Jimmy Núñez Pérez [2], Marco Vinicio Lara Fiallos [2], Ileana Pereda Reyes [3] and José Manuel Pais-Chanfrau [2,*]

1 Comunidad Santa Bárbara, Cotacachi 100350, Ecuador; sunnibaarotingo@gmail.com
2 Carrera de Agroindustria, FICAYA, Universidad Técnica del Norte, Ave. 17 de julio 5-21 y Gral. José Ma. De Córdova, Ibarra 100105, Ecuador; rcespin@utn.edu.ec (R.d.C.E.V.); jnunez@utn.edu.ec (J.N.P.); mvlara@utn.edu.ec (M.V.L.F.)
3 Process Engineering Centre (CIPRO), Universidad Tecnológica de La Habana "José Antonio Echeverría", Marianao 19390, Cuba; ileana@quimica.cujae.edu.cu
* Correspondence: jmpais@utn.edu.ec

Abstract: The biomethane accumulation of several combinations of whey and sugarcane molasses, inoculated with sludge from a treatment facility of one of the dairy enterprises of the Imbabura province in Ecuador, was assessed in the current experiment at a constant COD_0/VS_{in} ratio of 0.5. The whey/molasses (W:M) ratios for each treatment were (in % (m/m)) 0:100, 25:75, 50:50, 75:25, and 100:0, with a constant temperature of 37 °C and an initial pH adjustment of 7.5. Half a litre of total mixes was used for each treatment in duplicate. Six kinetic models were evaluated to account biomethane accumulation in anaerobic co-digestion processes in batch of whey and sugarcane molasses. Five of these have been tested by other researchers, and one was developed by modifying a first-order model to consider changes in the biomethane accumulation profile. This proposed model, along with the modified two-phase Gompertz model, resulted in the ones that were best able to adjust the experimental data, obtaining in all cases an $R^2 \geq 0.949$, indicating the accuracy of both models. In addition, the proposed here model has five parameters, one less than the modified two-phase Gompertz model, making it more straightforward and robust.

Keywords: anaerobic co-digestion; kinetics of biomethane accumulation; kinetic modelling; sugar-cane molasses; whey

Citation: Arotingo Guandinango, H.P.; Espín Valladares, R.d.C.; Núñez Pérez, J.; Lara Fiallos, M.V.; Pereda Reyes, I.; Pais-Chanfrau, J.M. Modelisation of the Biomethane Accumulation in Anaerobic Co-Digestion of Whey and Sugarcane Molasse Mixtures. *Fermentation* **2023**, *9*, 834. https://doi.org/10.3390/fermentation9090834

Academic Editors: Jose Luis García-Morales and Francisco Jesús Fernández Morales

Received: 12 August 2023
Revised: 4 September 2023
Accepted: 6 September 2023
Published: 13 September 2023

1. Introduction

The world's population has grown steadily over the past centuries, reaching 7.9 billion inhabitants today (Population Growth—Our World in Data). Alongside this, the demand for food and energy is also growing, and pressure on arable land and ecosystems is increasing [1]. With the increase in food production, there is also, logically, an increase in the waste generated. In contrast, the incessant increase in energy demand makes it essential to explore other renewable sources of energy to provide a future response to the depletion of traditional non-renewable sources, which will inexorably occur in the not-too-distant future.

In this sense, Ecuador, and specifically Zone 1 (formed by the provinces of Imbabura, Carchi, Esmeraldas, and Sucumbíos), is characterised by an active agricultural economy, which includes the daily production of more than 50% of Ecuador's milk production [2]. An essential part of this production is destined for cheese production, which generates significant quantities of cow-whey. In 2017, it was estimated that in the provinces of Imbabura and Carchi alone, more than 120 m^3 of whey was generated daily [3]. About 70% of this whey is used for pig feed, but the rest must be treated in treatment plants due to its high polluting power [3].

On the other hand, one of the central sugar mills in the country is in the province of Imbabura, which generates sugar cane molasses as waste [4]. In this sense, agro-industrial waste could be studied as a possible source of raw material for biogas generation [5], an alternative to the circular economy for local industries.

Finally, modelling the complex fermentative processes that take place within anaerobic co-digestion [6–8], mediated by complex consortia of bacteria and yeast, including acetogenic and methanogenic bacteria and diverse sources of carbon and nitrogen, is of the utmost importance for the design of treatment processes for bioremediation and as a source of renewable biomethane from these agriculture or agro-industrial wastes [9–11].

The present work aims to evaluate the anaerobic digestion of cow's whey and sugarcane molasse, alone or formed by different mixtures, and to fit different kinetics models for the biomethane accumulation reported by other authors. A modified first-order model in two stages, not reported before, has also been evaluated.

2. Materials and Methods

2.1. Raw Materials Used

The whey used in this study came from the Ibarra branch of the company Floralp S.A. (Princesa Paccha 5-163, Caranqui, Ibarra, Imbabura, Ecuador, https://floralp-sa.com (accessed on 30 July 2023)). The company's waste treatment plant supplied the sludge. The sugar cane molasses was purchased on the local market from the *Ingenio Azucarero del Norte* (Panamericana Norte, km 25 vía Tulcán, Imbabura, Ecuador, http://www.tababuela.com (accessed on 30 July 2023).

2.2. Physico-Chemical Characterisation

The total and volatile solids were determined according to the methods described in APHA 2540 B and APHA 2540 E, respectively [12]. For the determination of COD, the method described in APHA 5520 D was used [12].

A known volume was weighed to determine the density and pH, and the pH was measured in a conventional pH meter, previously adjusted between pH 4 and pH 10.

2.3. Experimental Procedure

The experimental units were prepared with a constant COD_0/VS_{in} ratio of 0.5, where COD_0 is the amount (in grams) of COD at the start of the anaerobic co-digestion (AcD) process and contains the COD inputs from both the inoculum (activated sludge) and the substrates (milk whey and sugarcane molasse) and VS_{in} represents the quantity (in grams) of volatile solids present in the inoculum (activate sludge). The working volumes of activated sludge, whey, and sugarcane molasse mixtures were between 66–452 mL. For treatments that did not reach 452 mL, a volume of sterile deionised water was added until all the volume (including sterile deionised water) reached 452 mL. After that, the bottle caps were worn, and neoprene caps were placed to connect the pipes for the exit of the gases produced by anaerobic digestion.

Five of the six vials were inoculated with 10 mL of activated sludge, while to the sixth, with a "50:50" mixture, was added 10 mL sterile water and served as a "negative control" of the process. This last bottle did not produce gases practically.

The flasks were placed in a thermostatically controlled bath, maintaining the temperature at 37 ± 1 °C. The experimental setup consists of six 500 mL flasks, where anaerobic digestion occurs discontinuously, which are connected to six 250 mL flasks, which act as a trap to capture the CO_2 produced. Each trap flask was connected to 250 mL test tubes, inverted, and filled with the same solution as the traps (0.375 M NaOH + phenolphthalein), allowing the measurement of methane gas by liquid displacement, as described by other authors [7,13,14]. All test tubes were placed in a cuvette, partially filled with the same alkaline solution (Figure 1).

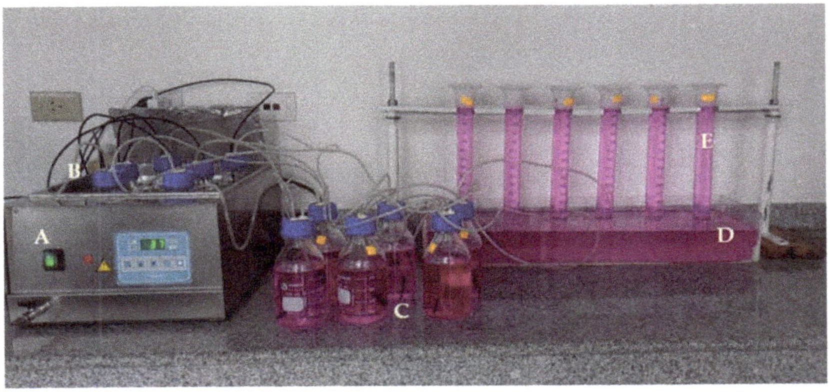

Figure 1. The experimental facility used in the research. (A) Thermostated water-bath with recirculation. (B) Anaerobic digester flasks. (C) Bubbling traps for CO_2 capture. (D) Immersion cuvette. (E) Inverted test tubes for methane measurement.

Before this, the sludge was adapted for 15 days, with similar amounts as in the whey and cane molasses mixture evaluation being supplied every 2–3 days, and when an appreciable decrease in gas bubbling was observed. The sludge was inoculated into the reactors once the fizzing had ceased after the last addition of the substrate.

Experimental blocks with six variants in each (five treatments + "negative control") maintained a constant ratio of COD_0/VS_{in} equal to 0.5 and were performed twice.

2.4. Kinetic Model for the Anaerobic Co-Digestion Mixes of Whey and Molasses

To kinetically characterise the process and model the generation of the primary metabolite, methane, the modified first-order in two-stage model (Equation (2)) was used together with other traditional models described by other authors [8], like the modified two-phase Gompertz model (Equation (3)), the multi-stage first-order model (Equation (4)), all conceived to describe the accumulative biomethane production obtained from complex substrates in which the diauxic growth has been observed.

Additionally, the three simplest models with three parameters each were also evaluated. The Fitzhugh model (Equation (5)), the transference-function model (Equation (6)), and Cone's model (Equation (7)), despite their simplicity, in most cases, as will demonstrate further, adjust the experimental values accurately.

The model used here is based on the first-order model and was conceived for anaerobic digestions of substrate mixtures and where the phenomenon of diauxic is observed. For this, we should estimate t_{d_i} when a change in the methane accumulation profile is observed. Therefore, it is a **modified first-order model** for mixtures of many substrates and multi-stages are available.

$$G = \begin{cases} G_{m1}\left[1 - e^{-k_{01}t}\right] & for\ 0 \le t < t_{d1} \\ G_{m2}\left[1 - e^{-k_{02}(t-t_{d1})}\right] & for\ t_{d1} \le t < t_{d2} \\ G_{m3}\left[1 - e^{-k_{03}(t-t_{d2})}\right] & for\ t_{d2} \le t \le t_{d3} \\ \vdots & \\ G_{mn}\left[1 - e^{-k_{0n}(t-t_{dn-1})}\right] & for\ t_{dn-1} \le t \le t_f \end{cases} \quad (1)$$

For two stages, the above model will transform into

$$G = \begin{cases} G_{m1}\left[1 - e^{-k_{01}t}\right] & for\ 0 \le t < t_{d1} \\ G_{m2}\left[1 - e^{-k_{02}(t-t_{d1})}\right] & for\ t_{d1} \le t < t_f \end{cases} \quad (2)$$

where Gm_1 and Gm_2 are the maximum accumulated value of methane in each stage, in Nml CH_4; k_{01} and k_{02} are the first-order constants of the kinetics of biomethane accumulation, in d^{-1}; t_d and t_f are the times where diauxic phenomenon and end of the AcD process are observed in days.

The **two-phase modified Gompertz model** was suggested to represent the accumulation of biomethane in AcD processes, where the phenomenon of diauxic growth is observed [14]. This model is based on six parameters (Gm_1, Gm_2, Rm_1, Rm_2, λ_1 and λ_2) (f = 6).

$$G = G_{m1}\, e^{\left\{-e^{([R_{m1}\cdot e\cdot(\lambda_1 - t)/G_{m1}]+1)}\right\}} + G_{m2}\, e^{\left\{-e^{([R_{m2}\cdot e\cdot(\lambda_2 - t)/G_{m2}]+1)}\right\}} \quad (3)$$

The Gm_1, Gm_2, Rm_1, Rm_2, λ_1 and λ_2 parameters that can be obtained, like that of the rest of the models, experimentally from having experimental data relating to G vs. t, and employing a non-linear regression analysis, represent the maximum values of biomethane accumulation (Gm_1 and Gm_2, in Nml CH_4), biomethane generation rate (Rm_1 and Rm_2, in Nml CH_4/d) and the duration of the lag phase (λ_1 and λ_2, in days), for each of the two phases of diauxic growth.

The **multi-stage first-order model** was conceived to model the production of biomethane in the presence of complex substrates formed by various sources of carbon, and their interactions, which lead to anaerobic digestion passing through different stages [15].

$$G = G_{m1}\left[1 - e^{-k_{01}t}\right] + G_{m2}\left[1 - e^{-k_{02}t}\right] + G_{m12}\left[1 - \frac{k_{02}\cdot e^{-k_{01}t}}{k_{02} - k_{01}} - \frac{k_{01}\cdot e^{-k_{02}t}}{k_{01} - k_{02}}\right] \quad (4)$$

It is a five-factor (f = 5) model (Gm_1, Gm_2, Gm_{12}, k_{01} and k_{02}), where Gm_1, Gm_2 and Gm_{12} represent the maximum accumulation of biomethane (Nml CH_4) in the stages "1", "2" and during the interaction of both substrates ("12"), whereas k_{01} and k_{02}, represent the first-order kinetic constants in the states "1" and "2", respectively.

The last three models to be analysed are simple models formed by only three factors (f = 3).

The **Fitzhugh model**, initially developed to monitor the production of biomethane by the action of microorganisms present in livestock rumen [16,17], has also been successfully used by other researchers to co-digest food waste with activated sludge [18]. It is a simple three-factor model (Gm, k_0 and n, f = 3), where n represents the presence (if $n \geq 1$) or the absence (If $n < 1$) of a lag phase in the anaerobic process.

$$G = G_m\left[1 - e^{(-k_0 t)^n}\right] \quad (5)$$

Gm, k_0 and n (f = 3), represent the maximum accumulation of biomethane (in Nml CH_4), the first order kinetic constant (in d^{-1}), and a dimensional constant, related to the existence or not of a lag phase in the AcD process, respectively.

Additionally, the **transference function model** was also assessed (Equation (6)). In some cases, this model has been used to describe anaerobic digestion [19].

$$G = G_m\left[1 - e^{(-(R_m/G_m)\cdot(t-\lambda))}\right] \quad (6)$$

Cone's empirical model, like others here, was initially developed to quantify methane production by the rumen microorganisms by metabolizing the grass [20].

$$G = \frac{G_m}{1 + (kt)^{-n}} \tag{7}$$

The values that need to be adjusted are Gm, k and n, representing the maximum cumulative amount of methane (in Nml CH_4), the first-order kinetic constant (d^{-1}), and a nondimensional number, respectively.

The experimental data (N = 19) for each mix were fitted by the least squares method and using the generalized reduced gradient (GRG) method [21], a nonlinear numerical optimization algorithm provided by the MS Office-365 Excel *Solver* tool.

2.5. Statistical Comparison of Models

Three known formulas will be used to judge whether the models represent the observed experimental data sufficiently well: the square regression coefficient (R^2, Equation (8)), the normalized root mean square error ($NRMSE$, Equation (9)) and the corrected Akaike information criterion (AIC_C, Equation (10)) [15,18,22].

$$R^2 = 1 - \frac{\sum_{i=1}^{19} \left(G_{exp} - G_{model}\right)_i^2}{\sum_{i=1}^{19} \left(G_{exp} - \overline{G}_{exp}\right)_i^2} \tag{8}$$

And

$$NRMSE = \left(\frac{\sqrt{\frac{\sum_{i=1}^{19} \left(G_{exp} - G_{model}\right)_i^2}{N}}}{G_{exp_{max}} - G_{exp_{min}}} \right) \times 100 \tag{9}$$

The correction that is introduced in the nondimensional Akaike Information Criterion (the last term on right in Equation (10)) [23] is recommended when the values obtained from AIC are small, and the number N of experimental data is not too large, as is the present case [24].

$$AIC_C = N \cdot \ln\left(\frac{\sum_{i=1}^{19} \left(G_{exp} - G_{model}\right)_i^2}{N} \right) + 2f + \left(\frac{2f(f+1)}{N-f-1} \right) \tag{10}$$

where N represents the number of experimental points used to construct each model (N = 19), and f represents the number of factors the model possesses.

In this case, models with R^2 values closer to one and with lower $NRMSE$ and AIC_C values are considered the most appropriate models to represent the observed experimental data.

3. Results and Discussion

For whey, sugarcane molasse and activated sludge, the values of volatile solids were 164.24, 726.94, and 935.4 g VS/L, respectively. The total solids were 237.70, 824.70, and 12.96 g TS/L, respectively, while the COD reached values of 0.64, 8.14, and 1.56 g COD/L in the same order. Additionally, the density was 0.98, 1.20, and 0.98 g/mL, while the initial pH that was had was of 6.90, 5.60, and 3.90, respectively.

According to the characterisation of the substrates in terms of volatile solids, total solids, and COD, it can be concluded that molasses has 4.4, 3.4 and 12.7 times more, respectively, than whey, suggesting a priori that molasses have a higher potential than whey for methane production.

The methane yield values are low, so it is suggested in further studies to raise the COD_0/VS_{in} ratio to values ≥ 1.

The values of each two-mixture treatment were used to represent the models for accumulative methane production. The models were charted alongside the observed experimental data, separating the five- and six-parameter models (Figure 2 (a1–a5)) from the simpler three-factor models (Figure 2(b1–b5)).

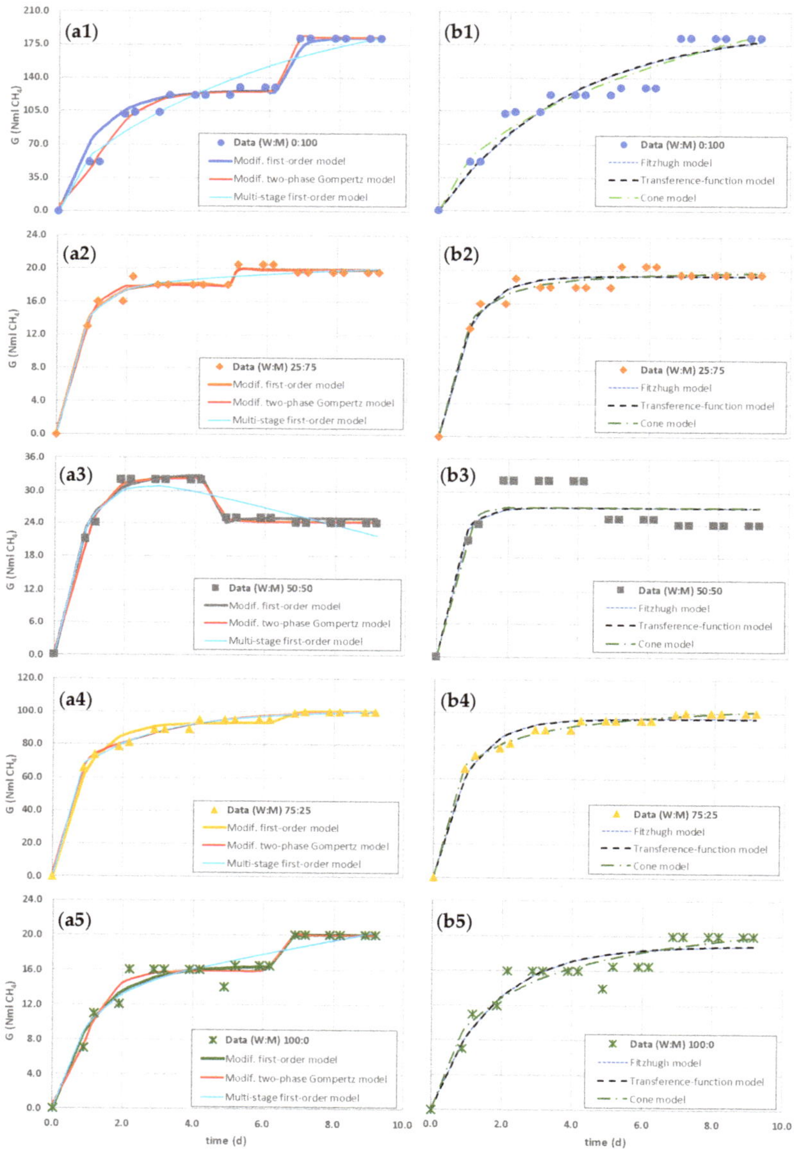

Figure 2. Biomethane accumulation kinetics for different mixtures of whey and sugarcane molasse. On the left side (**a1–a5**) are represented the three models that have between 5 and 6 factors, being from (**a1–a5**), the ratio of whey and molasse (W:M, in % (m/m)): 0:100; 25:75; 50:50; 75:25; and 100:0, respectively. On the right side (**b1–b5**) the three most straightforward, three-factor models were charted, being of (**b1–b5**), the ratios (W:M, in % (m/m)): 0:100; 25:75; 50:50; 75:25, and 100:0, respectively.

It is somewhat disconcerting to note the observation of the 50:50 mixture, where a decrease in the accumulation of biomethane is observed (Figure 2(a3,b3)). The causes of the biomethane reabsorption in the liquid phase, which leads to decreased accumulated volume, should be investigated. This phenomenon may be related, although additional experiments would be needed to prove it, to the temperature fluctuations in the lab between day and night, which can reach ≥ 15 °C.

The R^2, *NRMSE*, and AIC_C values of the five- and six-factor models exhibit better results, especially in those cases where changes in the methane accumulation profile are observed, and within these the modified first-order model and two-phase modified Gompertz model have shown higher performance than multi-stage first-order model (Table 1).

Table 1. Parameters of the kinetic models analysed in the present study and their respective statistical values of adjustment goodness.

Models	Parameters	Mix (W:M), % (m/m)				
		0:100	25:75	50:50	75:25	100:0
Modif. first-order (f = 5) Equation (2)	Gm_1, Nml CH_4	127.00	18.21	32.90	92.86	16.47
	k_{01}, d^{-1}	0.95	1.55	1.37	1.25	0.86
	Gm_2, Nml CH_4	182.00	19.83	24.90	100.00	20.00
	k_{02}, d^{-1}	3.92	32.18	4.72	12.53	12.65
	t_d, d	6.17	4.80	4.17	6.51	6.21
	R^2, -	0.949	0.983	0.979	0.986	0.966
	NRMSE, %	5.59%	2.84%	3.17%	2.62%	4.48%
	AIC_C, -	102.75	−5.95	15.12	51.24	10.44
Modif. two-phase Gompertz (f = 6) Equation (3)	Gm_1, Nml CH_4	125.71	1.90	32.30	66.89	15.90
	Rm_1, Nml $CH_4 \cdot d^{-1}$	59.39	35.66	28.95	159.11	10.80
	λ_1, d	0.13	4.88	0.16	0.39	0.20
	Gm_2, Nml CH_4	56.31	17.94	−8.02	33.45	4.10
	Rm_2, Nml $CH_4 \cdot d^{-1}$	251.23	20.22	−13.43	6.90	18.10
	λ_2, d	6.16	0.18	4.20	0.00	6.14
	R^2, -	0.990	0.983	0.993	0.995	0.975
	NRMSE, %	2.66%	2.81%	1.86%	1.48%	3.85%
	AIC_C, -	78.94	−2.02	−0.68	33.90	9.11
Multi-stage first-order (f = 5) Equation (4)	Gm_1, Nml CH_4	193.22	3.99	36.66	47.91	49.88
	k_{01}, d^{-1}	0.16	0.15	1.20	0.42	0.02
	Gm_2, Nml CH_4	1.10	10.22	1013.16	25.60	6.38
	k_{02}, d^{-1}	3.50	1.67	0.00	6.37	1.13
	Gm_{12}, Nml CH_4	32.5	6.70	1144.77	27.5	6.4
	R^2, -	0.926	0.968	0.921	0.996	0.932
	NRMSE, %	7.15%	3.84%	5.88%	1.46%	6.22%
	AIC_C, -	112.11	5.56	38.63	28.88	22.91
Fitzhugh (f = 3) Equation (5)	Gm, Nml CH_4	194.38	19.29	27.14	96.39	18.89
	k_0, d^{-1}	0.16	0.99	1.79	1.09	0.48
	n, -	1.65	1.28	1.19	0.99	1.21
	R^2, -	0.915	0.960	0.667	0.971	0.898
	NRMSE, %	8.09%	4.33%	10.93%	3.83%	7.75%
	AIC_C, -	109.80	3.05	55.18	58.63	24.25
Transference function (f = 3) Equation (6)	Gm, Nml CH_4	194.40	19.30	27.14	96.39	18.89
	Rm, Nml $CH_4 \cdot d^{-1}$	52.63	24.47	58.03	104.96	10.99
	λ, d	0.00	0.00	0.00	0.00	0.00
	R^2, -	0.915	0.960	0.672	0.971	0.898
	NRMSE, %	8.09%	4.33%	10.93%	3.83%	7.75%
	AIC_C, -	109.80	3.05	55.18	58.63	24.25
Cone (f = 3) Equation (7)	Gm, Nml CH_4	746.80	20.24	27.18	116.36	26.06
	k, d^{-1}	0.02	2.00	1.40	1.76	0.48
	n, -	0.61	1.26	5.78	0.67	0.77
	R^2, -	0.933	0.968	0.688	0.996	0.919
	NRMSE, %	6.91%	3.85%	10.72%	1.42%	6.81%
	AIC_C, -	103.79	−1.40	54.44	20.93	19.32

Except for mixing (W:M) 50:50 in three-parameter models (f = 3), for the rest of the cases, all models showed good performance, with $R^2 > 0.89$.

In the present study, only the modified first-order (f = 5) and the two-phase modified Gompertz (f = 6) models were able to represent all the experimental values of the mixtures adequately and in which notable results were consistently obtained as demonstrated by the $R^2 \geq 0.949$, and values of $NRMSE \leq 5.59\%$, and $AIC_C \leq 102.75$, for all the mixes.

It should be noted that the modified two-phase Gompertz model has been used successfully to represent the accumulation of methane and its yield, in numerous studies of anaerobic digestion [25–28], both in the single substrate and in mixtures, where the phenomenon of diauxic growth has often been observed [29].

Both models quite accurately represent the experimental data obtained. The modified first order model, however, does so with one factor less, which means that, for equal values of R^2, as is the case for mixing (W:M) 25:75 (see Figure 2(a2) and Table 1), the AIC_C value of the modified second order model is lower, and therefore better, than the one obtained for the two-phase modified Gompertz model.

4. Conclusions

In the study presented here, we managed to model the cumulative production of biomethane from various mixtures of whey and sugarcane molasse, using various models reported elsewhere. In addition, a new model was suggested that accurately predicts the observed experimental behaviour and is less complex than the best of the models used here. Further studies are in progress to evaluate and validate this model for other anaerobic co-digestion processes.

Author Contributions: Conceptualization. I.P.R. and J.M.P.-C.; methodology. I.P.R., J.M.P.-C. and H.P.A.G.; software. J.M.P.-C.; validation. R.d.C.E.V. and M.V.L.F.; formal analysis. J.N.P.; investigation. H.P.A.G. and J.M.P.-C.; resources. R.d.C.E.V.; data curation. J.N.P.; writing—original draft preparation. H.P.A.G.; writing—review and editing. J.M.P.-C.; visualization. I.P.R.; supervision. I.P.R.; project administration. M.V.L.F.; funding acquisition. R.d.C.E.V. All authors have read and agreed to the published version of the manuscript.

Funding: This research received no external funding.

Institutional Review Board Statement: Not applicable.

Informed Consent Statement: Not applicable.

Data Availability Statement: Not applicable.

Acknowledgments: The authors of the work wish to express their gratitude to the Marcelo Cevallos, of the FICAYA, for the support given to this research.

Conflicts of Interest: The authors declare no conflict of interest.

References

1. Araújo, R.G.; Chavez-Santoscoy, R.A.; Parra-Saldívar, R.; Melchor-Martínez, E.M.; Iqbal, H.M.N. Agro-food systems and environment: Sustaining the unsustainable. *Curr. Opin. Environ. Sci. Health* **2023**, *31*, 100413. [CrossRef]
2. Peralta, A. Eficiencia en la producción lechera. In Proceedings of the VIII Foro del Sector Lechero Ecuatoriano. Estrategias Para Mejorar el Consumo de Lácteos en la Población, Quito, Ecuador, 1 June 2017; Centro de la Industria Láctea (CIL); p. 20.
3. Pais Chanfrau, J.M.; Núñez Pérez, J.; Lara Fiallos, M.V.; Rivera Intriago, L.M.; Abril Porras, V.H.; Cuaran Guerrero, M.J.; Trujillo Toledo, L.E. Milk whey-from a problematic byproduct to a source of valuable products. *Prensa Med. Argent* **2017**, *103*, 1.
4. Sánches, A.; Vayas, T.; Mayorga, F.; Freire, C. Sector Azucarero del Ecuador. *Obs. Econ. Soc. Tungurагua* **2018**, *4*, 1–4.
5. Alghoul, O.; El-Hassan, Z.; Ramadan, M.; Olabi, A.G. Experimental investigation on the production of biogas from waste food. *Energy Sources Part A Recover. Util. Environ. Eff.* **2019**, *41*, 2051–2060. [CrossRef]
6. Acosta-Pavas, J.C.; Morchain, J.; Dumas, C.; Ngu, V.; Arnaud, C.; Aceves-Lara, C.A. Towards Anaerobic Digestion (ADM No.1) Model's Extensions and Reductions with In-situ Gas Injection for Biomethane Production. *IFAC-Pap.* **2022**, *55*, 635–640. [CrossRef]
7. López González, L.M.; Pereda Reyes, I.; Romero Romero, O. Anaerobic co-digestion of sugarcane press mud with vinasse on methane yield. *Waste Manag.* **2017**, *68*, 139–145. [CrossRef] [PubMed]
8. Zeller, V.; Lavigne, C.; D'Ans, P.; Towa, E.; Achten, W.M.J. Assessing the environmental performance for more local and more circular biowaste management options at city-region level. *Sci. Total Environ.* **2020**, *745*, 140690. [CrossRef]

9. Acosta-Pavas, J.C.; Robles-Rodríguez, C.E.; Morchain, J.; Dumas, C.; Cockx, A.; Aceves-Lara, C.A. Dynamic modeling of biological methanation for different reactor configurations: An extension of the anaerobic digestion model No. 1. *Fuel* **2023**, *344*, 128106. [CrossRef]
10. Ardolino, F.; Parrillo, F.; Arena, U. Biowaste-to-biomethane or biowaste-to-energy? An LCA study on anaerobic digestion of organic waste. *J. Clean. Prod.* **2018**, *174*, 462–476. [CrossRef]
11. Rao, P.V.; Baral, S.S.; Dey, R.; Mutnuri, S. Biogas generation potential by anaerobic digestion for sustainable energy development in India. *Renew. Sustain. Energy Rev.* **2010**, *14*, 2086–2094. [CrossRef]
12. Rice, E.; Baird, R.; Eaton, A.; Clesceri, L. *Standard Methods for the Examination of Water and Wastewater*; Standard Methods: Denver, CO, USA, 2012.
13. Pagés Díaz, J.; Pereda Reyes, I.; Lundin, M.; Sárvári Horváth, I. Co-digestion of different waste mixtures from agro-industrial activities: Kinetic evaluation and synergetic effects. *Bioresour. Technol.* **2011**, *102*, 10834–10840. [CrossRef] [PubMed]
14. Montesdeoca-Pichucho, N.B.; Garibaldi-Alcívar, K.; Baquerizo-Crespo, R.J.; Gómez-Salcedo, Y.; Pérez-Ones, O.; Pereda-Reyes, I. Synergistic and antagonistic effects in anaerobic co-digestion. Analysis of the methane yield kinetics. *Rev. Fac. Ing. Univ. Antioq.* **2022**, *107*, 80–87. [CrossRef]
15. Lima, D.R.S.; Adarme, O.F.H.; Baêta, B.E.L.; Gurgel, L.V.A.; de Aquino, S.F. Influence of different thermal pretreatments and inoculum selection on the biomethanation of sugarcane bagasse by solid-state anaerobic digestion: A kinetic analysis. *Ind. Crop. Prod.* **2018**, *111*, 684–693. [CrossRef]
16. Pitt, R.E.; Cross, T.L.; Pell, A.N.; Schofield, P.; Doane, P.H. Use of in vitro gas production models in ruminal kinetics. *Math. Biosci.* **1999**, *159*, 145–163. [CrossRef] [PubMed]
17. Venkateshkumar, R.; Shanmugam, S.; Veerappan, A. Anaerobic co-digestion of cow dung and cotton seed hull with different blend ratio: Experimental and kinetic study. *Biomass Convers. Biorefinery* **2022**, *12*, 5635–5645. [CrossRef]
18. Jhunjhunwala, U.; Padhi, S.K.; Pattanaik, L.; Sharma, D.; Kumar, A.; Chaudhary, P.; Saxena, V. Anaerobic co-digestion of food waste and waste activated sludge for methane production: Evaluation of optimum ratio, microbial analysis, and kinetic modeling. *Biomass Convers. Biorefinery* **2023**. [CrossRef]
19. Redzwan, G.; Banks, C. The use of a specific function to estimate maximum methane production in a batch-fed anaerobic reactor. *J. Chem. Technol. Biotechnol.* **2004**, *79*, 1174–1178. [CrossRef]
20. Groot, J.C.J.; Cone, J.W.; Williams, B.A.; Debersaques, F.M.A.; Lantinga, E.A. Multiphasic analysis of gas production kinetics for in vitro fermentation of ruminant feeds. *Anim. Feed Sci. Technol.* **1996**, *64*, 77–89. [CrossRef]
21. Lasdon, L.S.; Fox, R.L.; Ratner, M.W. Nonlinear optimization using the generalized reduced gradient method. *RAIRO Oper. Res.* **1974**, *8*, 73–103. [CrossRef]
22. Morais, N.W.S.; Coelho, M.M.H.; de Oliveira, M.G.; Mourão, J.M.M.; Pereira, E.L.; dos Santos, A.B. Kinetic Study of Methanization Process Through Mathematical Modeling in Biochemical Methane Potential Assays from Four Different Inoculants. *Water. Air. Soil Pollut.* **2021**, *232*, 423. [CrossRef]
23. Akaike, H. A New Look at the Statistical Model Identification. *IEEE Trans. Automat. Contr.* **1974**, *19*, 716–723. [CrossRef]
24. Chai, A.; Wong, Y.S.; Ong, S.A.; Lutpi, N.A.; Sam, S.T.; Kee, W.C.; Eng, K.M. Kinetic model discrimination on the biogas production in thermophilic co-digestion of sugarcane vinasse and water hyacinth. *Environ. Sci. Pollut. Res.* **2022**, *29*, 61298–61306. [CrossRef] [PubMed]
25. Roubinek, O.K.; Wilinska-Lisowska, A.; Jasinska, M.; Chmielewski, A.G.; Czerwionka, K. Production of Biogas from Distillation Residue as a Waste Material from the Distillery Industry in Poland. *Energies* **2023**, *16*, 3063. [CrossRef]
26. AL-Huqail, A.A.; Kumar, V.; Kumar, R.; Eid, E.M.; Taher, M.A.; Adelodun, B.; Abou Fayssal, S.; Mioč, B.; Držaić, V.; Goala, M.; et al. Sustainable Valorization of Four Types of Fruit Peel Waste for Biogas Recovery and Use of Digestate for Radish (*Raphanus sativus* L. cv. Pusa Himani) Cultivation. *Sustainability* **2022**, *14*, 10224. [CrossRef]
27. Hassan, M.; Anwar, M.; Sarup Singh, R.; Zhao, C.; Mehryar, E. Co-digestion of chicken manure with goose manure and thermo-oxidative-treated wheat straw in CSTR: Co-digestion synergistics and OLR optimization through kinetic modeling. *Biomass Convers. Biorefinery* **2022**. [CrossRef]
28. Khatun, M.L.; Nime, J.; Nandi, R.; Alam, M.M.; Saha, C.K. Co-digestion of poultry droppings and banana waste for maximizing biogas production in Bangladesh. *Fuel* **2023**, *346*, 128346. [CrossRef]
29. Gomes, C.S.; Strangfeld, M.; Meyer, M. Diauxie studies in biogas production from gelatin and adaptation of the modified gompertz model: Two-phase gompertz model. *Appl. Sci.* **2021**, *11*, 1067. [CrossRef]

fermentation

Article

Biorefinery Approach for H_2 and Acids Production Based on Uncontrolled pH Fermentation of an Industrial Effluent

María Eugenia Ibañez-López [1], Encarnación Díaz-Domínguez [1], Miguel Suffo [2], Jacek Makinia [3], Jose Luis García-Morales [1] and Francisco Jesús Fernández-Morales [4,*]

1 Department of Environmental Technologies, Faculty of Marine and Environmental Sciences, IVA-GRO-Wine and Agrifood Research Institute, University of Cadiz, 11510 Cadiz, Spain; encarnacion.diaz@uca.es (E.D.-D.); joseluis.garcia@uca.es (J.L.G.-M.)

2 Department of Mechanical Engineering and Industrial Design, High Engineering School, Universidad de Cádiz, Campus Río San Pedro S/N, Puerto Real, 11510 Cadiz, Spain

3 Faculty of Civil and Environmental Engineering, Gdansk University of Technology, 80-233 Gdansk, Poland; jmakinia@pg.edu.pl

4 Department of Chemical Engineering, University of Castilla-La Mancha, Avda. Camilo José Cela S/N, 13071 Ciudad Real, Spain

* Correspondence: fcojesus.fmorales@uclm.es; Tel.: +34-926-05-21-79

Abstract: In this work, the feasibility of uncontrolled pH acidogenic fermentation of industrial organic effluent from corn-bioethanol production was studied and modelled by using a Monod-based mathematical model. In order to do that, several tests were carried out at different initial pH values, ranging from 4 to 6. The experimental data showed a pH reduction during the fermentation process due to the generation of short-chain acids. When starting at initial pH of 5.0 and 6.0, the substrates were fully fermented reaching final pH s over 4 units in both cases and a final undissociated fatty acid concentration of about 80 ($mmol \cdot L^{-1}$) in both cases. Regarding fermentation at an initial pH of 4, the pH decreased to 3.5 units, and the organic substrates were not fully fermented due to the stoppage of the fermentation. The stoppage was caused by the very acidic pH conditions. The biomass showed an uncoupled growth as the operating conditions became more acidic, and, finally, the biomass growth was zero. Regarding the generation of fermentation products, in general terms, the highest economical value of products was obtained when fermenting at an initial pH of 5. More specifically, acetic acid was the acid that presented the highest yield at an initial pH value of 4. Butyric yield showed the highest values at initial pH values of 5 and 6. The highest H_2 yield (1.1 mol $H_2 \cdot mol^{-1}$ dextrose) was achieved at an initial pH value of 5. Finally, the experimental data were modelled using a Monod-based model. From this model, the value of the main kinetics and stoichiometric parameters were determined.

Keywords: corn-bioethanol effluent; acidogenic fermentation; uncontrolled pH; modelling

Citation: Ibañez-López, M.E.; Díaz-Domínguez, E.; Suffo, M.; Makinia, J.; García-Morales, J.L.; Fernández-Morales, F.J. Biorefinery Approach for H_2 and Acids Production Based on Uncontrolled pH Fermentation of an Industrial Effluent. *Fermentation* **2023**, *9*, 937. https://doi.org/10.3390/fermentation9110937

Academic Editor: Miguel Ladero

Received: 29 September 2023
Revised: 26 October 2023
Accepted: 26 October 2023
Published: 28 October 2023

1. Introduction

The increasing consumption of energy and chemicals has caused undesirable effects on the planet. Regarding the energy aspects, future energy sources should not only be renewable and sustainable, but also versatile [1]. Hydrogen (H_2) has emerged as a potential candidate for energy supply over the past few decades. The advantages of using H_2 include its clean consumption, lack of polluting emissions, and high energy output, i.e., 123 MJ/kg, which is approximately three times that of fossil fuels: gasoline is 47.3 MJ/kg, kerosene is 46.2 MJ/kg, and diesel fuel is 44.8 MJ/kg [2]. The market price of hydrogen depends on its precedence, with the prices of grey, blue, and green H_2 being about 1.5, 2.5, and 5 kEUR/ton, respectively [3,4]. In addition, a significant number of chemicals are produced daily by the industry, among which Volatile Fatty Acids (VFA) is significant. The most known and widely used VFA are acetic, propionic, and butyric acids, which represent the whole production of close to 4000 kt/year, with a market price of 0.4–0.8 kEUR/ton for acetic acid, 1.7–1.9 for propionic acid, and 1.8–2.0 for butyric acid [5]. These chemicals are usually produced from fossil fuels, with a significant environmental impact, making it necessary to search for more sustainable options [6].

To ensure the sustainability of energy and chemical production, eco-friendly production technologies and sustainable raw materials should be employed. Amongst the sustainable production technologies, biochemical processes stand out. There are different biochemical processes for H_2 and chemical generation, and acidogenic fermentation is one of the most attractive options due to its ability to convert organic waste effluents into H_2 and VFA [7].

Organic-rich effluents can serve as an efficient substrate for H_2 and VFA production, reducing the overall treatment costs [8]. Furthermore, using waste materials as raw materials aligns with the principles of circular economy, providing additional value. In the literature, the generation of industrial wastewater has been described as containing high concentrations of easily degradable organic substrates, which can ensure high material and energy yields. Common examples include the sugar industry [9], the food processing industry [10,11], distilleries [12], chemical industries [13], and pharmaceuticals [14]. Even the energy industry from biofuel production results in the excessive generation of byproducts, which can be effectively utilized as a substrate for H_2 and VFA production [15,16].

One of the most interesting bioenergy industry effluents is corn syrup containing wastewater, which is generated during corn-bioethanol production and typically involves the fermentation and distillation of corn starch to produce bioethanol [17,18]. During this process, enzymes break down corn starch into simpler sugars, primarily dextrose. The dextrose is fermented and then distilled to separate the ethanol from the non-fermentable components, resulting in corn syrup, which is essentially a concentrated solution of dextrose [19]. The wastewaters containing corn syrup can be fermented with the aim of producing chemicals and energy.

Acidogenic fermentation can provide an attractive and efficient process for converting corn syrup substrates into H_2 and VFA, which have broad applications in chemical synthesis [20,21]. In the literature, the influence of the main operational variables on the acidogenic fermentation process has been described, including the influence of the operational pH, temperature, substrate concentration, etc. The pH has been identified as one of the most important variables influencing the metabolic pathways, and therefore, the generation of fermentation products [1,20,22–31]. It is known that there is an optimal operational pH for acidogenic fermentation in which the highest fermentation rate and yield are obtained [32]. This behaviour can be explained by the inhibitory effects of the accumulation of acids generated during fermentation [33]. Acid accumulation reduces pH, which influences the intracellular pH of microorganisms and leads to metabolic disruptions [34]. When the pH falls below the pKa value of the acids, they become protonated and can easily permeate via passive diffusion inside the microorganisms [35]. Once inside the cells, these protonated acids accumulate, resulting in a decrease in the intracellular pH [28]. Therefore, to ensure optimal operational conditions, most fermentation processes are conducted under controlled pH. However, this enhancement of productivity using pH control entails the utilization of control systems and the consumption of acids or alkalis to maintain the pH near the set point. On the one hand, from an environmental perspective, the doses of acids and alkalis lead to an increase in the effluent salinity due to the Cl^-, Na^+, etc. dosage, reducing the sustainability of the generated H_2 and VFA. On the other hand, from an economical perspective, the necessity of control systems and the consumption of acids and alkalis increase the operational costs of the process. Regarding the costs, it must be stated that while pH-controlled fermentative processes could be viable when operated at a laboratory or bench scale, they could become economically challenging when scaled up to full production. The increasing operational costs associated with larger-scale operations can render pH-controlled full-scale fermentative processes economically unsustainable. In any case, the increase in operational cost due to the acid and alkali consumption of the controlled pH acidogenic fermentation should also be balanced with the lower economic cost of the fermentation product separation due to the higher concentration reached.

Because of this economic drawback, in the literature, some research focused on uncontrolled pH acidogenic fermentation [1,36,37]. In these studies, which were carried out without pH control, the initial pH value is of crucial importance in order to ensure optimum performance. In the literature, different initial pH values have been studied, observing an optimum pH range even broader than that shown by the studies carried out with pH control. For example, the optimal initial pH observed by Khanal et al. [38] and Lin et al. [39] were 4.5 and 5.5, respectively, whereas the optimal initial pH reported by Lee et al. [40] was 9.

In this context, the present work focused on the study of the influence of the initial pH on the uncontrolled pH, mixed culture, and acidogenic fermentation of an industrial organic effluent from corn-bioethanol production. Additionally, a Monod-based mathematical model was proposed with the aim of determining the main kinetic and stoichiometric parameters of substrate fermentation, product generation, and biomass growth.

2. Materials and Methods

2.1. Innoculum and Synthetic Corn-Bioethanol Effluent

The inoculum used in this study was taken from the activated sludge reactor of a conventional municipal wastewater treatment plant. Further details about this facility can be found in the literature [41,42]. To facilitate the acidogenic fermentation of the corn-bioethanol effluent, the inoculum underwent an acclimatization process, following a previously described procedure in the literature [43]. This process involved the utilization of a Sequential Batch Reactor (SBR) operating under strict anaerobic conditions at 35 °C.

In order to ensure the reproducibility of the experiments, corn-bioethanol effluent was used during the batch experiments without pH control, and synthetic corn-bioethanol wastewater was used. According to the literature, the effluent mainly consists of dextrose 80–300 g/L and presents a pH ranging from 4 to 6 [19]. Because of that, a synthetic corn-bioethanol effluent consisting primarily of dextrose, with an initial dextrose concentration of 150 $g \cdot L^{-1}$, and variable initial pH ranging from 4 to 6 was synthesized in the laboratory. The synthetic corn-bioethanol effluent was supplemented with inorganic and trace minerals, as outlined in the literature [19,44]. A detailed composition of the inorganic and trace minerals used in the study is presented in Table 1.

Table 1. Inorganic components and trace minerals used in this study.

Compound	Concentration ($g \cdot L^{-1}$)	Compound	Concentration ($g \cdot L^{-1}$)
$(NH_4)Cl$	4.89	$FeSO_4\ 7H_2O$	11.3×10^{-3}
KH_2PO_4	2.85	$MnCl_2\ 4H_2O$	9.1×10^{-3}
NaCl	1.07	$CuCl\ 2H_2O$	8.0×10^{-3}
Na_2SO_4	0.21	$CoCl_2\ 6H_2O$	3.5×10^{-3}
$MgCl_2\ 6H_2O$	0.44	$CaCl_2$	2.2×10^{-3}
EDTA	0.18	$NiCl_2\ 6H_2O$	1.8×10^{-3}
$ZnSO_4\ 7H_2O$	11.7×10^{-3}		

2.2. Experimental Set-Up

To measure the hydrogen production, batch tests were carried out using Oxitop® reactors. In parallel, liquid samples were extracted from water-sealed serum bottles. Each serum bottle was discarded after sampling, so the number of serum bottles used matched the number of samples taken.

Before the start of the experiments, 1.2 L of acclimatized culture was mixed with 0.2 L of synthetic corn-bioethanol effluent and 1.6 L of demineralised water, resulting in a total mixture volume of 3 L. This mixture yielded an initial dextrose concentration of about 10 $g \cdot L^{-1}$, 50 $mmol \cdot L^{-1}$. Subsequently, the pH of the mixtures was adjusted to the desired initial pH levels of 4, 5, and 6 units using HCl and NaOH. These mixtures were then distributed among the serum bottles and the Oxitop® reactors. The Oxitop® reactors of 1 L of volume contained 0.3 L of the mixture, whereas the serum bottles of 0.1 L contained 0.03 L of the mixture as they were identical to the ratio liquid/gas.

Once the Oxitop® reactors and the serum bottles were filled, N_2 gas at a flow rate of 5 $L \cdot min^{-1}$ was used to remove the oxygen from the liquid and gas phases. Then, the reactors were hermetically sealed. Throughout the fermentation process, the temperature was maintained at 35 °C using a thermostatic chamber. To ensure homogeneity during the experiments, magnetic stirring was induced in all the Oxitop® reactors and serum bottles.

2.3. Analytical Methods

Total Suspended Solids (TSS) were determined by filtering the samples through a Millipore 0.7-μm membrane filter, followed by overnight drying at 105 °C. To measure Volatile Suspended Solids (VSS), the same sample was subsequently ignited for 2 h at 550 °C. Both TSS and VSS were determined following standard methods [45]. The pH measurements were conducted using a pH probe (Mettler-Toledo, Worthington, EEUU).

Gas production was determined by means of the barometric sensors of the Oxitop® reactors. Continuous recordings of the gas pressure inside these systems were obtained using the sensors located in the Oxitop® reactor heads. To eliminate the contribution of CO_2 to the total pressure, NaOH pearls were placed in the rubber quivers of the Oxitop® reactors.

In order to characterize the substrate and fermentation product concentrations during the fermentative processes, liquid samples were collected and subjected to centrifugation at 12,000 rpm. Then, the supernatant was filtered through a 0.45 μm glass fiber filter. The dextrose concentration was determined from the filtrate using High-Performance Liquid Chromatography (HPLC) with an Agilent refractive index detector (series 1200). A Zorbax Carbohydrate Column (4.6 × 150 mm, 5-micron) was employed to separate the components at a temperature of 35 °C. The mobile phase consisted of 75 vol. % acetonitrile and 25 vol. % water, with a flow rate of 1.5 $mL \cdot min^{-1}$. The VFA (including acetic, propionic, and butyric acids) were determined from the filtrate samples using a Perkin Elmer flame ionization detector (FID) gas chromatograph. A Crossbond Carbowax Column (15 m × 0.32 mmID, 0.25 mm df) was used. The oven temperature was initially set to 140 °C for 1.5 min, followed by an increase at a rate of 25 °C/min until it reached 190 °C, where it was held for 2 min. The injector and detector temperatures were 200 °C and 230 °C, respectively. Nitrogen was used as the carrier gas. The lactic acid concentration was measured from the filtrate samples by using an HPLC system with an Agilent ultraviolet diode array detection (UV-DAD) detector and a Zorbax SB-Aq column (4.6 × 150 mm, 5 μm). The mobile phase was a pH 2 buffer (0.05 M phosphate) consisting of 99% water and 1% acetonitrile.

2.4. Mathematical Model

In order to model the behaviour of the corn-bioethanol effluent fermentation without pH control, a mathematical model was developed and fitted to the experimental data. The model developed was based on the Monod kinetics and takes into account the inhibition caused by the undissociated acid accumulation described in the literature [46,47]. This kind of inhibition has been described as one of the most relevant when performing fermentation processes [47]. The experimental results of biomass growth and decay, substrate fermentation, and fermentation product generation are described by means of the model presented in the Petersen Matrix of Table 2.

Table 2. Petersen matrix of the processes taking place in this study.

Process \ Components	Dextrose	Acetic	Lactic	Butyric	Propionic	Biomass	CO_2	H_2	*Process Rate*
Acetic production	$-1/Y_A$	1							$q_a \frac{S}{K_s+S} \cdot \frac{1}{1+1.2^{(C_{HA}-K_{substrate})}} X$
Lactic production	$-1/Y_L$		1						$q_l \frac{S}{K_s+S} \cdot \frac{1}{1+1.2^{(C_{HA}-K_{substrate})}} X$
Butyric production	$-1/Y_B$			1					$q_b \frac{S}{K_s+S} \cdot \frac{1}{1+1.2^{(C_{HA}-K_{substrate})}} X$
Propionic production	$-1/Y_P$				1				$q_p \frac{S}{K_s+S} \cdot \frac{1}{1+1.2^{(C_{HA}-K_{substrate})}} X$
Biomass growth	$-1/Y_{obs}$					1			$\mu_{max} \frac{S}{K_s+S} \cdot \frac{1}{1+1.2^{(C_{HA}-K_{biomass})}} X - DX$
Carbon dioxide production	$-1/Y_{CO2}$						1		$q_{CO_2} \frac{S}{K_s+S} \cdot \frac{1}{1+1.2^{(C_{HA}-K_{substrate})}} X$
Hydrogen production	$-1/Y_{H2}$							1	$q_{H_2} \frac{S}{K_s+S} \cdot \frac{1}{1+1.2^{(C_{HA}-K_{substrate})}} X$
Lactic consumption		1	−2		1		1		$q_{lc} \frac{S}{K_s+S} \cdot \frac{1}{1+1.2^{(C_{HA}-K_{substrate})}} X$
Lysis of biomass						−1			$D \cdot X$

In Table 2, X is the biomass concentration (g SSV·L^{-1}), μ_{max} is the maximum specific growth rate (h^{-1}), S is substrate concentration (mmol·L^{-1}), K_s is substrate half-saturation constant (mmol·L^{-1}), D is the decay constant (h^{-1}), and $K_{biomass}$ and $K_{sustrate}$ are the undissociated acid threshold concentrations (mmol·L^{-1}), which caused an important shift in biomass growth and substrate fermentation when those concentrations were reached. Y_{obs} is the observed biomass yield for each experiment (g SSV·$mmol^{-1}$ dextrose). The Y_{obs} can be determined by using Equation (1), where Y_x^{max} (g SSV·$mmol^{-1}$ dextrose) is the maximum biomass growth yield and Y_{maint} (g SSV·$mmol^{-1}$ dextrose) is the biomass maintenance.

$$Y_{obs} = Y_x{}^{max} - Y_{maint} \tag{1}$$

The C_{HA}, total undissociated acid concentration (mmol·L^{-1}), can be determined by taking into account that the acids generated can remain dissociated or undissociated as a function of their pKa and the pH of the medium. The pKa values of the obtained acids are as follows: 3.86 (lactic acid); 4.75 (acetic acid); 4.83 (butyric acid); 4.87 (propionic acid). The undissociated acid concentrations were calculated by using Equation (2).

$$C_{HA_i} = \frac{C_{Total}}{1 + 10^{(pH - pKa)}} \tag{2}$$

Therefore, the total concentration of undissociated acids can be calculated using Equation (3).

$$C_{HA} = \sum C_{HA_i} \tag{3}$$

Because of the continuous modification of the pH, the values of the kinetic and stoichiometric parameters changed during the experiment. In this work, according to the experimental results, linear functions were used to describe the pH dependence of the μ_{max} and the Y_{maint}. In the case of the K_s, a potential function was used. In the literature, similar dependencies have been described for these parameters [47,48].

$$\mu_{max} = a \cdot pH + b \tag{4}$$

$$Y_{maint} = c \cdot pH + d \tag{5}$$

$$k_s = e \cdot pH^2 + f \cdot pH + g \tag{6}$$

In Equations (4)–(6), the parameters a–g are the fitting values. Once the set of equations describing the process was defined, the experimental results of biomass growth, dextrose consumption, and fermentation product generation were used to fit the mathematical model.

In order to fit the model to the experimental results, the set of equations presented was solved simultaneously using the Gauss–Newton algorithm. An initial set of values was assigned to the model parameters, and after several iterations, the values of the parameters that minimised the sum of squared errors (SSE) were chosen as the best estimate. The SSE expression used to estimate the values of the model parameters was as follows:

$$\chi(p) = \frac{\sqrt{\sum_{i=1}^{n} (z_{meas,i} - z_i(p))^2}}{z} + \frac{\sqrt{\sum_{i=1}^{n} (y_{meas,i} - y_i(p))^2}}{y} + \frac{\sqrt{\sum_{i=1}^{n} (x_{meas,i} - x_i(p))^2}}{x} \tag{7}$$

where x, y, and z are the model variables: dextrose, biomass, and fermentation product concentration, $x_{meas,i}$, $y_{meas,i}$ and $z_{meas,i}$ are the ith measurement of the dextrose, biomass, and fermentation products concentration, and $x_i(p)$, $y_i(p)$, and $z_i(p)$ are the calculated values of the model variables corresponding to the ith measurement. x, y, and z are the mean value of the measurements and n is the number of data points.

3. Results

3.1. Experimental Results

The first effects observed during the non-controlled pH corn-bioethanol effluent acidogenic fermentation were the dextrose consumption and the pH reduction. The results obtained are presented in Figure 1. Regarding the substrate consumption rate, it was observed that the higher the initial pH, the higher the substrate consumption rate. The time required to complete the fermentation was around 15 and 20 h when the initial pH values were 6 and 5 units, respectively. In the case of the initial pH of 4 units, the substrate was not completely fermented. In the latter case, the substrate concentration only decreased about 10 mmol·L^{-1}, indicating the existence of an inhibition event that stopped the fermentation process when the initial pH was 4.

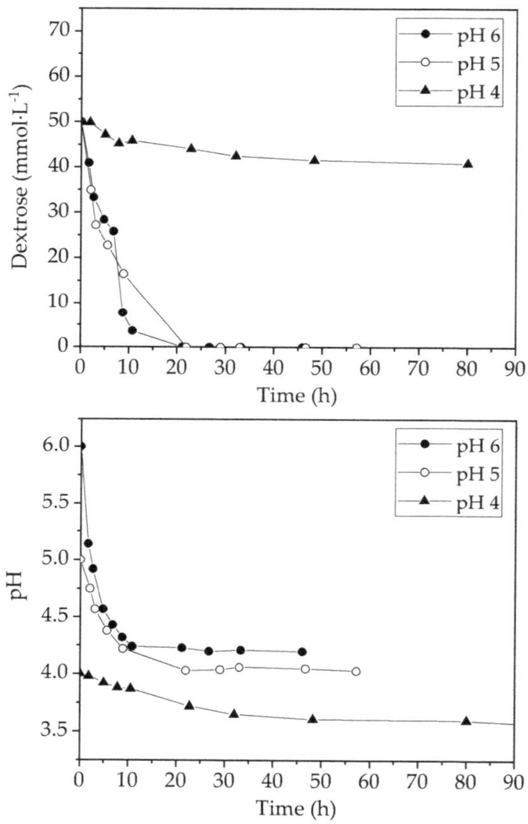

Figure 1. Dextrose and pH evolution during the uncontrolled pH fermentation.

The pH reduction can be explained by the pH values reached and the accumulation of the acids produced as a result of acidogenic fermentation. When the initial pH was 5 or 6, the dextrose was fully consumed and the pH decreased to values close to 4.1 and 4.3 units, respectively. These values remained constant due to the stopping of the fermentation process caused by the full consumption of the dextrose. The higher pH drop that was observed when operating at an initial pH of 6 units could be explained by the effect of the pH on acid dissociation. When operating at a higher pH, a lower fraction of the acids is undissociated due to the relationship between the pKa of the acid and the pH. Taking into account that the undissociated acids are responsible for inhibition events, a lower undissociated acid concentration leads to a lower inhibitory effect [34]. This lower inhibition could explain the full consumption of the dextrose previously described when fermenting at initial pH values of 5 and 6. When the initial pH was 4, the pH reached a value of 3.5 units, which was a value lower than those obtained at the initial pH of 5 or 6. Once, the system reached the pH of 3.5 the value remained stable. It must be highlighted that the lower pH reached when starting at a pH of 4 could be explained by the inhibition caused by the combined effect of the pH and the undissociated acids accumulated in the medium. Starting at pH 4, a negligible acid concentration was present in the medium; therefore, the concentration of the undissociated acid was also very low. This allowed the fermentation of some substrates in spite of the low operational pH.

The trends of pH and dextrose influenced the kinetic and stoichiometric parameters and, therefore, affected other variables, such as biomass growth and fermentation product generation. Biomass growth, as well as the production of every fermentation product, is defined by the metabolic pathway followed. In this case, due to the evolution of the operational conditions (mainly the pH and the undissociated acid concentration), the biomass growth and fermentation product generation changed during the experiment. Figure 2 presents the experimental results of biomass growth. By comparing the trend of biomass growth with that of dextrose, important differences can be identified.

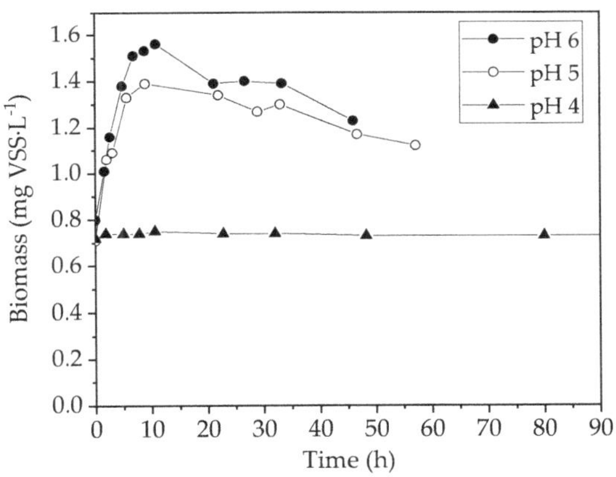

Figure 2. Biomass evolution during uncontrolled pH fermentation.

In Figure 2, it can be observed that when the initial pH was 4 units, the biomass concentration remained stabilized during the experiment. However, the biomass increased during the first hours of fermentation when the initial pH values were 5 or 6 units. At the initial pH values of 5 and 6, it can be seen that the biomass concentration starting at an initial pH of 6 was higher than that achieved at an initial pH of 5. In principle, taking into account that in both cases the substrate was fully consumed, the biomass growth should be identical. However, by comparing these biomass growth results with the substrate consumption trends shown previously, it can be observed that the biomass growth decreased and even stopped before the full consumption of the substrate. This behaviour indicates the difficulties faced by the microorganisms when exposed to the operational conditions reached at the end of fermentation. This biomass growth deviation from theoretical behaviour is called uncoupled growth [35]. Uncoupled growth is caused by the increasing maintenance energy requirements of the cells when developing under hostile conditions [49]. This increase in energy maintenance requirements leads to a decrease in the energy available for biomass growth [50]. As previously stated, hostile conditions could be caused by the accumulation of undissociated acids when operating at a lower pH. The dissociated and undissociated fractions can be determined by means of the acid dissociation constant (K_a) defined by Equation (8).

$$K_a = \frac{[A^-]\cdot[H^+]}{[AH]} \quad (8)$$

Based on this equation, K_a expresses the strength of an acid (in other words, how easily the acid releases a proton). In addition, the equation shows how the dissociation state of weak acids varies according to the proton's concentration in the solution.

With the aim of investigating the effect of acid accumulation on the behaviour of the system, dissociated and undissociated acid concentrations were determined. Figure 3 shows the total undissociated acid concentrations during the fermentation at initial pH values of 4, 5, and 6. As can be seen at the beginning of fermentation, the undissociated acid concentration decreased as the initial pH increased due to the pH dependence of the acid dissociation. In the fermentations carried out with initial pH values of 5 and 6, the differences between their undissociated acid concentrations decreased during the experiment, and finally, both fermentations achieved a very similar final concentration of undissociated acid, around 80 mmol·L^{-1}. On the other hand, when fermentation was carried out at an initial pH of 4, the total undissociated acid concentration increased slowly due to the very low fermentation rate and conversion caused by the inhibition.

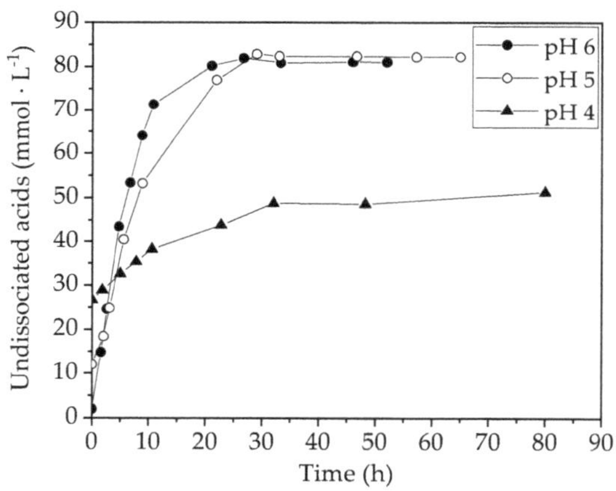

Figure 3. Undissociated acid evolution during uncontrolled pH fermentation.

The total undissociated acid concentration presented a trend different from that of the biomass growth in all the experiments. By comparing these trends, it can be seen that the undissociated acid concentration kept increasing even when the biomass growth stopped. Thus, by comparing Figures 2 and 3, it can be seen that the values of undissociated acid concentration increased the energy maintenance requirements and avoided biomass growth. These values were around 65 mmol·L^{-1} when fermenting at the initial pH of 5 and 6. This value is very similar to that reported in the literature [47,51].

The substrate was completely consumed when the fermentative processes were carried out at the initial pH values of 5 and 6. In both cases, the total undissociated acid concentration reached about 80 mmol·L^{-1} when the metabolism ceased. Regarding the experimental results at an initial pH of 4, it was observed that the total undissociated acid concentrations were significantly lower than 65 mmol·L^{-1} when the metabolism ceased due to the inhibition caused by the undissociated acids. Because of this, it was concluded that a high undissociated acid concentration was not the sole mechanism that stopped the biomass growth and substrate fermentation. By comparing the results obtained at an initial pH of 4 with those obtained at an initial pH of 5 and 6, it was observed that the only variable that significantly changed was the pH, which reached a value of 3.5 units.

In order to investigate the effect of the initial pH on the generation of fermentation products, the specific production yield of each acid produced was determined. These yields were calculated according to Equation (9), where *P* refers to the product generated and *S* refers to the substrate consumed.

$$Y_X = \frac{P_{Final} - P_{Initial}}{S_{Initial} - S_{Final}} \quad (9)$$

The obtained average fermentation product yields are presented in Figure 4.

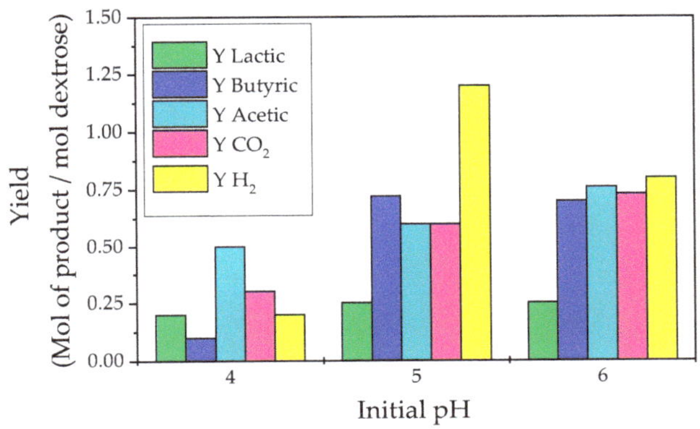

Figure 4. Average experimental yields of the main fermentation products generated.

In Figure 4, it can be observed that the fermentation product yields and distribution changed with the initial pH. In this Figure, it can be seen that acetic and lactic acid yields were higher at initial pH 4. However, the butyric acid showed an opposite trend, achieving the highest yields at the initial pH values of 5 and 6. The highest hydrogen yield (1.1 mol H_2·mol dextrose^{-1}) was achieved at the initial pH of 5, showing a slight decrease at the initial pH of 6 and a significant fall at the initial pH of 4. This H_2 trend is similar to that reported in the literature [27,46]. H_2 production, which is one of the most interesting fermentation products obtained from acidogenic fermentation, has been researched previously by many researchers [3,21]. The H_2 yield results obtained in this work were similar to other results published in the literature. In the literature, similar research studying mixed culture acidogenic fermentation with pH control showed H_2 yields ranging from 1.1 mol H_2·mol dextrose^{-1} [52] to 1.46 mol·H_2·mol dextrose^{-1} [53], and our yield was 1.1 mol·H_2·mol dextrose^{-1}, which is a typical value.

Considering the different substrate conversions obtained when operating at different initial pH values, the net productivities of the acids and H_2 as well as the estimated economic value were determined and are presented in Table 3.

Table 3. Average productivities of the different fermentation products generated at each initial pH.

Fermentation Product	pH 4	pH 5	pH 6
Acetic acid (mmol·L^{-1})	0.52	0.63	0.80
Butyric acid (mmol·L^{-1})	0.07	0.58	0.50
Propionic acid (mmol·L^{-1})	0.03	0.02	0.03
H_2 (mmol·L^{-1})	0.12	0.89	0.56
CO_2 (mmol·L^{-1})	0.23	0.50	0.61
Estimated economic value (EUR·L^{-1})	0.03	0.12	0.11

According to Table 3, the best initial pH from an economic point of view was pH 5 and 6, with the accumulated economic value being approximately the same in both cases. In the cases of initial pH 5 and 6, the main contribution to the economic value was butyric acid due to its higher specific yield and market price. Hydrogen has an even higher market price; however, its lower mass productivity reduced its contribution to the accumulated economic value. Based on these results, it must be stated that from a global economic point of view, it is indifferent to start fermentation at an initial pH of 5 or 6. For hydrogen production, it would be better to ferment the corn-bioethanol effluent at pH 5, and for VFA production, it would be better to ferment at pH 6.

3.2. *Mathematical Modelling*

With the objective of accurately quantifying all the fermentation products generated during the corn-bioethanol effluent fermentations at different initial pH values, modelling studies were carried out. Before the modelling studies, the quality of the data set obtained was verified by means of a carbon balance, as shown in Figure 5. From the carbon balance, a final carbon recovery of over 80% was obtained in all cases.

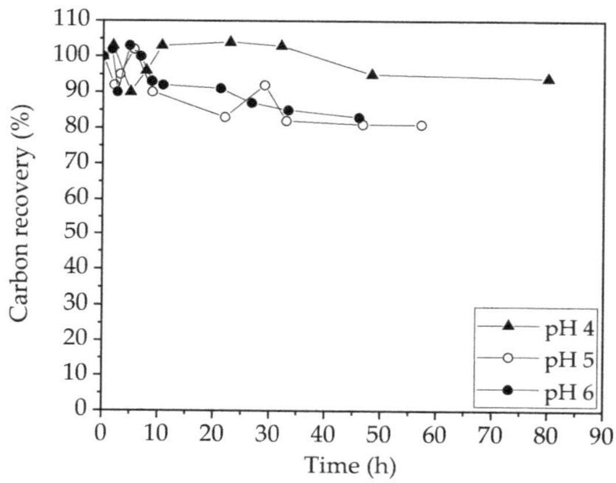

Figure 5. Carbon recovery during uncontrolled pH fermentation.

The lower carbon recovery experienced at the end of the fermentation process can be explained by the loss of carbon compounds such as in the biomass growth. Moreover, the error observed in the carbon balance can be explained by the existence of traces of undetected fermentation products in the liquid bulk. It is remarkable that higher carbon recovery occurs during fermentation at an initial pH of 4. This behaviour could be explained by the lower substrate consumption, and therefore, the lower biomass and fermentation product generation and its associated error. Regarding the carbon balance, it must be highlighted that the SBR operation involves maintaining 1.2 L from the previous cycle to the subsequent ones as an inoculum. This volume contained fermentation products from the previous fermentation cycle; therefore, its carbon content was added to the carbon content of the dextrose added at the beginning of every cycle, 10 $g \cdot L^{-1}$.

Once the quality of the data set was ensured, the main kinetic and stoichiometric parameters when operating at different initial pH values were determined. To do this, the mathematical Monod-based model previously described was fitted to the obtained results.

As previously shown, the pH decreased during the fermentation process due to the generation of short-chain fatty acids. This pH change during the fermentation process could affect the metabolic pathways of the mixed microbial culture used, thereby affecting the kinetic and stoichiometric parameters. In order to take this effect into account, pH-dependent functions were used to describe several kinetic and stoichiometric parameters. In the literature, the mathematical functions describing the stoichiometric and kinetic parameters are presented as the functions of the pH value [47,48]. In this work, linear and exponential functions were proposed according to Equations (4)–(6).

Linear functions were used to describe the dependence of μ_{max} and Y_{maint} on the pH value. The μ_{max} increased when the pH increased within the range studied in this work, whereas the Y_{maint} decreased as the pH increased. A second-order function was used to describe the dependence of Ks on the pH. This function was selected due to the quick increase in Ks when the pH changed from 5 to 4. This behaviour was also previously described in the literature [48].

To begin the iterative determination of the pH dependence of the model parameters, the values of Y_x^{max} (0.023 g SSV·mmol^{-1} dextrose) and the decay coefficient (0.03 h^{-1}, 0.07 h^{-1}, 0.13 h^{-1} at pH 4, 5 and 6) were taken from the literature as initial values [47]. The results obtained from the modelling are presented in Table 4.

Table 4. Modelling values of the kinetic and stoichiometric parameters.

Parameter		pH 4	pH 5	pH 6
Y_x^{max} (g VSS·mmol S^{-1})		0.023	0.023	0.023
Decay (h^{-1})		0.001	0.005	0.007
μ_{max} (h^{-1})	a	0.041	00.49	0.046
	b	−0.142	−0.130	−0.133
Y_{maint} (g VSS·mmol S^{-1})	c	0.0040	0.0041	0.0041
	d	0.0271	0.0251	0.0251
K_s (mmol·L^{-1})	e	43.7	40.3	40.3
	f	−472	−472	−472
	g	1295	1285	1288
$K_{susbtrate}$ (mmol·L^{-1})		90	90	90
$K_{biomass}$ (mmol·L^{-1})		70	70	70
Y_a (mol·mol^{-1})		0.50	0.80	0.96
Y_b (mol·mol^{-1})		0.10	0.60	0.60
Y_l (mol·mol^{-1})		0.80	0.20	0.20
Y_p (mol·mol^{-1})		0.01	0.02	0.04
Y_{H2} (mol·mol^{-1})		0.10	1.10	0.69
Y_{CO2} (mol·mol^{-1})		0.32	0.60	0.73

The undissociated acid inhibition parameters, $K_{biomass}$ and $K_{substrate}$, were determined, with their values being 70 and 90 mmol·L^{-1}, respectively. The $K_{biomass}$ value was very similar to the concentration of undissociated acids, resulting in biomass growth detention. The $K_{susbstrate}$ value was higher than the experimental concentration of the achieved undissociated acids. For this reason, it did not cause the detention of the fermentative process, as previously stated. Regarding the parameter values presented in Table 4, they accurately described the process behaviour in spite of the different initial pH values. This demonstrates that the values shown in this Table can be used to describe the corn syrup acidogenic fermentation at any pH value within the range studied.

In the literature, the changes in parameter values when the pH of the medium changes have also been presented. The researchers found that μ_{max} increased from a pH of 6.4 to a pH of 7.8, where it achieved the highest value, after which it started to decrease. In previous studies, the influence of pH on biomass yield in fermentations of rich protein wastewater was carried out by a culture taken from a methanogenic reactor [54]. This study showed that the biomass yield increased when the pH increased from 4 to 6; at this pH, the biomass yield started to decrease. The optimal values of the kinetic parameters for lactose fermentation and lactic acid production were also estimated by using *Lactobacillus plantarum*. The obtained values at different pH values showed an important increase in Ks when the pH changed from 5 to 4, whereas Ks was almost constant in the pH range of 5–7. Similar trends to those obtained in this work for fermenting synthetic corn-bioethanol effluent by using a mixed culture have been described in the literature [48]. This coincidence indicates that these trends are general when fermentation is carried out without pH control. For example, Figure 5 presents the experimental data and the fitted model. As can be seen in Figure 6, the model accurately predicts the evolution of the main variables, indicating that the fitting values of the kinetic and stoichiometric parameters were adequate for predicting the behaviour during corn syrup fermentation with uncontrolled pH.

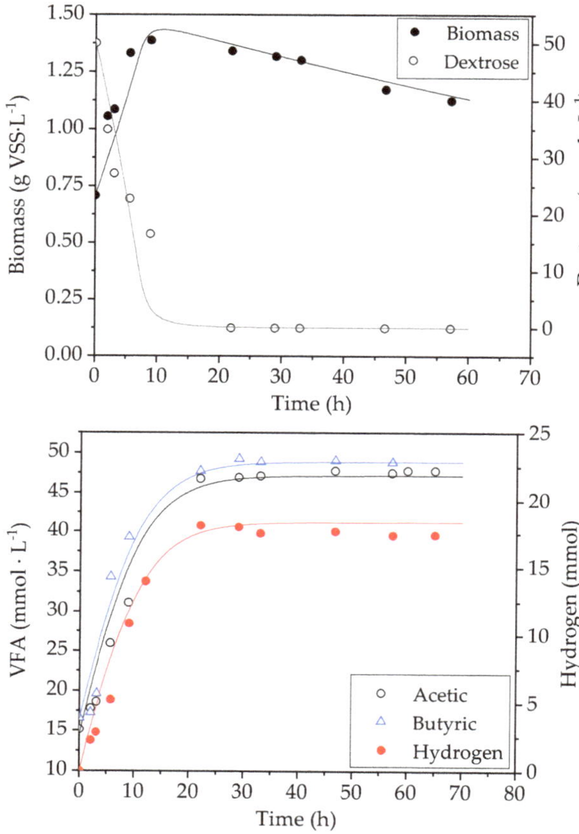

Figure 6. Evolution of the main variables during uncontrolled pH fermentation. Lines correspond to the modelling fit.

4. Conclusions

The fermentation of the substrates contained in the corn-bioethanol effluent without pH control generates a wide spectrum of VFAs, H_2, and CO_2. The decreased pH facilitated acid dissociation, thereby inhibiting the fermentative process. The significance of this inhibition increased as the initial pH decreased and even stopped the fermentation process when the initial pH was 4. The high dissociated acid concentration caused an uncoupled biomass growth when the undissociated acid concentration reached approximately 65 mmol·L^{-1}. The influence of the initial pH and the subsequent evolution of the operational conditions affected the fermentation product distribution. In general, fermentation product yields increased when the initial pH was higher. However, butyric acid presented a maximum yield when fermenting at an initial pH of 5. H_2 and CO_2 were the only gaseous products generated during fermentation. The CO_2 production increased as the pH increased, and H_2 achieved the highest production when fermenting at an initial pH of 5. Based on these results, it can be concluded that corn-bioethanol effluent can be valorised by means of an uncontrolled pH acidogenic fermentation process, especially when starting at an initial pH of 5 or 6. Finally, a mathematical model was fitted to the experimental data, and the values of the main kinetic and stoichiometric parameters were determined. The parameter values obtained were able to accurately predict the uncontrolled pH of acidogenic fermentation of corn-bioethanol effluent.

Author Contributions: Conceptualization, F.J.F.-M. and J.L.G.-M.; methodology, F.J.F.-M. and J.L.G.-M.; software, M.S., J.M. and F.J.F.-M.; validation, J.M. and F.J.F.-M.; formal analysis, M.S., J.M. and F.J.F.-M.; investigation, M.E.I.-L. and E.D.-D.; resources, J.L.G.-M., M.S. and F.J.F.-M.; data curation, M.E.I.-L. and E.D.-D.; writing—original draft preparation, F.J.F.-M. and J.L.G.-M.; writing—review and editing, F.J.F.-M. and J.M.; visualization, M.S., F.J.F.-M. and J.L.G.-M.; supervision, J.L.G.-M., J.M. and F.J.F.-M.; project administration, J.L.G.-M.; funding acquisition, J.L.G.-M. All authors have read and agreed to the published version of the manuscript.

Funding: This research was funded by the project FEDER-UCA-18-107460 entitled "Integration of the anaerobic bio refinery in the co-digestion treatment of biosolids and agri-food waste—Integracion de la bio refinería anaerobia en el tratamiento de co-digestión de biosólidos y residuos agroalimentarios".

Institutional Review Board Statement: Not applicable.

Informed Consent Statement: Not applicable.

Data Availability Statement: The data presented in this study are available on request from the corresponding author.

Conflicts of Interest: The authors declare no conflict of interest. The funders had no role in the design of the study; in the collection, analyses, or interpretation of data; in the writing of the manuscript; or in the decision to publish the results.

References

1. Sarkar, O.; Rova, U.; Christakopoulos, P.; Matsakas, L. Influence of initial uncontrolled pH on acidogenic fermentation of brewery spent grains to biohydrogen and volatile fatty acids production: Optimization and scale-up. *Bioresour. Technol.* **2021**, *319*, 124233. [CrossRef] [PubMed]
2. Baeyens, J.; Zhang, H.; Nie, J.; Appels, L.; Dewil, R.; Ansart, R.; Deng, Y. Reviewing the potential of bio-hydrogen production by fermentation. *Renew. Sustain. Energy Rev.* **2020**, *131*, 110023. [CrossRef]
3. Al-Qahtani, A.; Parkinson, B.; Hellgardt, K.; Shah, N.; Guillen-Gosalbez, G. Uncovering the true cost of hydrogen production routes using life cycle monetisation. *Appl. Energy* **2021**, *281*, 115958. [CrossRef]
4. Spazzafumo, G.; Raimondi, G. Economic assessment of hydrogen production in a Renewable Energy Community in Italy. *e-Prime—Adv. Electr. Eng. Electron. Energy* **2023**, *4*, 100131. [CrossRef]
5. Jankowska, E.; Chwialkowska, J.; Stodolny, M.; Oleskowicz-Popiel, P. Volatile fatty acids production during mixed culture fermentation—The impact of substrate complexity and pH. *Chem. Eng. J.* **2017**, *326*, 901–910. [CrossRef]
6. Pandey, A.K.; Pilli, S.; Bhunia, P.; Tyagi, R.D.; Surampalli, Y.R.; Zhang, C.T.; Kim, S.-H.; Pandey, A. Dark fermentation: Production and utilization of volatile fatty acid from different wastes- A review. *Chemosphere* **2022**, *288*, 132444. [CrossRef] [PubMed]
7. Sarkar, O.; Modestra, J.A.; Rova, U.; Christakopoulos, P.; Matsakas, L. Waste-Derived Renewable Hydrogen and Methane: Towards a Potential Energy Transition Solution. *Fermentation* **2023**, *9*, 368. [CrossRef]
8. Sivagurunathan, P.; Kumar, G.; Mudhoo, A.; Rene, E.R.; Saratale, G.D.; Kobayashi, T.; Xu, K.; Kim, S.H.; Kim, D.H. Fermentative hydrogen production using lignocellulose biomass: An overview of pre-treatment methods, inhibitor effects and detoxification experiences. *Renew. Sustain. Energy Rev.* **2017**, *77*, 28–42. [CrossRef]
9. Jayabalan, T.; Matheswaran, M.; Naina Mohammed, S. Biohydrogen production from sugar industry effluents using nickel based electrode materials in microbial electrolysis cell. *Int. J. Hydrogen Energy* **2019**, *44*, 17381–17388. [CrossRef]
10. Kumar, G.; Bakonyi, P.; Sivagurunathan, P.; Nemestóthy, N.; Bélafi-Bakó, K.; Lin, C.Y. Improved microbial conversion of de-oiled Jatropha waste into biohydrogen via inoculum pretreatment: Process optimization by experimental design approach. *Biofuel Res. J.* **2015**, *2*, 209–214. [CrossRef]
11. Dahiya, S.; Sarkar, O.; Swamy, Y.V.; Venkata Mohan, S. Acidogenic fermentation of food waste for volatile fatty acid production with co-generation of biohydrogen. *Bioresour. Technol.* **2015**, *182*, 103–113. [CrossRef] [PubMed]
12. Wicher, E.; Seifert, K.; Zagrodnik, R.; Pietrzyk, B.; Laniecki, M. Hydrogen gas production from distillery wastewater by dark fermentation. *Int. J. Hydrogen Energy* **2013**, *38*, 7767–7773. [CrossRef]
13. Venkata Mohan, S.; Vijaya Bhaskar, Y.; Sarma, P.N. Biohydrogen production from chemical wastewater treatment in biofilm configured reactor operated in periodic discontinuous batch mode by selectively enriched anaerobic mixed consortia. *Water Res.* **2007**, *41*, 2652–2664. [CrossRef] [PubMed]
14. Yang, G.; Wang, J.; Shen, Y. Antibiotic fermentation residue for biohydrogen production using different pretreated cultures: Performance evaluation and microbial community analysis. *Bioresour. Technol.* **2019**, *292*, 122012. [CrossRef] [PubMed]

15. Mahapatra, D.M.; Chanakya, H.N.; Ramachandra, T.V. *Euglena* sp. as a suitable source of lipids for potential use as biofuel and sustainable wastewater treatment. *J. Appl. Phycol.* **2013**, *25*, 855–865. [CrossRef]
16. Rossi, D.M.; da Costa, J.B.; de Souza, E.A.; Peralba, M.D.C.R.; Ayub, M.A.Z. Bioconversion of residual glycerol from biodiesel synthesis into 1,3-propanediol and ethanol by isolated bacteria from environmental consortia. *Renew. Energy* **2012**, *39*, 223–227. [CrossRef]
17. Lu, M.; Li, J.; Han, L.; Xiao, W. High-solids enzymatic hydrolysis of ball-milled corn stover with reduced slurry viscosity and improved sugar yields. *Biotechnol. Biofuels* **2020**, *13*, 77. [CrossRef]
18. Chen, M.H.; Kaur, P.; Dien, B.; Below, F.; Vincent, M.L.; Singh, V. Use of tropical maize for bioethanol production. *World J. Microbiol. Biotechnol.* **2013**, *29*, 1509–1515. [CrossRef]
19. Hafez, H.; Nakhla, G.; El Naggar, H. Biological Hydrogen Production from Corn-Syrup Waste Using a Novel System. *Energies* **2009**, *2*, 445–455. [CrossRef]
20. Eryildiz, B.; Lukitawesa; Taherzadeh, M.J. Effect of pH, substrate loading, oxygen, and methanogens inhibitors on volatile fatty acid (VFA) production from citrus waste by anaerobic digestion. *Bioresour. Technol.* **2020**, *302*, 122800. [CrossRef]
21. Atilano-Camino, M.M.; Luévano-Montaño, C.D.; García-González, A.; Olivo-Alanis, D.S.; Álvarez-Valencia, L.H.; García-Reyes, R.B. Evaluation of dissolved and immobilized redox mediators on dark fermentation: Driving to hydrogen or solventogenic pathway. *Bioresour. Technol.* **2020**, *317*, 123981. [CrossRef] [PubMed]
22. Slezak, R.; Grzelak, J.; Krzystek, L.; Ledakowicz, S. Influence of initial pH on the production of volatile fatty acids and hydrogen during dark fermentation of kitchen waste. *Environ. Technol.* **2020**, *42*, 4269–4278. [CrossRef] [PubMed]
23. Mota, V.T.; Ferraz Júnior, A.D.N.; Trably, E.; Zaiat, M. Biohydrogen production at pH below 3.0: Is it possible? *Water Res.* **2018**, *128*, 350–361. [CrossRef] [PubMed]
24. Ma, H.; Chen, X.; Liu, H.; Fu, B. Improved volatile fatty acids anaerobic production from waste activated sludge by pH regulation: Alkaline or neutral pH? *Waste Manag.* **2016**, *48*, 397–403. [CrossRef] [PubMed]
25. Jiang, J.; Zhang, Y.; Li, K.; Wang, Q.; Gong, C.; Li, M. Volatile fatty acids production from food waste: Effects of pH, temperature, and organic loading rate. *Bioresour. Technol.* **2013**, *143*, 525–530. [CrossRef] [PubMed]
26. Cubillos, G.; Arrué, R.; Jeison, D.; Chamy, R.; Tapia, E.; Rodríguez, J.; Ruiz-Filippi, G. Simultaneous effects of pH and substrate concentration on hydrogen production by acidogenic fermentation. *Electron. J. Biotechnol.* **2010**, *13*, 11–12. [CrossRef]
27. Tang, G.L.; Huang, J.; Sun, Z.J.; Tang, Q.Q.; Yan, C.H.; Liu, G.Q. Biohydrogen production from cattle wastewater by enriched anaerobic mixed consortia: Influence of fermentation temperature and pH. *J. Biosci. Bioeng.* **2008**, *106*, 80–87. [CrossRef] [PubMed]
28. Temudo, M.F.; Kleerebezem, R.; Van Loosdrecht, M. Influence of the pH on (Open) mixed culture fermentation of glucose: A chemostat study. *Biotechnol. Bioeng.* **2007**, *98*, 69–79. [CrossRef]
29. Fang, H.H.P.; Liu, H. Effect of pH on hydrogen production from glucose by a mixed culture. *Bioresour. Technol.* **2002**, *82*, 87–93. [CrossRef]
30. Yu, H.G.; Fang, H.H. Acidogenesis of dairy wastewater at various pH levels. *Water Sci. Technol. A J. Int. Assoc. Water Pollut. Res.* **2002**, *45*, 201–206. [CrossRef]
31. Veeken, A.; Kalyuzhnyi, S.; Scharff, H.; Hamelers, B. Effect of pH and VFA on hydrolysis of organic solid waste. *J. Environ. Eng.* **2000**, *126*, 1076–1081. [CrossRef]
32. Wang, J.; Wan, W. Factors influencing fermentative hydrogen production: A review. *Int. J. Hydrogen Energy* **2009**, *34*, 799–811. [CrossRef]
33. Castro-Villalobos, M.C.; Garcia-Morales, J.L.; Fernandez, F.J. By-products inhibition effects on bio-hydrogen production. *Int. J. Hydrogen Energy* **2012**, *37*, 7077–7083. [CrossRef]
34. Rodríguez, J.; Kleerebezem, R.; Lema, J.M.; Van Loosdrecht, M.C.M. Modeling product formation in anaerobic mixed culture fermentations. *Biotechnol. Bioeng.* **2006**, *93*, 592–606. [CrossRef] [PubMed]
35. Kleerebezem, R.; Rodríguez, J.; Temudo, M.F.; van Loosdrecht, M.C.M. Modeling mixed culture fermentations; the role of different electron carriers. *Water Sci. Technol.* **2008**, *57*, 493–497. [CrossRef] [PubMed]
36. Ribeiro, J.C.; Mota, V.T.; de Oliveira, V.M.; Zaiat, M. Hydrogen and organic acid production from dark fermentation of cheese whey without buffers under mesophilic condition. *J. Environ. Manag.* **2022**, *304*, 114253. [CrossRef] [PubMed]
37. Yuan, Q.; Lou, Y.; Wu, J.; Sun, Y. Long-term semi-continuous acidogenic fermentation for food wastes treatment: Effect of high organic loading rates at low hydraulic retention times and uncontrolled pH conditions. *Bioresour. Technol.* **2022**, *357*, 127356. [CrossRef]
38. Khanal, S.K.; Chen, W.H.; Li, L.; Sung, S. Biological hydrogen production: Effects of pH and intermediate products. *Int. J. Hydrogen Energy* **2004**, *29*, 1123–1131. [CrossRef]
39. Lin, C.-Y.; Chang, R.-C. Fermentative hydrogen production at ambient temperature. *Int. J. Hydrogen Energy* **2004**, *29*, 715–720. [CrossRef]
40. Lee, Y.J.; Miyahara, T.; Noike, T. Effect of pH on microbial hydrogen fermentation. *J. Chem. Technol. Biotechnol.* **2002**, *77*, 694–698. [CrossRef]
41. De Lucas, A.; Rodriguez, L.; Villasenor, J.; Fernandez, F.J. Fermentation of agro-food wastewaters by activated sludge. *Water Res.* **2007**, *41*, 1635–1644. [CrossRef]
42. Rodríguez Mayor, L.; Villaseñor Camacho, J.; Fernández Morales, F.J. Operational optimisation of pilot scale biological nutrient removal at the Ciudad Real (Spain) domestic wastewater treatment plant. *Water Air Soil Pollut.* **2004**, *152*, 279–296. [CrossRef]

43. Fernandez-Morales, F.J.; Villasenor, J.; Infantes, D. Modeling and monitoring of the acclimatization of conventional activated sludge to a biohydrogen producing culture by biokinetic control. *Int. J. Hydrogen Energy* **2010**, *35*, 10927–10933. [CrossRef]
44. Temudo, M.F.; Muyzer, G.; Kleerebezem, R.; Van Loosdrecht, M.C.M. Diversity of microbial communities in open mixed culture fermentations: Impact of the pH and carbon source. *Appl. Microbiol. Biotechnol.* **2008**, *80*, 1121–1130. [CrossRef] [PubMed]
45. American Public Health Association; American Water Works Association; Water Environment Federation. *Standard Methods for the Examination of Water and Wastewater*; APHA-AWWA-WEF: Washington, DC, USA, 2005. (In English)
46. Infantes, D.; Gonzalez del Campo, A.; Villasenor, J.; Fernandez, F.J. Influence of pH, temperature and volatile fatty acids on hydrogen production by acidogenic fermentation. *Int. J. Hydrogen Energy* **2011**, *36*, 15595–15601. [CrossRef]
47. Infantes, D.; González del Campo, A.; Villaseñor, J.; Fernández, F.J. Kinetic model and study of the influence of pH, temperature and undissociated acids on acidogenic fermentation. *Biochem. Eng. J.* **2012**, *66*, 66–72. [CrossRef]
48. Fu, W.; Mathews, A.P. Lactic acid production from lactose by Lactobacillus plantarum: Kinetic model and effects of pH, substrate, and oxygen. *Biochem. Eng. J.* **1999**, *3*, 163–170. [CrossRef]
49. Rodriguez, J.; Premier, G.C.; Guwy, A.J.; Dinsdale, R.; Kleerebezem, R. Metabolic models to investigate energy limited anaerobic ecosystems. *Water Sci. Technol.* **2009**, *60*, 1669–1675. [CrossRef]
50. Elbeshbishy, E.; Dhar, B.R.; Nakhla, G.; Lee, H.S. A critical review on inhibition of dark biohydrogen fermentation. *Renew. Sustain. Energy Rev.* **2017**, *79*, 656–668. [CrossRef]
51. van den Heuvel, J.C.; Verschuren, P.G.; Beeftink, H.H.; de Beer, D. Determination of the critical concentration of inhibitory products in a repeated fed-batch culture. *Biotechnol. Tech.* **1992**, *6*, 33–38. [CrossRef]
52. Lin, C.Y.; Chang, C.C.; Hung, C.H. Fermentative hydrogen production from starch using natural mixed cultures. *Int. J. Hydrogen Energy* **2008**, *33*, 2445–2453. [CrossRef]
53. Davila-Vazquez, G.; Alatriste-Mondragón, F.; de León-Rodríguez, A.; Razo-Flores, E. Fermentative hydrogen production in batch experiments using lactose, cheese whey and glucose: Influence of initial substrate concentration and pH. *Int. J. Hydrogen Energy* **2008**, *33*, 4989–4997. [CrossRef]
54. Yu, H.Q.; Fang, H.H.P. Acidogenesis of gelatin-rich wastewater in an upflow anaerobic reactor: Influence of pH and temperature. *Water Res.* **2003**, *37*, 55–66. [CrossRef]

fermentation

MDPI

Article

Recovery of Energy and Carbon Dioxide from Craft Brewery Wastes for Onsite Use

Dhanashree Rawalgaonkar [1], Yan Zhang [1], Selina Walker [2], Paul Kirchman [2], Qiong Zhang [1] and Sarina J. Ergas [1,*]

[1] Department of Civil & Environmental Engineering, University of South Florida, Tampa, FL 33620, USA; dhanashree@usf.edu (D.R.); yanzhang1@usf.edu (Y.Z.); qiongzhang@usf.edu (Q.Z.)

[2] Department of Integrative Biology, University of South Florida, Sarasota, FL 34243, USA; selina5@usf.edu (S.W.); pkirchman@usf.edu (P.K.)

* Correspondence: sergas@usf.edu; Tel.: +1-813-974-1119

Abstract: Interest in craft beers is increasing worldwide due to their flavor and variety. However, craft breweries have high water, energy, and carbon dioxide (CO_2) demands and generate large quantities of high-strength waste and greenhouse gases. While many large breweries recover energy using anaerobic digestion (AD) and recapture CO_2 from beer fermentation, little is known about the economic feasibility of applying these technologies at the scale of small craft breweries. In addition, compounds in hops (*Humulus lupulus*), which are commonly added to craft beer to provide a bitter or "hoppy" flavor, have been shown to adversely affect anaerobic microbes in ruminant studies. In this study, biochemical methane potential (BMP) assays and anaerobic sequencing batch reactor (ASBR) studies were used to investigate biomethane production from high-strength craft brewery waste, with and without hop addition. A spreadsheet tool was developed to evaluate the economic feasibility of bioenergy and CO_2 recovery depending on the brewery's location, production volume, waste management, CO_2 requirement, energy costs, and hop waste addition. The results showed that co-digestion of yeast waste with 20% hops (based on chemical oxygen demand (COD)) resulted in slightly lower methane yields compared with mono-digestion of yeast; however, it did not significantly impact the economic feasibility of AD in craft breweries. The use of AD and CO_2 recovery was found to be economically feasible if the brewery's annual beer production is >50,000 barrels/year.

Keywords: anaerobic digestion; carbon dioxide recovery; craft breweries; economic sustainability; hops; methanogenesis inhibition; yeast waste

Citation: Rawalgaonkar, D.; Zhang, Y.; Walker, S.; Kirchman, P.; Zhang, Q.; Ergas, S.J. Recovery of Energy and Carbon Dioxide from Craft Brewery Wastes for Onsite Use. *Fermentation* **2023**, *9*, 831. https://doi.org/10.3390/fermentation9090831

Academic Editors: Jose Luis García-Morales and Francisco Jesús Fernández Morales

Received: 1 August 2023
Revised: 8 September 2023
Accepted: 9 September 2023
Published: 12 September 2023

1. Introduction

Interest in craft beers is increasing worldwide due to their flavor, variety, and artisanal approach to brewing. Craft breweries are typically defined as those with an annual production of 0.7 million m^3 (6 million barrels) of beer or less [1,2]. Craft breweries have high water, energy and carbon dioxide (CO_2) demands, and generate large quantities of solid and liquid wastes and greenhouse gases. Spent grains account for up to 85% of the solid waste generated in craft breweries [3] and are typically sent to farmers for use as animal feed. Beer brewing requires 4 to 20 m^3 of water to produce each m^3 of beer. Wastewater is generated from various processes, including low-strength wastewater from cleaning operations and high-strength wastewater, including trub, spent yeast, and hops. Spent yeast, which makes up the largest fraction of high-strength liquid waste, has high chemical oxygen demand (COD) concentrations ranging from 100,000 to 300,000 mg/L [4]. While some of the yeast can be recycled within the brewery or directed for use as animal feed, most craft breweries direct this wastewater to local treatment plants, which often impose high waste surcharges [5,6]. Craft beer brewing is also energy intensive, with approximately 240 to 280 kWh of thermal energy and 75 to 138 kWh of electrical energy consumed per m^3 of beer produced [7,8].

Anaerobic digestion (AD) is a biological process that converts organic wastes into biogas, which is a mixture of methane (CH_4) and CO_2. Biogas can be further processed into renewable natural gas (RNG) and used onsite to meet a brewery's thermal energy needs or processed into compressed natural gas (CNG) and liquefied natural gas (LNG) for offsite use. Alternatively, it can be utilized for generating electricity and heat through combined heat and power (CHP) systems to offset a brewery's electrical and thermal demand [9]. Many large breweries employ AD for both wastewater treatment and energy cost reduction [10]. For example, Sierra Nevada Brewing Company (Chico, CA, USA) reported annual energy and waste management savings of >USD 500,000 after implementing AD [11].

Beers containing large quantities of hops (*Humulus lupulus*), such as India Pale Ales (IPAs), are a trademark of craft brewing. Spent hops have a bitter flavor and a lower nutritional value than spent grain. Hence, only a small portion of hop waste can be directed to animal feed [4]. In addition, hop metabolites include alpha acids, beta acids, and Xanthohumol, which have antimicrobial properties that aid in beer preservation [12,13]. These compounds have been shown to inhibit CH_4 production in ruminant animals, which has been proposed as a way to increase the nutritive value of feeds while reducing greenhouse gas emissions from cattle [14–16]. Two mechanisms have been identified for CH_4 inhibition in ruminants: (a) inhibition of Gram-positive bacteria in the acetogenic and acidogenic stage [17,18] and (b) inhibition of methanogenic archaea [16].

Although it is evident that hop metabolites inhibit CH_4 production in cattle, the effect of hop metabolites on the AD of brewery waste has not previously been investigated. Sosa-Hernandez and colleagues conducted biomethane potential (BMP) assays with spent yeast from different sources and reported low CH_4 yields from hoppy beers (28 $mLCH_4/gCOD$) compared with less hoppy beers (42 and 68 $mLCH_4/gCOD$), suggesting potential inhibition by hop metabolites [19].

Carbon dioxide (CO_2) is a by-product of beer fermentation and is also used in the brewing process for bottling, flushing, and carbonation. Prior studies have shown that CO_2 can be recovered from fermentation, scrubbed, and compressed for in-process recycling and reducing costs and greenhouse gas emissions [20]. CO_2 that is recovered from fermenters is also a high-quality product without industrial contaminants that may be present when by-product CO_2 is purchased from ammonia and urea facilities. Recovered CO_2 can be further processed into dry ice and compressed or liquefied CO_2 for offsite applications. CO_2 recovery units are available as modular skid-mounted systems [21]. Considering its economic and environmental benefits, CO_2 recovery could improve the sustainability of small craft breweries.

Several spreadsheet tools have been developed to aid in the economic and environmental assessment of AD systems. However, most of these tools focus on livestock manure as the primary AD substrate. For example, the US Environmental Protection Agency (US EPA) has developed a Co-Digestion Economic Analysis Tool (CoEAT) to evaluate the economic feasibility of AD co-digestion of manure with food waste, fats, oils, and grease [22]. Astill and colleagues developed a tool to aid farmers in AD adoption decision-making. The tool is designed to assess the economic feasibility of AD using farm-derived feedstocks, including manure and crop residues [23]. Therefore, the existing tools are not directly applicable to craft brewery waste. Furthermore, no prior study examines the economic tradeoffs of CO_2 recovery systems for small craft breweries.

The overall goal of this study was to improve the environmental and economic sustainability of small craft breweries by recovering bioenergy and CO_2 for onsite use. The specific objectives were to: (1) investigate the effect of hops on spent yeast waste AD through BMP assays, (2) conduct bench-scale anaerobic sequencing batch reactor (ASBR) studies without and with hops addition to provide data for full-scale economic analysis, and (3) develop a tool to evaluate the feasibility of bioenergy and CO_2 recovery at craft breweries depending on factors such as production volume, location, waste surcharges, CO_2 costs, energy costs, and hop waste addition.

2. Materials and Methods

2.1. Bench Scale Experiments

2.1.1. Materials

Characteristics of spent yeast, hops, and inoculum are shown in Table 1. Spent yeast was obtained from a small craft brewery in Sarasota, FL, USA. Hops, with an alpha acid content of 7.3%, were obtained from Yakima Chief Hops HBC 692 (Yakima, WA, USA). AD inoculum was obtained from a mesophilic AD that was used to process waste-activated sludge at the South Cross Bayou Advanced Wastewater Treatment Facility in Pinellas County, FL, USA. Fresh AD inoculum was obtained for each phase of the study. Magnesium Carbonate ($MgCO_3$), which was used as an alkalinity source, was obtained from Thermo-Scientific (Haverhill, MA, USA). Note that $MgCO_3$ was used instead of $NaHCO_3$ due to the high Na^+ concentration of spent yeast, which can be toxic to anaerobic microbes [24–26]. Well water was sourced from Botanical Gardens located at the University of South Florida.

Table 1. Average characteristics of spent yeast, hops, and inoculum used in this study.

	Yeast	Hops	Inoculum
pH	4.5	4.6 *	8
Alkalinity (mg/L)	NA	45 *	5900
VSS (mg/L)	46,497		21,000
COD (mg/L)	231,280	1 **	38,215

* Hop alkalinity and pH were measured on stock solutions prepared with 1 g dry hops in 15 mL DI. ** Units of mg COD/mg hops.

2.1.2. Biomethane Potential Assays (BMPs)

Mesophilic (35 °C) BMP assays were conducted in two phases (Table 2). BMPs were set up in 200 mL glass serum bottles with crimp caps and septum seals. In Phase I, the substrate to inoculum ratio (S/I) was set at 2.5 g COD/g VSS based on prior studies [19,27,28]. The yeast-only system in Phase 1 soured due to volatile fatty acid (VFA) accumulation. Hence, based on the results from Phase 1, another round of BMPs (Phase 2) was conducted at a lower S/I ratio of 1.7, a higher initial alkalinity, and with fresh inoculum. In both phases, digestion sets were set up with hop concentrations of 0% hops (yeast only), 20% hops, and 40% hops (based on total COD supplied by the substrate). These hop percentages were based on estimates of relative hop and yeast waste production rates at the craft breweries we partnered with in Sarasota and Tampa (FL, USA). Additional digestion sets were used as inoculum-only controls in both phases. Biogas and methane contents were determined on duplicate bottles. Duplicate bottles were sacrificed for chemical analysis (described below) on days 0, 42, and 58 during Phase I and days 0, 38, and 60 during Phase II. Additional details can be found in [29].

Table 2. Summary of conditions for BMP phases.

	BMP Phase 1	BMP Phase 2
S/I (mg COD/mg VSS) *	2.5	1.7
Substrates used	Spent yeast, Hops	Spent Yeast, Hops
Hop Dosages (g-hopCOD/g-totalCOD)	0, 20%, 40%	0, 20%, 40%
Alkalinity Addition (mg/L as $CaCO_3$)	1000	1500

* mg of COD in substrate/mg VSS in inoculum.

2.1.3. Anaerobic Sequencing Batch Reactor (ASBR) Studies

Two bench-scale ASBRs were created from glass bottles (1.6 L working volume) with screw caps drilled with three holes for tubing reaching the: (a) head space, which was connected to a biogas collection system; (b) supernatant, for feeding the reactor and wasting effluent; and (c) settled solids, for solids wasting. Preliminary studies were carried out in duplicate mesophilic ASBRs for 2 months with yeast waste only at varying hydraulic residence times (HRTs), organic loading rates (OLRs), solids residence times (SRTs), and $MgCO_3$ dosing to determine optimal ASBR operating conditions [29]. Subsequently, one of the duplicate ASBRs continued to be fed with yeast waste only (Y), and the second ASBR was set up with 80% yeast waste and 20% hops based on COD (YH). The OLR and HRT were maintained at 720 mg COD/L/day and 20 days, respectively, by wasting 240 mL of supernatant every 3 days and feeding fresh influent diluted with well water. The SRT was maintained at 190 days by wasting settled solids every 6 days. In addition, 0.25 g $MgCO_3$ was added as an alkalinity source on sludge-wasting days.

2.1.4. Analytical Methods

In the BMP assays, biogas volume was measured using a frictionless syringe (Cadence Inc., Staunton, VA, USA). In the ASBR studies, biogas flowrate was measured using a gas flow meter (Wet Tip Gas Meter, Nashville, TN, USA). Methane content of the biogas was measured using a Gas Chromatograph (GOW MAC, Bethlehem, PA, USA) equipped with a Hay Sep Column and Thermal Conductivity Detector. The detector, column, and injector temperatures were 100 °C, 60 °C, and 100 °C, respectively. A current of 200 mV and high-purity helium (Airgas, Inc., Radnor, PA, USA) at a flow rate of 32 mL/min were used. All chemical characteristics were measured using *Standard Methods* [30] for pH (4500), alkalinity (2320), volatile suspended solids (VSS; 1648), COD (5220), and VFA (5560). Test kits were used to measure VFAs (Hach, Loveland, CO, USA) and COD (Lovibond, Sarasota, FL, USA) concentrations. Ammonium concentrations were measured using a Timberline TL-2800 Ammonia Analyzer (Timberline Instrument, Boulder, CO, USA).

2.1.5. Data Analysis

Gas volumes were adjusted to standard temperature (273 °C) and pressure (1 atm) using the ideal gas law. Paired T-tests with a p-value of 0.05 were used to evaluate statistical significance between chemical characteristics data for BMP assays and ASBR studies. The Modified Gompertz Equation (Equation (1); [31]) was used to determine the methane rate constant for the BMP assays. Excel was used to minimize the sum of absolute errors between experimental and model methane volumes.

$$M_P = P_M \cdot exp\left\{-exp\left[\frac{R_{exp}}{P_M}(X_O - X) + 1\right]\right\} \quad (1)$$

where:

M_p = cumulative methane production at time X (mL);
P_M = methane production potential (mL);
R_{exp} = maximum methane production rate (mL/day);
X_O = lag period (days);
X = time (days).

2.2. Decision Support Tool

A custom Excel spreadsheet tool was created specifically for craft brewers to analyze the cost and benefits of incorporating AD and CO_2 recovery systems at varying scales of craft breweries. Figure 1 illustrates the system boundary of the tool, covering inputs, analysis, and outputs. The tool's input interface includes essential details about the brewery, waste generation and management, energy consumption, and CO_2 consumption. General information includes location, beer production rate, and available space. Waste generation and management are determined either through user input or calculations based on annual production, considering factors such as waste characteristics, transportation distance, and existing waste disposal methods. Similarly, energy consumption is obtained either through user input or by performing calculations based on the annual production rate, which includes factors such as electricity and/or natural gas consumption. The tool's output interface provided AD plant sizing, predicted biogas production, capital and operation and maintenance (O&M) costs, CO_2 offset potential, and economic performance (net present value (NPV) and payback period).

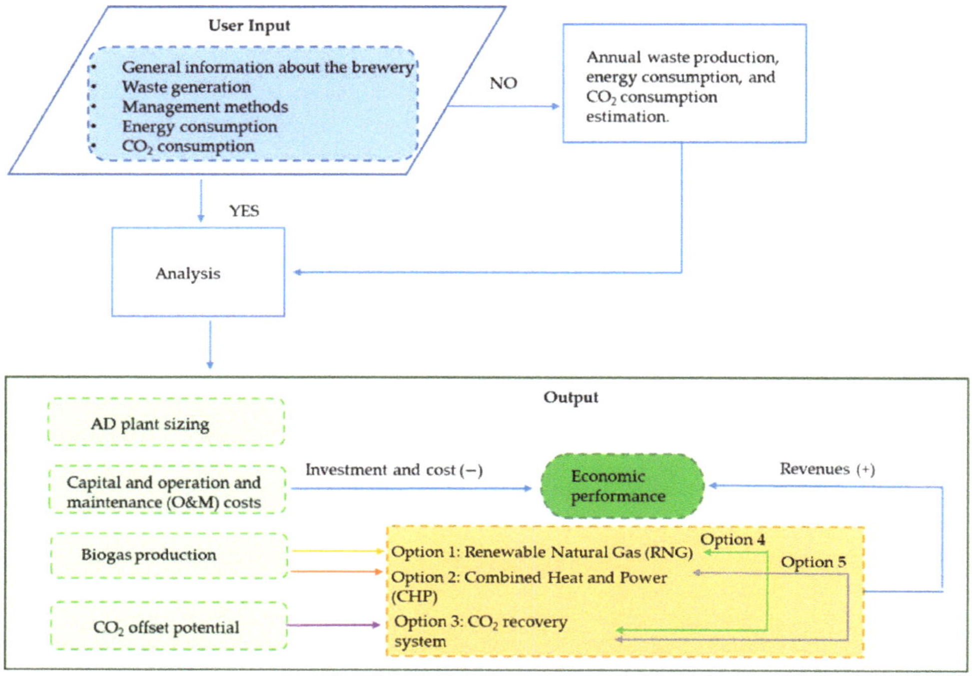

Figure 1. Decision support tool diagram with inputs, analysis, and outputs.

2.2.1. Anaerobic Digester Sizing

The tool employs user inputs and assumed constants to estimate an AD plant and determine the suitable size of the AD reactor for processing the high-strength fraction of the brewery wastewater. The digester size is a function of the flow rate, influent substrate concentration, and OLR. In addition, a safety factor of 20% was applied to the digester's headspace to account for gas storage and variations in wastewater characteristics [32,33]. This safety factor is flexible and can be adapted to the unique annual production of each brewery. The optimal size of the digester could be determined using Equation (2):

$$V_D = \frac{Q \times C_{0,COD}}{OLR} \times (1 + H_D) \quad (2)$$

where:

Q = feedstock flowrate (m^3/s);
$C_{0,COD}$ = influent substrate concentration (kg/m^3);
OLR = COD loading rate ($kg/m^3/s$);
H_D = head space of the digester (%);
V_D = volume of the AD (m^3).

2.2.2. Biogas Production and Utilization

The amount of methane produced (m^3/d) was estimated using Equation (3).

$$Methane\ production\ rate = (COD_{Total} \times \alpha) \quad (3)$$

where:
COD_{Total} = COD of waste generated (kg/day);
α = the methane yield (m^3 CH_4/kg COD).

Note that the methane yield used in the model was based on the experimental data from the ASBR studies of 0.3 m^3 CH_4/kg COD for yeast waste alone and 0.23 m^3 CH_4/kg COD for co-digestion of yeast waste with 20% hops.

Based on discussions with our collaborating breweries, two different onsite uses for biogas were considered: CHP and RNG. Each method involves slightly different processes and equipment. CHP produces heat and power by combustion of biogas generated during AD. The most common application of biogas in AD facilities is for generating electricity and heat [34]. The CHP process includes gas cleaning, combustion, generator driving, and heat exchange. On the other hand, RNG systems purify the biogas by removing nearly all non-methane components, making it meet natural gas standards and suitable for use in conventional natural gas applications. In this study, the biogas would be upgraded to meet the natural gas quality for onsite utilization within the craft brewery.

2.2.3. Carbon Offset Potential

By capturing and reusing by-product CO_2 from beer fermentation, breweries can reduce their environmental impact, save money, and contribute to a more sustainable future. Typically, during beer production, approximately 1–1.5 kg CO_2/hL of beer is utilized for bottling and pre-pressurizing tanks. At current levels of recovery technology, it is possible to recover up to 2 kg CO_2 per hectoliter (hL) of fermented beer. However, it is important to note that any excess CO_2 generated during the pressurization of filtered beer tanks is reclaimed and reintroduced into the CO_2 recovery system [35]. Since this study focused on the onsite use of recovered CO_2, other potential products from CO_2, such as dry ice, were not considered.

2.2.4. Economic Analysis

A customized economic analysis tool was created to evaluate the viability of bioenergy and CO_2 recovery in craft breweries. This tool allows for the assessment of plant revenues, expenditures, and economic indicators such as the payback period and NPV. The economic factors influencing bioenergy and CO_2 recovery systems encompass capital costs, O&M costs, the benefits derived from the produced biogas, income from the digestate, and savings from avoided waste disposal costs. These costs depend on local costs as well as the region's political and economic policies.

The cost associated with AD depends on the facility's processes, design, and size (Table S1). In this study, the capital costs of biogas facilities were obtained from the EPA CoEAT model [22]. In order to validate the feasibility of using the EPA CoEAT model to estimate the installation costs of AD plants in craft breweries, the capital costs obtained from the model were compared with several real case studies (Table S2) [36–39].

Currently, there is a lack of available data on the O&M costs of AD plants. In this study, the average O&M costs of AD plants were estimated to be \$10/m^3 of AD plant capacity [23]. O&M costs of RNG, CHP, and CO_2 recovery systems were calculated based on 3%, 1.5%, and 1% of capital costs, respectively [23].

The revenue generated by the AD plant is obtained through the sale of the liquid digestate, which serves as a fertilizer. This liquid digestate contains a substantial concentration of nitrogen and phosphorus, comparable to that found in industrial fertilizers. Other savings include avoided electricity costs, natural gas costs, and CO_2 costs, depending on the alternative chosen. The economic analysis conducted by the tool excluded tax credits as these factors are contingent upon the particular region or country in which the craft brewery is located.

The tool incorporates various economic indicators to enable users to effectively evaluate the economic feasibility of their chosen alternative, including the payback period and

NPV, as described in Equations (4) and (5). Alternatives with a shorter payback period and a positive NPV are preferred.

$$\text{Payback period} = \frac{Inital\ investment}{Cash\ flow} \tag{4}$$

$$NPV = \sum \frac{CF_n}{(1+R)^n} - I_0 \tag{5}$$

where:

n = the period which takes values from 0 to the nth period till the cash flow ending period;
CF_n = the cash flow in the nth period (USD);
R = the discount rate;
I_0 = the initial investment (USD).

3. Results

3.1. Bench Scale Experiments

3.1.1. Biomethane Potential Assays (BMPs)

Methane production data for Phase I BMPs are shown in Figure 2. During Phase I, yeast waste-only BMPs produced almost no methane, while BMPs with added hops had maximum methane yields of 0.10 mL CH_4/mg COD for 20% hops and 0.076 mL CH_4/mg COD for 40% hops (Table 3). VFA concentrations during the second sacrifice on day 42 (Table 4) were much higher in the yeast-only system (7500 mg/L) compared with the 20% hops (300 mg/L) and 40% hops (600 mg/L). The sudden release of VFAs in the yeast-only system consumed available alkalinity, resulting in the pH dropping below the conducive range for anaerobic digestion [40], which soured the system, resulting in little to no methane generation. While alkalinity concentrations in the yeast-only system dropped below the conducive limit of 2000 mg/L [41], the 20% hops and 40% hops BMPs had adequate alkalinity. Hops have a high crude fiber content [4], which is not readily bioavailable for anaerobic microbes; therefore, hop addition may have prevented souring due to the more distributed release of VFAs during fermentation. The higher bioavailability of yeast is further evident as the yeast-only system had the highest COD degradation compared to the 20% and 40% hops assays (Table 3).

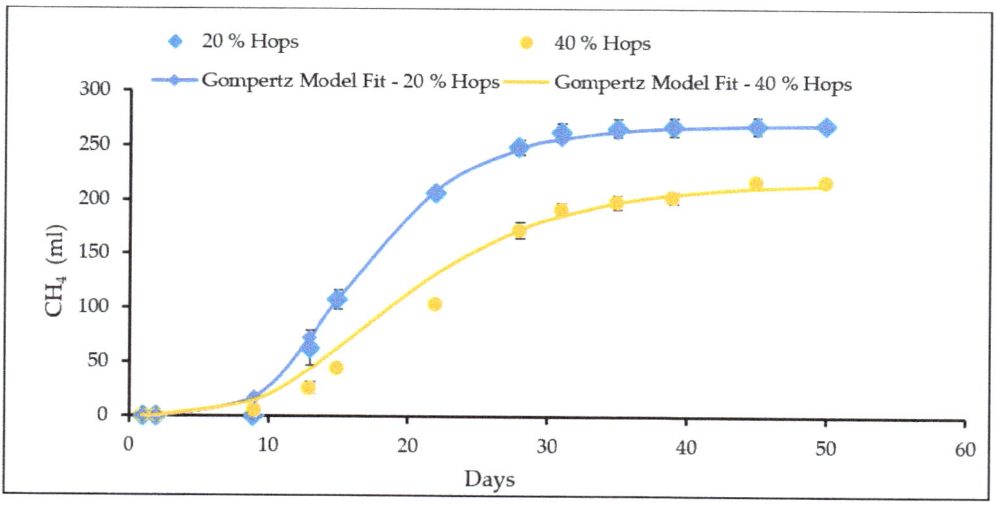

Figure 2. Cumulative methane volumes and Modified Gompertz model fit for Phase I (error bars show standard deviations between duplicate BMPs).

Table 3. Summary of methane data obtained from BMPs and Gompertz analysis.

	BMP Phase I			BMP Phase II		
	Yeast	20% Hops	40% Hops	Yeast	20% Hops	40% Hops
Methane yield (mL CH_4/mg COD)	NA *	0.10	0.076	0.17	0.15	0.11
Cumulative methane (mL)	4.0	269	216	338	309	227
Lag period (days)	NA *	9	9	17	11	11
R_{max} (mL CH_4/day)	NA *	18	10	27	28	8
COD degradation (%)	58	51	37	53	44	36

* Yield, lag period, and R_{max} values for yeast are not reported during Phase I due to souring.

Table 4. Summary of chemical analysis from BMPs during Phase I (standard deviations shown in parentheses).

	Yeast			20% Hops			40% Hops		
	Day 0	Day 42	Day 58	Day 0	Day 42	Day 58	Day 0	Day 42	Day 58
VFA (mg/L)	216(8)	7500(181)	3423(1)	236(8)	287(145)	134(11)	264(7)	584(367)	170(1)
Alkalinity ** (mg/L)	2167(28)	1659(141)	NA *	2017(104)	4100(141)	4275(35)	2200(343)	3275(388)	3850(40)
Ammonium (mg/L)	236(15)	720(80)	624(17)	217(2)	633(12)	654(8)	252(21)	549(12)	570(42)
VSS	7722(308)	5495(321)	5760(56)	7713(245)	5390(181)	6036(90)	8762(439)	7221(240)	6939(196)

* Value not reported due to the yeast-only assay souring during Phase I, ** as $CaCO_3$.

During Phase II, the S/I was decreased from 2.4 to 1.7 g COD/gVS, and the initial alkalinity was increased to prevent souring observed in Phase I. Maximum methane yields of 0.17, 0.15, and 0.11 mL CH_4/mg COD were observed for yeast waste alone, 20% hops, and 40% hops, respectively (Figure 3). Methane yields obtained during Phase II were similar to values reported in the prior literature of 0.025–0.24 mL CH_4/mg COD [19,42], indicating that a lower S/I ratio and the addition of alkalinity avoided VFA accumulation, reactor souring and methanogenesis inhibition, as shown in Table 3. Similar to Phase I, for the BMPs with added hops, the lower hop percentage resulted in a higher methane yield, suggesting that hop dosage affects their inhibitory effects. Methane yield was significantly lower for 40% hops compared with yeast only or 20% hops; however, differences between yeast only and 20% hops were not significant. Similar to Phase I, assays with higher hop concentrations had lower COD degradation during Phase II (Table 3).

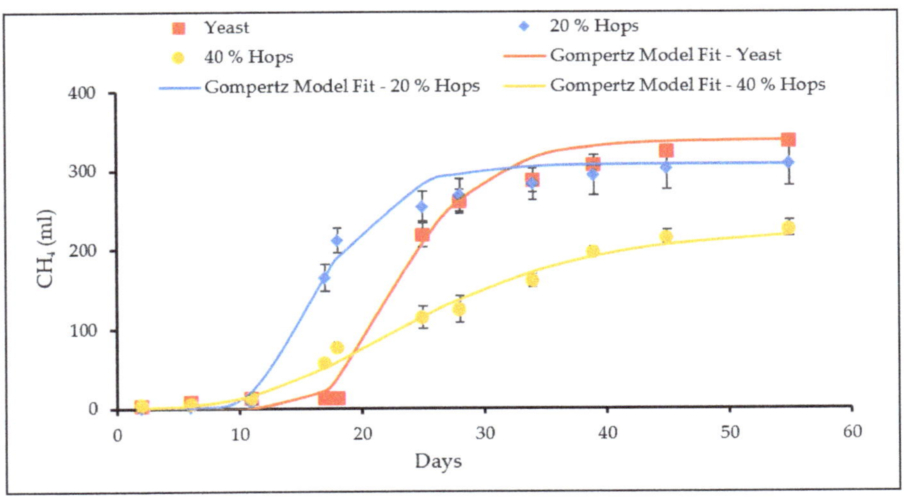

Figure 3. Cumulative methane volumes and Modified Gompertz model fit for Phase II (error bars show standard deviations between duplicate BMPs).

As shown in Tables 4 and 5, ammonium concentrations increased over time, with the yeast-only system having the highest concentrations on days 58 and 60 during Phases I and II, respectively. This was likely due to high protein compositions typically found in spent yeast [4]. VSS concentrations decreased after the first sacrifice in both phases as the volatile solids were consumed over time. The increase in VSS between the second and third sacrifices in the yeast only and 20% hops assays could have been due to the growth of microbial biomass.

Table 5. Summary of chemical analysis from BMPs during Phase II (standard deviations shown in parentheses).

	Yeast			20% Hops			40% Hops		
	Day 0	Day 38	Day 60	Day 0	Day 38	Day 60	Day 0	Day 38	Day 60
VFA (mg/L)	138(3)	254(23)	90(9)	194(4)	118(2.8)	120(56)	229(7)	589(148)	177(4.20)
Alkalinity * (mg/L)	2725(35)	3775(35)	4075(35)	2650(70)	3850(70)	4175(35)	3050(280)	3500(70)	3875(176)
Ammonium (mg/L)	114(4)	461(1)	516(8)	114(8)	432(16)	498(8)	114(4)	384(6)	462(8)
VSS (mg/L)	6880(170)	5430(183)	5265(487)	7590(70)	5812(34)	5297(349)	7500(340)	3534(190)	6971(72)

* as $CaCO_3$.

Gompertz analysis of the BMP data (Table 3) shows that a greater hop content in the feed led to lower methane yields in both BMP Phases. The results are consistent with AD studies by Sosa-Hernandez et al. [19], who found that spent yeast from hoppy beers had lower methane yields than less hoppy beers. As mentioned previously, prior studies with ruminant microbial communities showed that hop metabolites have antimicrobial properties that inhibit methanogenesis [16–18]. Concentrations of VFAs during the second and third sacrifices of both BMP phases were higher, with 40% hops compared with 20% hops (Tables 4 and 5). This suggests that VFAs produced during fermentation in high hop dosage assays were consumed by methanogens at a slower rate compared to lower dosages. This is further supported by the Gompertz rate constants (Table 3), which showed lower methane production rates at higher hop dosages. Surprisingly, the lag period for the yeast-only BMP in Phase II was longer than for the digesters containing hops in both Phases (Table 3). This may have been due to initial reactor souring followed by recovery in the yeast waste-only BMPs; however, chemical analysis was not conducted until day 38.

3.1.2. Anaerobic Sequencing Batch Reactors (ASBRs)

Bench-scale ASBRs were set up with yeast waste alone and with 20% hops (based on COD) and operated at an OLR of 720 mg COD/L/day, an HRT of 20 days, and an SRT of 190 days. Biogas and methane yields in the yeast-only ASBR were similar to results from the preliminary 2-month study performed with duplicate ASBRs operated with yeast waste alone [29]. In both ASBRs, methane yields (Figure 4) were comparable to those reported for co-digestion of spent yeast with brewery wastewater, which ranges 0.20–0.35 mL CH_4/mg COD [42–45].

Consistent with Phase II BMPs, the mono-digestion of yeast resulted in higher methane yields than the co-digestion of yeast and hops (Figure 4). Inhibition increased over approximately one HRT as hops from the feed accumulated in the system. Hop addition resulted in both lower biogas production and lower biogas methane content (Table 6); however, the lower methane yields in the ASBR with hops were largely a result of lower biogas production. Lower methane yield in the ASBR with hops may have been due to: (1) direct inhibition of methanogenesis due to the accumulation of hop metabolites, such as alpha and beta acids, and/or (2) slower VFA release during fermentation since hops are more difficult to break down by hydrolytic bacteria. COD degradation in the ASBR with hops was lower than in the ASBR without hops (Table 6), which is similar to results found in the BMP studies (Table 3). The mean alkalinity concentration in the Y digester was higher than the YH digester. Although the VFA concentrations were not measured during the ASBR studies, lower alkalinity concentrations might suggest VFA accumulation due to methanogenesis inhibition by hops in the YH digester, which likely consumed the alkalinity. The mean ammonium concentrations in digester Y were lower than the YH digester.

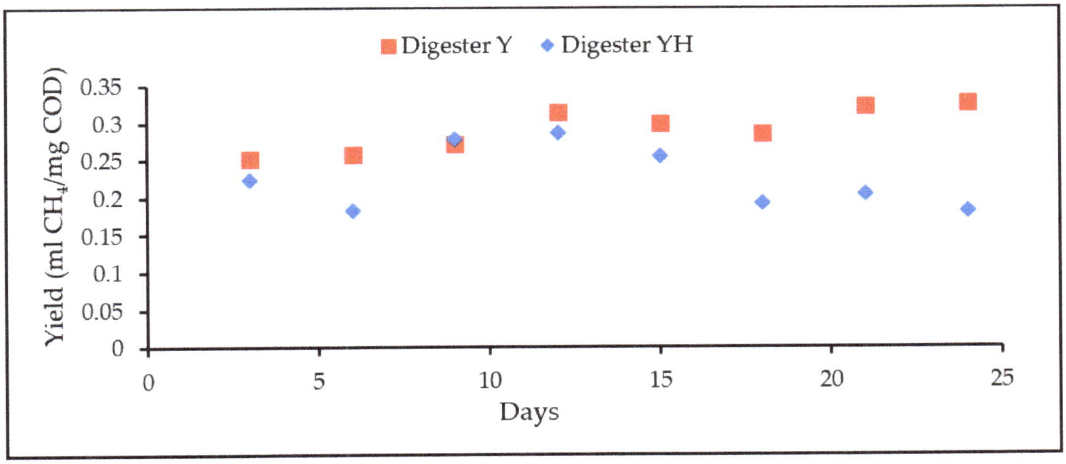

Figure 4. Methane yield for ASBR study. Y = yeast only, YH = yeast + hops.

Table 6. Mean values for ASBR performance (standard deviations shown in parentheses).

	Digester Y	Digester YH
Biogas volume (mL/d)	538 (44)	466 (36)
Biogas methane content (%)	76 (3.3)	73 (5.4)
Methane volume (mL/d)	330 (35)	256 (47)
COD degradation (%)	96 (2.1)	93 (1.5)
Alkalinity (mg/L) *	1918 (94)	1878 (157)
Ammonium (mg/L)	438 (6)	470 (7)

Note: Biogas and methane volumes adjusted to STP. Data averaged over the final HRT. * as $CaCO_3$.

Due to the short operating time of the ASBR studies (25 days), steady-state operations were not established, which is a limitation of this study. In a prior study, Blaxland et al. [16] observed acclimation of the microbial community against inhibitory hop substances over time. Therefore, longer studies should be carried out to determine whether acclimation of microbial communities to hops might result in increased methane yields.

3.2. Economic Analysis

This study examined potential biogas production for craft breweries, considering various annual production levels: 50,000, 500,000, 1 million, 2 million, 4 million, and 6 million barrels, based on typical production rates for U.S. craft breweries [2]. AD capital costs, with additional investments required for RNG or CHP with and without CO_2 recovery, are shown in Figure 5. AD capital costs included tanks, mixers, inlet and outlet pumps, and piping (listed in Table S2). Small-scale systems that can recover CO_2 from beer fermentation gases are generally affordable, with a current capital cost of approximately USD 150,000 [46].

When evaluating total capacity costs, the combination of AD with RNG results in AD accounting for >90% of the total capital cost. The relative percentage of AD cost increases as the annual production increases (Figure 6). However, in the combination of AD with a CHP system, the relative percentage of AD capital cost decreases as annual production increases, indicating that AD + CHP is more economically viable for large-scale breweries, which have more organic matter available for CH_4 production. When considering an annual production of 50,000 barrels using an AD with RNG, the capital cost of CO_2 recovery accounts for up to 5.7%. As the annual production increases, the difference between options with and without CO_2 recovery is negligible.

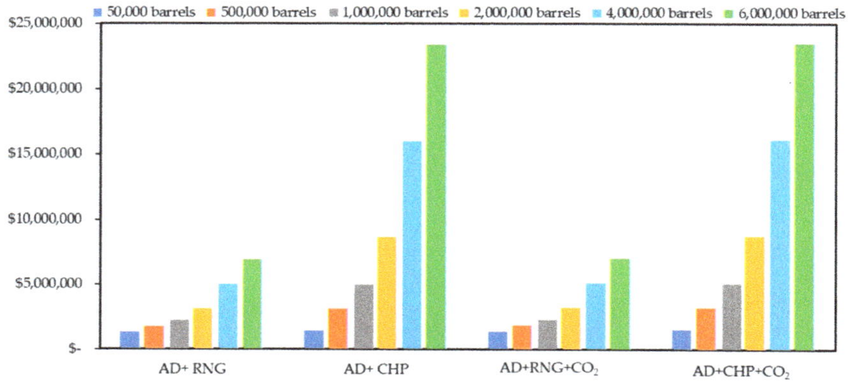

Figure 5. Capital cost for different options and annual production levels. Data from EPA CoEAT model (additional details are provided in Table S2).

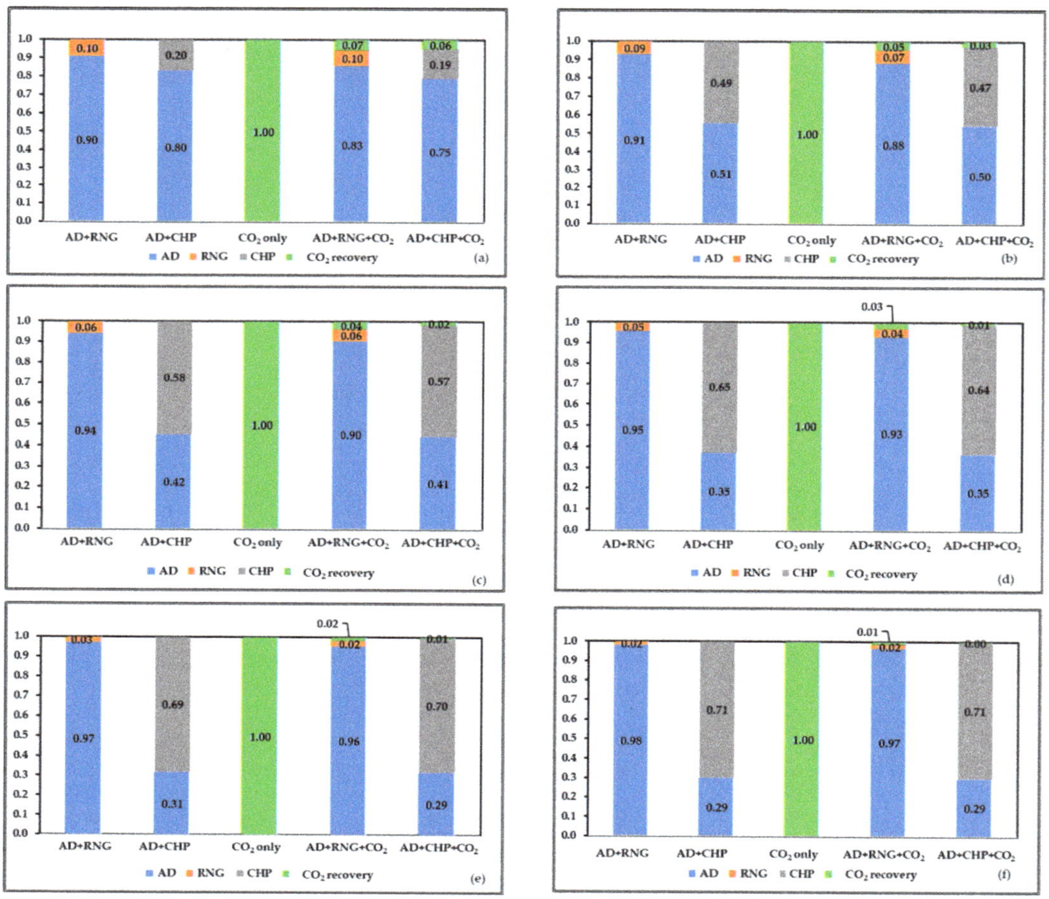

Figure 6. Relative percentage of total capital cost for different annual production levels (**a**) 50,000, (**b**) 500,000, (**c**) 1 million, (**d**) 2 million, (**e**) 4 million, and (**f**) 6 million barrels.

Annual income primarily comes from cost savings on natural gas, electricity, and CO_2, as well as tax credits (see Table S3 for details). Figure 7 illustrates the payback period for different options. For an annual production of 50,000 barrels, the payback period is 43.0 years for AD with RNG and 45.4 years for AD with CHP. However, when considering the implementation of a CO_2 recovery system, the payback period significantly decreases to 3.5 years for the combination of AD and RNG and 3.7 years for the combination of AD and CHP. Without CO_2 recovery, both AD + RNG and AD + CHP become economically feasible for craft breweries with annual production >500,000 barrels, with a payback period of <10 years.

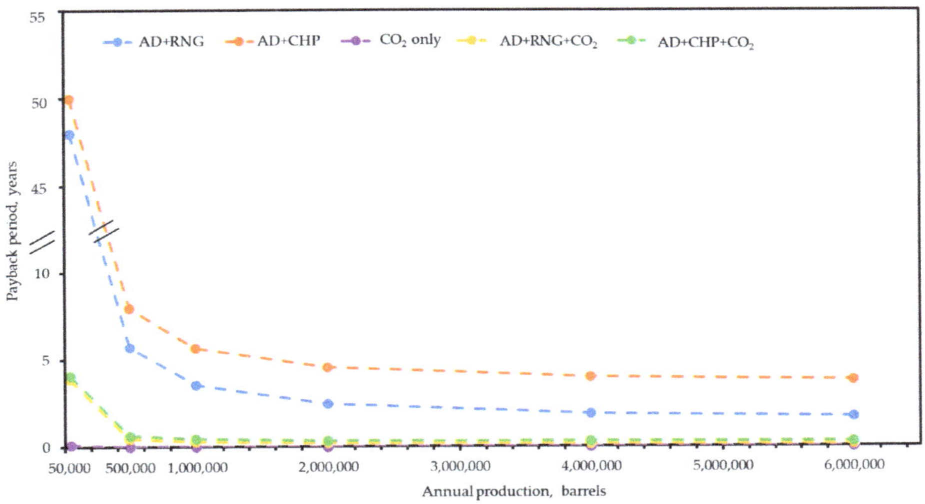

Figure 7. Payback period in terms of different annual production without hops.

Implementing a CO_2 recovery system can significantly reduce the payback period for both the AD + RNG option and the AD + CHP option. Without the CO_2 recovery system, the payback periods for both options are quite long, indicating a slower return on investment. However, when the CO_2 recovery system is included, the payback periods decrease significantly, making both alternatives economically feasible. This is because recovered CO_2 is a high-value product compared with electricity and natural gas. Overall, the information emphasizes the potential economic benefits of implementing AD and RNG or AD and CHP systems, especially when coupled with CO_2 recovery, and provides valuable insights for decision making in the context of craft breweries.

The co-digestion of yeast waste with 20% hops decreased methane yield from 0.3 m^3 CH_4/kg COD to 0.23 m^3 CH_4/kg COD. Despite this reduction, adding 20% of hops waste had minimal impact on the payback period. Across various scenarios with annual production levels ranging from 50,000 to 6,000,000 barrels, the payback period decreased by 0–4.2% (as shown in Table S4). The results of this study highlight the potential benefits of co-digestion with 20% of hops waste, not only from an environmental perspective but also from an economic standpoint.

4. Conclusions

This study evaluated the effects of AD of spent brewery yeast, co-digestion of spent yeast with hops, and the economic feasibility of AD and CO_2 recovery systems at craft breweries. Bench-scale experiments showed that the AD of yeast alone requires dilution with lower-strength waste, such as wastewater from cleaning operations, to avoid reactor overload since yeast has an acidic pH and high concentrations of readily bioavailable COD.

During co-digestion, a 20% hop dosage resulted in little to no inhibition of methanogenesis, whereas a 40% hop dosage led to significantly lower methane yields. Future studies should consider pilot-scale AD studies with varying hop dosages.

An economic analysis tool was used to evaluate the feasibility of bioenergy and CO_2 recovery at craft breweries. The findings indicated that AD and CO_2 recovery were economically viable for breweries producing over 50,000 barrels annually. The analysis demonstrated that the AD + RNG option is more financially viable than the AD + CHP option. Implementation of CO_2 recovery significantly reduced payback periods for AD plants. Although co-digestion with 20% hops waste led to a slight decrease in methane yield, it did not significantly impact the economic feasibility of the AD plant. Future studies should explore the economics of other pathways for resource recovery from craft breweries, including CO_2 recovery from biogas and production of compressed natural gas (CNG), liquefied natural gas (LNG), dry ice, and compressed or liquefied CO_2. In addition, the Excel tool should be compared with results from real-world breweries at different scales to enhance its usability.

Supplementary Materials: The following supporting information can be downloaded at: https://www.mdpi.com/article/10.3390/fermentation9090831/s1, Table S1. Capital costs for AD plant. Table S2. Comparison of capital costs using the CoEAT model and the actual construction costs from real case studies. Table S3. Annual income and avoided costs for varying production levels. Table S4. Comparison of payback period with 20% hops waste and without hops.

Author Contributions: Conceptualization, S.J.E., Q.Z. and P.K.; methodology, D.R. and Y.Z.; software, Q.Z. and Y.Z.; validation, Y.Z. and D.R.; formal analysis, Y.Z. and D.R.; investigation, Y.Z., D.R. and S.W.; resources, S.J.E., P.K. and Q.Z.; data curation, Y.Z. and D.R.; writing—original draft preparation, Y.Z. and D.R.; writing—review and editing, S.J.E., Q.Z., P.K. and S.W.; visualization, Y.Z. and D.R.; supervision, S.J.E., Q.Z. and P.K.; project administration, S.J.E.; funding acquisition, S.J.E., Q.Z. and P.K. All authors have read and agreed to the published version of the manuscript.

Funding: This research is based upon work supported by the University of South Florida Interdisciplinary Research Grant program and by the US National Science Foundation under Grant No. 1930451. Dhanashree Rawalgaonkar was partially supported by the USF Trailblazers Scholarship program.

Institutional Review Board Statement: Not applicable.

Informed Consent Statement: Not applicable.

Data Availability Statement: The data presented in this study are available on request from the corresponding author (Sarina J. Ergas: sergas@usf.edu).

Acknowledgments: The authors would like to thank the breweries who collaborated with us on this research, including Motorworks Brewery (Bradenton, FL, USA), Calusa Brewery (Sarasota, FL, USA), and Cigar City Brewery (Tampa, FL, USA). Many thanks to Xia Yang for her mentorship and guidance throughout the project. We would also like to thank Yaritza Vargas and Dahlia Martinez for their help with preliminary experiments.

Conflicts of Interest: The authors declare no conflict of interest.

References

1. Brewers Association. Craft Brewery Definition. Available online: https://www.brewersassociation.org/statistics-and-data/craft-brewer-definition/ (accessed on 31 July 2023).
2. Brewers Association. State Craft Beer Sales & Production Statistics. Available online: https://www.brewersassociation.org/statistics-and-data/state-craft-beer-stats/ (accessed on 31 July 2023).
3. Swart, L.J.; Bedzo, O.K.K.; van Rensburg, E.; Görgens, J.F. Pilot-Scale Xylooligosaccharide Production through Steam Explosion of Screw Press–Dried Brewers' Spent Grains. *Biomass Convers. Biorefin.* **2020**, *12*, 1295–1309. [CrossRef]
4. Karlović, A.; Jurić, A.; Ćorić, N.; Habschied, K.; Krstanović, V.; Mastanjević, K. By-Products in the Malting and Brewing Industries—Re-Usage Possibilities. *Fermentation* **2020**, *6*, 82. [CrossRef]
5. Kerby, C.; Vriesekoop, F. An Overview of the Utilisation of Brewery By-Products as Generated by British Craft Breweries. *Beverages* **2017**, *3*, 24. [CrossRef]

6. Walker, M.; Kruger, P.; Mercer, J.; Webster, T.; Swersey, C.; Skypeck, C. *Wastewater Management Guidance Manual*; Brewers Association: Boulder, CO, USA, 2015. Available online: https://www.brewersassociation.org/educational-publications/wastewater-management-guidance-manual/ (accessed on 31 July 2023).
7. Brewers Association. *Brewers Association Energy Usage, GHG Reduction, Efficiency and Load Management Manual*; Brewers Association: Boulder, CO, USA, 2014. Available online: https://www.brewersassociation.org/educational-publications/energy-sustainability-manual/ (accessed on 31 July 2023).
8. Olajire, A.A. The Brewing Industry and Environmental Challenges. *J. Clean. Prod.* **2012**, *256*, 102817. [CrossRef]
9. Baiano, A. Craft Beer: An Overview. *Comp. Rev. Food Sci. Food Saf.* **2020**, *20*, 1829–1856. [CrossRef] [PubMed]
10. Mainardis, M.; Buttazzoni, M.; Gievers, F.; Vance, C.; Magnolo, F.; Murphy, F.; Goi, D. Life Cycle Assessment of Sewage Sludge Pretreatment for Biogas Production: From Laboratory Tests to Full-Scale Applicability. *J. Clean. Prod.* **2021**, *322*, 129056. [CrossRef]
11. Reid, N.; Gatrell, J. Brewing Growth Regional Craft Breweries and Emerging Economic Development Opportunities. *Econ. Dev. J.* **2015**, *14*, 5–12.
12. Steenackers, B.; De Cooman, L.; De Vos, D. Chemical Transformations of Characteristic Hop Secondary Metabolites in Relation to Beer Properties and the Brewing Process: A Review. *Food Chem.* **2015**, *172*, 742–756. [CrossRef]
13. Preedy, V.R.; Watson, R.R. *The Mediterranean Diet: An Evidence-Based Approach*; Academic Press: Cambridge, MA, USA, 2020.
14. Bryant, R.W.; Burns, E.E.R.; Feidler-Cree, C.; Carlton, D.; Flythe, M.D.; Martin, L.J. Spent Craft Brewer's Yeast Reduces Production of Methane and Ammonia by Bovine Rumen Microbes. *Front. Animal Sci.* **2021**, *2*, 720646. [CrossRef]
15. Bryant, R.W.; Cohen, S.D. Characterization of Hop Acids in Spent Brewer's Yeast from Craft and Multinational Sources. *J. Am. Soc. Brew. Chem.* **2015**, *73*, 159–164. [CrossRef]
16. Blaxland, J.A.; Watkins, A.J.; Baillie, L.W.J. The Ability of Hop Extracts to Reduce the Methane Production of Methanobrevibacter Ruminantium. *Archaea* **2021**, *2021*, 5510063. [CrossRef] [PubMed]
17. Flythe, M.D.; Aiken, G.E. Effects of Hops (*Humulus lupulus* L.) Extract on Volatile Fatty Acid Production by Rumen Bacteria. *J. Appl. Microbiol.* **2010**, *109*, 1169–1176. [CrossRef] [PubMed]
18. Pszczolkowski, V.L.; Bryant, R.W.; Harlow, B.E.; Aiken, G.E.; Martin, L.J.; Flythe, M.D. Effects of Spent Craft Brewers' Yeast on Fermentation and Methane Production by Rumen Microorganisms. *Adv. Microbiol.* **2016**, *6*, 716–723. [CrossRef]
19. Sosa-Hernández, O.; Parameswaran, P.; Alemán-Nava, G.S.; Torres, C.I.; Parra-Saldívar, R. Evaluating Biochemical Methane Production from Brewer's Spent Yeast. *J. Ind. Microbiol. Biotechnol.* **2016**, *43*, 1195–1204. [CrossRef] [PubMed]
20. Copco, A. CO_2 Recovery from Fermentation in Breweries. Available online: https://www.atlascopco.com/en-us/compressors/industry-solutions/brewery-air-compressor/co2-recovery-brewery (accessed on 31 July 2023).
21. Gribbins, K. The CO_2 Shortage: Pros and Cons of Craft Brewery CO_2 Recapture. Available online: https://www.craftbrewingbusiness.com/ingredients/the-co2-shortage-brewers-can-produce-carbon-dioxide-onsite-with-a-co2-recapture-unit/ (accessed on 28 April 2023).
22. United States Environmental Protection Agency (USEPA). User's Manual: Co-Digestion Economic Analysis Tool. Available online: https://www.epa.gov/sites/default/files/2017-09/documents/co-eat_users_manual_fin_sept_2017.pdf (accessed on 20 May 2022).
23. Astill, J.; Dara, R.A.; Fraser, E.D.G.; Roberts, B.; Sharif, S. Smart Poultry Management: Smart Sensors, Big Data, and the Internet of Things. *Comput. Electron. Agric.* **2020**, *170*, 105291. [CrossRef]
24. Rakowska; Sadowska, A.; Dybkowska, E.; Swiderski, F. Spent Yeast as Natural Source of Functional Food Additives. *Rocz. Państwowego Zakładu Hig.* **2017**, *68*, 115–121.
25. Matin, A.; Bashir, B.H. Sodium Toxicity Control by the Use of Magnesium in an Anaerobic Reactor. *J. Appl. Sci. Environ. Manag.* **2005**, *8*, 17–21. [CrossRef]
26. Jaeger, A.; Arendt, E.K.; Zannini, E.; Sahin, A.W. Brewer's Spent Yeast (BSY), an Underutilized Brewing By-Product. *Fermentation* **2020**, *6*, 123. [CrossRef]
27. Ahnert, M.; Schalk, T.; Brückner, H.; Effenberger, J.; Kuehn, V.; Krebs, P. Organic Matter Parameters in WWTP—A Critical Review and Recommendations for Application in Activated Sludge Modelling. *Water Sci. Technol.* **2021**, *84*, 2093–2112. [CrossRef]
28. Holliger, C.; Astals, S.; de Laclos, H.F.; Hafner, S.D.; Koch, K.; Weinrich, S. Towards a Standardization of Biomethane Potential Tests: A Commentary. *Water Sci. Technol.* **2020**, *83*, 247–250. [CrossRef]
29. Rawalgaonkar, D. Anaerobic Digestion of Brewery Waste Including Spent Yeast and Hops. Master's Thesis, Department of Civil & Environmental Engineering, University of South Florida, Tampa, FL, USA, 2023.
30. American Public Health Association; American Water Works Association; Water Environment Federation. *Standard Methods for Examination of Water and Wastewater 2012*; Rice, E.W., Baird, R.B., Eaton, A.D., Clesceri, L.S., Eds.; American Public Health Assn: Washington, DC, USA, 2012.
31. Etuwe, C.N.; Momoh, Y.O.L.; Iyagba, E.T. Development of Mathematical Models and Application of the Modified Gompertz Model for Designing Batch Biogas Reactors. *Waste Biomass Valorization* **2016**, *7*, 543–550. [CrossRef]
32. Singh, A.K.; Kaushal, R.K. Design of small scale anaerobic digester for application in Indian village: A review. *Int. J. Eng. Appl. Sci.* **2016**, *3*, 257612.
33. Sheffler, K. Anaerobic Digestion and Biogas Production Feasibility Study. Master's Thesis, University of Idaho, Moscow, ID, USA, 23 April 2018.

34. Huttunen, S.; Manninen, K.; Leskinen, P. Combining Biogas LCA Reviews with Stakeholder Interviews to Analyse Life Cycle Impacts at a Practical Level. *J. Clean. Prod.* **2014**, *80*, 5–16. [CrossRef]
35. Titu, A.M.; Simonffy, A. Contributions Regarding the Reduction of Production Costs for Brewing by Recovering and Reusing the Carbon Dioxide. *Procedia Econ. Financ.* **2014**, *16*, 141–148. [CrossRef]
36. Christiansen, R. Barrels of Biogas. Available online: https://biomassmagazine.com/articles/2540/barrels-of-biogas/#:~:text=Blossman%20says%20the%20anaerobic%20digester (accessed on 22 May 2022).
37. Crubaugh, L. Brewery's Anaerobic Digester System Reduces Loading to Municipal Wastewater. Available online: https://www.tpomag.com/online_exclusives/2013/02/brewerys_anaerobic_digester_system_reduces_loading_to_municipal_wastewater (accessed on 2 November 2022).
38. Tucker, M. Digester in Magic Hat's Sustainability Mix—BioCycle. Available online: https://www.biocycle.net/digester-in-magic-hats-sustainability-mix/ (accessed on 15 October 2022).
39. Nagelkirk, J. Bell's Brewery Gives Tour Highlighting Benefits of Advanced Energy, Energy Efficiency. Available online: https://mieibc.org/bells-brewery-gives-tour-highlighting-benefits-of-advanced-energy-energy-efficiency/ (accessed on 28 October 2022).
40. Gerardi, M.H. *The Microbiology of Anaerobic Digesters*; John Wiley: Hoboken, NJ, USA, 2003.
41. Speece, R.E. *Anaerobic Biotechnology for Industrial Wastewaters*; ACS Press: Nashville, TN, USA, 1996.
42. Zupančič, G.D.; Škrjanec, I.; Marinšek Logar, R. Anaerobic Co-Digestion of Excess Brewery Yeast in a Granular Biomass Reactor to Enhance the Production of Biomethane. *Bioresour. Technol.* **2012**, *124*, 328–337. [CrossRef]
43. Zupančič, G.D.; Panjičko, M.; Zelić, B. Biogas Production from Brewer's Yeast Using an Anaerobic Sequencing Batch Reactor. *Food Technol. Biotechnol.* **2017**, *55*, 187–196. [CrossRef] [PubMed]
44. Neira, K.; Jeison, D. Anaerobic Co-Digestion of Surplus Yeast and Wastewater to Increase Energy Recovery in Breweries. *Water Sci. Technol.* **2010**, *61*, 1129–1135. [CrossRef]
45. Connaughton, S.; Collins, G.; O'Flaherty, V. Psychrophilic and Mesophilic Anaerobic Digestion of Brewery Effluent: A Comparative Study. *Water Res.* **2006**, *40*, 2503–2510. [CrossRef]
46. Verive, J. Recapturing CO_2: It's a Gas. Brewing Industry Guide. Available online: https://brewingindustryguide.com/recapturing-co2-its-a-gas (accessed on 12 February 2022).

fermentation

MDPI

Article

Modelling of Amino Acid Fermentations and Stabilization of Anaerobic Digestates by Extracting Ammonium Bicarbonate

Alejandro Moure Abelenda *, George Aggidis and Farid Aiouache

School of Engineering, Lancaster University, Lancaster LA1 4YW, UK; g.aggidis@lancaster.ac.uk (G.A.); f.aiouache@lancaster.ac.uk (F.A.)

* Correspondence: a.moureabelenda@lancaster.ac.uk; Tel.: +44-79-33-712-762; Fax: +34-650-222-342

Abstract: With the current increase in demand for animal and agricultural products, management of agrowaste has become critical to avoid greenhouse gas emissions. The present article investigates the applicability of ammonium bicarbonate synthesis via flash distillation to valorize and stabilize several types of anaerobic digestates which are produced from individual fermentations of amino acids. The content of CO_2 in the digestate was found to be responsible for the OH alkalinity (0.4 equivalents of acid/kg digestate), while the partial and total alkalinities (0.8 eq/kg digestate) were essentially derived from the content of NH_3. The most suitable conditions for the flash distillation were 95 °C and 1 bar with the condensation occurring at 25 °C. However, in order to attain the precipitation of NH_4HCO_3 in the distillate, it was necessary to consider digestates with a moisture content of 50 wt.%, since saturation levels of inorganic nitrogen and inorganic carbon were not attained otherwise. Even under these conditions, few amino acids (i.e., arginine, glycine, and histidine) were able to provide an anaerobic digestate upon fermentation that would be suitable for NH_4HCO_3 stabilization. The process of stabilization with a capacity of a t of digestate per h was improved by adding hydrochloric acid or sodium hydroxide at a rate of 44 kg/h, leading to production of 34 kg NH_4HCO_3/h. Given the role of the volatile elements of the biogas as endogenous stripping agents, it is recommended to use a fresh and saturated digestate as feed for the flash distillation.

Keywords: circular economy; greenhouse gas; carbon capture; endogenous striping agents; biogas upgrading; slow-release fertilizer

Citation: Moure Abelenda, A.; Aggidis, G.; Aiouache, F. Modelling of Amino Acid Fermentations and Stabilization of Anaerobic Digestates by Extracting Ammonium Bicarbonate. *Fermentation* **2023**, *9*, 750. https://doi.org/10.3390/fermentation9080750

Academic Editors: Jose Luis García-Morales and Francisco Jesús Fernández Morales

Received: 4 July 2023
Revised: 30 July 2023
Accepted: 9 August 2023
Published: 12 August 2023

1. Introduction

Anaerobic digestion (AD) is one the most promising technologies for reliable production of clean energy [1]. According to Patel et al. [2], AD has the potential to convert up to 95 wt.% of organic matter to biogas. The valorization of waste by means of this fermentative process enhances the circular economy because most of the nutrients required for cultivation of crops end up in the organic residue (i.e., anaerobic digestate), which is mainly applied to land as organic soil amendment and reduces the consumption of industry-based inorganic fertilizers [3]. The mineralization of organic nutrients during AD implies that these become more readily available to plants but also that they can be easily lost via volatilization and leaching during the management of organic manures. Current regulations in force prescribe closing slurry stores and the use of low-emission spreading techniques at the time of land application, which should be done at specific locations and during the right season [4]. It is also possible to carry out the isolation of the nutrients, for example, by precipitating the struvite (i.e., magnesium ammonium phosphate). This open-loop management strategy requires prior solid–liquid separation of the anaerobic digestate, an external source of magnesium [5], and often additional phosphate to be able to deplete the ammoniacal nitrogen (NH_4^+-N) initially contained in the aqueous solution [6,7]. The formation of ammonium carbonate in the anaerobic digestate has been traditionally reported as one of the phenomena that increase and regulate the pH during anaerobic fermentation [8]; therefore, this could be among the most efficient routes for the isolation of NH_4^+-N.

The ammonium bicarbonate process (known as ABC) [9–11] is, therefore, considered a more sustainable alternative for recovering ammonia, compared with other processes involving the use of exogenous scrubbing agents, such as sulfuric acid [12], and the operation of expensive equipment with high energy demand, such as an air blower [13]. Hence, the stabilization of anaerobic digestate by means of NH_4HCO_3 synthesis has gained attention as a promising profitable technology [13]. Unlike open-loop processes, the patented ABC [9] offers a synergistic approach in which biogas with a composition of 65 vol.% CH_4 and 35 vol.% CO_2 [14] is efficiently used as a scrubbing agent of the stripped gas. In this process, the pH of the liquid fraction of the anaerobic digestate is adjusted to 11.5 via addition of NaOH to reduce 10 times its content of total ammonia (from 800 to 80 mg/L), using the NH_3&CO_2-depleted biogas (i.e., biomethane grade or 99 vol.% CH_4) as stripping gas. Subsequently, the resulting stripped gas is mixed with 58 vol.% of the fresh biogas, originally produced in the AD, to promote the precipitation of NH_4HCO_3 and produce the biomethane stream. In addition to the NH_3&CO_2-depleted/stabilized liquid digestate and the solid crystals of NH_4HCO_3, the purges of 42 vol.% fresh biogas and 1 vol.% biomethane (these percentages refer to the total volume of the streams generated in the steps of anaerobic digestion and NH_4HCO_3 precipitation, respectively) could be regarded as valuable outputs of the ABC. The 99 vol.% of the biomethane generated in the step of NH_4HCO_3 precipitation is recirculated and employed as stripping gas. The role of the CH_4 in the ABC is not clear [9], nor whether the use of the biomethane as stripping gas affects (a) the volatilization of NH_3 from the alkalinized liquid digestate and (b) the subsequent NH_4HCO_3 precipitation during the contact of the stripped gas with the fresh biogas.

In the distillation process engineered by Drapanauskaite et al. [13] for the production of NH_4HCO_3 from the liquid fraction of an anaerobic digestate, the release of CH_4 was not accounted in the liquid distillate stream obtained at the top of the column at 3 bar and 49 °C (89.0 vol.% H_2O, 5.0 vol.% HCO_3^-, 5.0 vol.% NH_4^+, 0.3 vol.% CO_2, 0.1 vol.% NH_3, 0.3 vol.% NH_2COO^-, and 0.1 vol.% CO_3^{2-}). It is noteworthy that the Henry's law volatility constant of methane ($71.43 \cdot 10^3$ $Pa \cdot m^3 \cdot mol^{-1}$) is greater than those of ammonia (1.69 $Pa \cdot m^3 \cdot mol^{-1}$) and carbon dioxide ($3.03 \cdot 10^3$ $Pa \cdot m^3 \cdot mol^{-1}$) in an aqueous solution [15]. It should be noted that since NH_3 and CO_2 behave like non-condensable gases under the normal operating conditions of the distillation, Henry's law is preferred to Raoult's law to describe this two-step system: degasification and absorption. This agrees with the fact that less than 1% of the NH_4^+-N is volatilized as part of the biogas release during the AD [8]. In the condensed distillate at 3 bar and 49 °C and in the uncondensed purge there were shares of 0.2 vol.% and 0.3 vol.%, respectively, that were not identified and could be attributed to methane [13]. It is important to account for the release of any gas during the NH_4HCO_3 synthesis because this can have a significant effect on the residual biogas potential of the stabilized digestate, for which an upper limit of 250–450 mL/g volatile solids is considered [16–18].

The present article investigates whether NH_4HCO_3 manufacturing technology via flash distillation is suitable for any composition of anaerobic digestate or whether some requirements need to be introduced at the stage of selecting the feedstock for AD. It is proposed to further expand the process simulation model by Rajendran et al. [19] for AD while minimizing the heat requirements for the flash distillation, in agreement with the approach followed by Centorcelli et al. [20,21]. Therefore, this investigation supports and advances the 7th (affordable and clean energy) and 13th (climate action) Sustainable Development Goals of the United Nations. The closed-loop process, where the CO_2 is valorized as absorbent of the NH_3 for synthesis of NH_4HCO_3, is relevant for agroindustry and the chemical fertilizer industry.

2. Materials and Methods

Following the methodology of Drapanauskaite et al. [13], prior to the development of the models in Aspen Plus® v12, the simulation was validated by comparing the empirical data available in the literature to the values provided by the software for the partial pressures of NH_3 and CO_2 in an aqueous system at the bubble point. Similarly, the supersaturation of the aqueous system in NH_4HCO_3, NH_4COONH_2, $NaHCO_3$, and NH_4Cl was also tested to ensure that the formation of solid phases was closely monitored in the flash distillation process. A model digestate was elaborated (Table A1) with the descriptions by Rajendran et al. [19] and Akhiar et al. [22] to verify the simulation of adding acids or bases to shift the chemical equilibria, following the titration methodology of Moure Abelenda et al. [23]. When it was confirmed that the system CO_2-NH_3-H_2O was modelled correctly, the minimum conditions for the production of NH_4HCO_3 were investigated with the calculation block of the flash tank and using the electrolyte non-random two-liquid (ELECNRTL) property method. The ELECNRT method is defined by Aspen Plus® as versatile and capable of handling both very low and high concentrations of solutes in aqueous systems and other solvents. The first step was to determine the optimum temperatures for flash separation and condensation, to handle a tonne of digestate per hour with typical composition of 2 g/L of NH_3 and 3 g/L of CO_2 [22]. This processing capacity was selected based on the subsidies available for covering 40% of the total cost of the equipment dealing with slurry separation [24]; hence, a wider adoption by the stakeholders of agroindustry can be expected. The synthesis of NH_4HCO_3 was subsequently confirmed by a parametric study and the minimum feasible conditions (i.e., concentration of NH_3 and CO_2 and temperature of the flash distillation) were readjusted. In order to select the most suitable types of anaerobic digestates for the manufacturing of NH_4HCO_3, the minimum concentrations of NH_3 and CO_2 were matched to streams coming out of the anaerobic digestion of amino acids. These amino acids were set as the lower limit for the composition of the feedstock employed for the anaerobic digestion, as the presence of any other type of molecule would result in a diluted stream with lower concentrations of inorganic nitrogen and carbon, which would constrain the production of NH_4HCO_3 by the flash distillation process. The stoichiometry and the kinetics for amino acid anaerobic fermentation were mostly taken from the previous investigation by Rajendran et al. [19], except for the fermentation of glutamine (Equation (1)) which was proposed in a similar fashion to the degradation of the other amino acids. The production of oxygen in some of these reactions implies that they are only spontaneous under strongly reducing conditions and would be very limited under aerobic conditions (Le Chatelier's principle). It is important to mention that the type of model that was used for the AD was unstructured and unsegregated, which implies that it did not involve the metabolism of the microorganisms nor differentiate the species that degraded the biomass [25].

$$C_5H_{10}N_2O_3 + 3H_2O \rightarrow 2C_2H_4O_2 + CO_2 + 2NH_3 + H_2 \quad (1)$$

$$C_2H_5NO_2 + 0.5H_2O \rightarrow 0.75C_2H_4O_2 + NH_3 + 0.5CO_2 \quad (2)$$

$$C_2H_5NO_2 + H_2O \rightarrow C_2H_4O_2 + NH_3 + 0.5O_2 \quad (3)$$

$$C_4H_9NO_3 + H_2O \rightarrow C_3H_6O_2 + NH_3 + H_2 + CO_2 \quad (4)$$

$$C_4H_9NO_3 + H_2O \rightarrow C_2H_4O_2 + 0.5C_4H_8O_2 + NH_3 + 0.5O_2 \quad (5)$$

$$C_6H_9N_3O_2 + 5H_2O \rightarrow CH_3NO + 2C_2H_4O_2 + 2NH_3 + CO_2 + H_2 \quad (6)$$

$$C_6H_9N_3O_2 + 5H_2O \rightarrow CH_3NO + C_2H_4O_2 + 0.5C_4H_8O_2 + 2NH_3 + CO_2 + 0.5O_2 \quad (7)$$

$$C_6H_{14}N_4O + 6H_2O \rightarrow 2C_2H_4O_2 + 4NH_3 + 2CO_2 + 3H_2 \quad (8)$$

$$C_6H_{14}N_4O + 4H_2O \rightarrow 0.5C_2H_4O_2 + 0.5C_3H_6O_2 + 0.5C_5H_{10}O_2 + 4NH_3 + CO_2 + 0.5O_2 \quad (9)$$

$$C_5H_{11}NO_2S + 2H_2O \rightarrow C_3H_6O_2 + CO_2 + NH_3 + H_2 + CH_4S \quad (10)$$

$$C_3H_7NO_3 + H_2O \rightarrow C_2H_4O_2 + CO_2 + NH_3 + H_2 \quad (11)$$

$$C_4H_7NO_4 + 2H_2O \rightarrow C_2H_4O_2 + 2CO_2 + NH_3 + 2H_2 \quad (12)$$

$$C_5H_9NO_4 + H_2O \rightarrow C_2H_4O_2 + 0.5C_4H_8O_2 + CO_2 + NH_3 \quad (13)$$

$$C_6H_{14}N_2O_2 + 2H_2O \rightarrow C_2H_4O_2 + C_4H_8O_2 + 2NH_3 \quad (14)$$

$$C_6H_{13}NO_2 + 2H_2O \rightarrow C_5H_{10}O_2 + CO_2 + NH_3 + H_2 \quad (15)$$

$$C_5H_{11}NO_2 + 2H_2O \rightarrow C_4H_8O_2 + CO_2 + NH_3 + 2H_2 \quad (16)$$

$$C_9H_{11}NO_2 + 2H_2O \rightarrow C_6H_6 + C_2H_4O_2 + CO_2 + NH_3 + H_2 \quad (17)$$

$$C_9H_{11}NO_3 + 2H_2O \rightarrow C_6H_6O + C_2H_4O_2 + CO_2 + NH_3 + H_2 \quad (18)$$

$$C_{11}H_{12}N_2O_2 + 2H_2O \rightarrow C_8H_7N + C_2H_4O_2 + CO_2 + NH_3 + H_2 \quad (19)$$

$$C_3H_7NO_2 + 2H_2O \rightarrow C_4H_8O_2 + CO_2 + NH_3 + 2H_2 \quad (20)$$

$$C_3H_6NO_2S + 2H_2O \rightarrow C_2H_4O_2 + CO_2 + NH_3 + 0.5H_2 + H_2S \quad (21)$$

$$C_5H_9NO_2 + 2H_2O \rightarrow 0.5C_2H_4O_2 + 0.5C_3H_6O_2 + 0.5C_5H_{10}O_2 + NH_3 + 0.5O_2 \quad (22)$$

As in the simulation by Rajendran et al. [19], thermophilic conditions (55 °C) were considered to simulate the AD of amino acids. The hydrolysis of all amino acids followed the first-order reaction, with the kinetic constant (s^{-1}) described in Equation (23) as function of the temperature (K) [19]:

$$k = 1.2753 \cdot 10^{-6} \cdot \frac{T}{328.15} \cdot e^{\frac{1.41437 \cdot 10^4}{8.3145} \cdot \left(\frac{1}{T} - \frac{1}{328.15}\right)} \quad (23)$$

The subsequent methanogenic stage [26] was also included as part of the anaerobic fermentation of all amino acids by considering the stoichiometry of Equations (24) and (25) and the first-order kinetic constant (s^{-1}) following Equation (26) with the temperature in K [19]. Equation (25) was only considered if there was production of hydrogen in the previous acetogenic stage of degradation of amino acids. The modelling of the fermentation was confirmed by monitoring a constant value for Henry's parameter of each volatile compound.

$$C_2H_4O_2 + 0.022NH_3 \rightarrow 0.022C_5H_7NO_2 + 0.945CH_4 + 0.066H_2O + 0.945CO_2 \quad (24)$$

$$14.4976H_2 + 3.8334CO_2 + 0.0836NH_3 \rightarrow 0.0836C_5H_7NO_2 + 3.4154CH_4 + 7.4996H_2O \quad (25)$$

$$k = 2.394 \cdot 10^{-7} \cdot \frac{T}{328.15} \quad (26)$$

Finally, a comprehensive model was developed to confirm the synthesis of NH_4HCO_3 with particular types of feedstocks for AD. The model included the stages of amino acid fermentation and extraction of NH_4HCO_3 for stabilization of the resulting anaerobic digestate (Figure 1). The process engineering software employed to develop the model allows consideration of a single stream with liquid and gas phases coming out of the fermenter. This concept is representative of real conditions when treating fresh anaerobic digestate saturated with biogas. The titrations of amino acid digestates were investigated by adding NaOH or HCl before the flash distillation, with the purpose of tuning the ratio of NH_4^+ to HCO_3^- in the distillate. The equilibrium-driven reactions that were considered for the modelling of the multiphase system NH_3-CO_2-H_2O are shown in Table 1. A discussion is offered on which other parameters would need to be considered, in addition to the molar ratio of inorganic nitrogen (NH_4^++NH_3+NH_2COO^-) to inorganic carbon (CO_2+HCO_3^-+CO_3^{2-}+NH_2COO^-) of the anaerobic digestate, to facilitate the stabilization of this organic material by means of isolating the NH_4HCO_3. The 3 streams coming out of this synergistic process (Figure 1) are the upgraded biogas, the distillate (i.e., saturated NH_4HCO_3 aqueous solution), and the stabilized digestate.

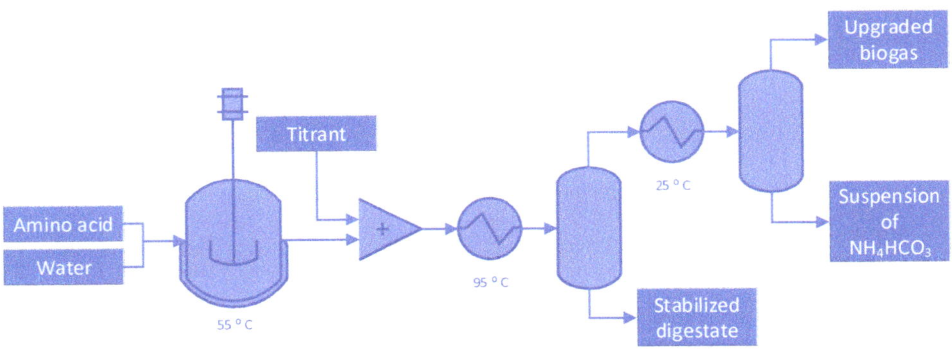

Figure 1. Process flow diagram with capacity to handle 1 t of feedstock (blend of amino acid and water) per h, to evaluate the feasibility of producing NH_4HCO_3 from different types of anaerobic digestates.

Table 1. Equilibrium-driven reactions for the system NH_3-CO_2-H_2O proposed by the property method ELECNRTL of the electrolyte template of Aspen Plus® v12.

$\ln(K_{eq})=A+\frac{B}{T}+C\cdot\ln(T)+D\cdot T$	A	B	C	D
$2H_2O + CO_2 \leftrightarrow HCO_3^- + H_3O^+$	231.465424	12092.099609	36.781601	0
$H_2O + HCO_3^- \leftrightarrow CO_3^{2-} + H_3O^+$	216.050446	12431.700195	35.481899	0
$NH_3 + H_2O \leftrightarrow OH^- + NH_4^+$	1.256563	3335.699951	1.4971	−0.037057
$NH_3 + HCO_3^- \leftrightarrow H_2O + NH_2COO^-$	4.583437	2900	0	0
$NH_4^+ + HCO_3^- \leftrightarrow NH_4HCO_3$	554.818115	22442.529297	89.006416	0.064732
$NH_4^+ + NH_2COO^- \leftrightarrow NH_4COONH_2$	4.289233	0	0	0

3. Results

3.1. Preliminary Validation of the Aspen Plus® v12 ELECNRTL Methodology

Figure 2a shows that the results of the simulation follow the trends of the experimental data of Goppert and Maurer (1988), which were reported by Darde [27] in Figures 5–12 and Figures 5–13. The greatest volatility of CO_2 corresponds to the greater partial pressure (at least two orders of magnitude greater) than that exerted by NH_3 for any of the compositions of the anaerobic digestate investigated at the bubble point, as shown in Figure 2a. The chemistry of this blend also plays a crucial role in determining the volatility of these compounds, as partial pressure of the CO_2 started to increase at a greater rate when the moles of this compound in the anaerobic digestate were greater than the moles of NH_3 (Figure 2a), which was set to 0.6 mol/kg H_2O for the whole test to comply with the experimental data available in Figures 5–12 and Figures 5–13 of Darde [27]. Typical concentrations of CO_2 and NH_3 in the anaerobic digestate are around 0.12 mol NH_3/kg H_2O and 0.07 mol CO_2/kg H_2O [22], which are milder concentrations than those tested by Darde et al. [28] for the development of a carbon capture process. However, given the wide scope of the ELECNRTL property method, the validation could be performed at the lower end of the concentration investigated by Darde et al. [28]. Figure 2b shows the calibration carried out with a concentration of 0.13 mol NH_3/kg H_2O with the experimental data of Pexton and Badger (1938), which is reported in Figures 2 and 3 of Darde et al. [28]. Figure 2c validates the precipitation of the NH_4HCO_3, which is less soluble than the NH_4COONH_2 and has similar solubility to NH_4Cl and higher solubility than $NaHCO_3$. Contrary to the explanation by Möller and Müller [8] on the formation of ammonium carbonate in the anaerobic digestate, Aspen Plus® v12 ELECNRTL does not consider the formation of this compound. Modelling the solubilities of NH_4Cl and $NaHCO_3$ was considered because the HCl and NaOH were tested as titrants of the anaerobic digestate to assist the flash distillation, tune the ratio of inorganic nitrogen to inorganic carbon in the distillate, and promote the precipitation of NH_4HCO_3. Therefore, NH_4Cl and $NaHCO_3$ might precipitate in the residual stabilized digestate and they are not expected in the distillate despite having solubility similar to or lower than NH_4HCO_3. The big difference in the solubility of NH_4HCO_3 modelled above 60 °C with regard to the experimental data available in the literature [29,30] is because this compound is not very stable and easily undergoes decomposition [31]. Figure 2d shows the modelling of the titration of the manure digestate resulting from the process described by Rajendran et al. [19] and considering some of the alkaline elements identified in the comprehensive characterization of various anaerobic digestates performed by Akhiar et al. [22]: $NaHCO_3$, $CaCl_2$, NaCl, and $KHCO_3$ (Table A1). The validation of the simulation of the effect of HCl and NaOH on the proposed anaerobic digestate was achieved by comparison with experimental data of titrations available in the literature, particularly the study by Vandré and Clemens [32] of raw animal slurry and the previous work of Moure Abelenda et al. [23] with agrowaste digestate and food waste digestate. The results showed that the amount of CO_2 and carbonates in the anaerobic digestate was responsible for the OH alkalinity at pH > 10. The content of ammonia was found to be the main cause of the partial (P) alkalinity and the total (M) alkalinity. This means that the volatile fatty acids (i.e., acetic acid and propionic acid) and other components (e.g., hydrogen sulfide; Table A1) previously reported with a role in regulating the pH of the anaerobic digestate [33] did not have a significant buffer effect in the present model. As can be seen in Figure A1, simply tuning of the ratio CO_2 to NH_3 in the digestate affects the pH, which shows the how the pH depends on the composition of the anaerobic digestates and the remaining stabilized digestates (Figure 1) after the flash separation indicated in Figure 2b.

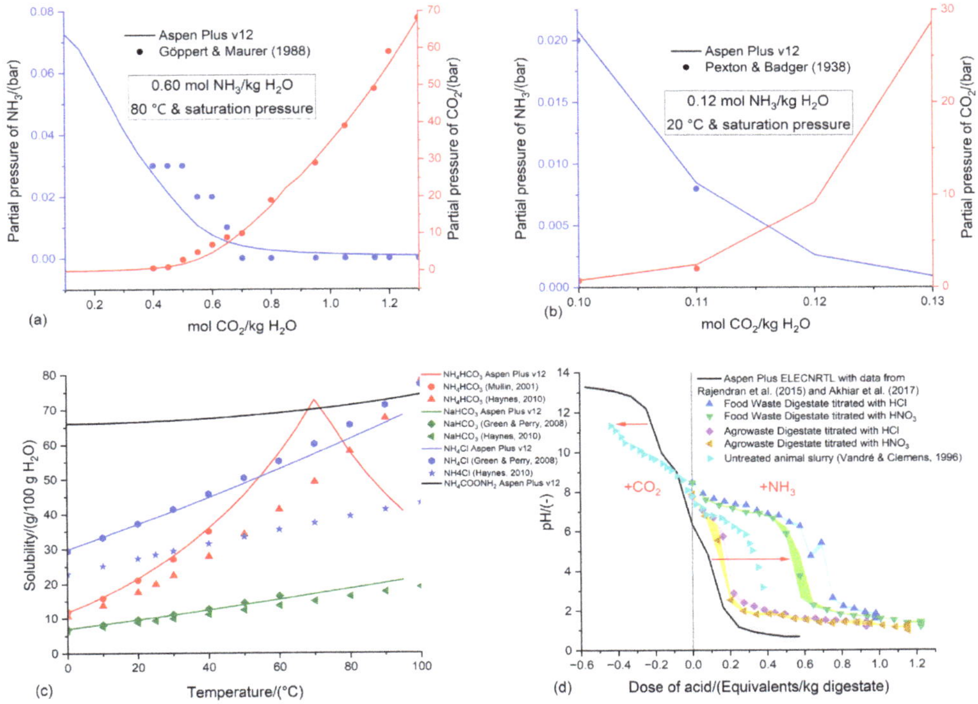

Figure 2. Validation of the modelling of the system NH_3-CO_2-H_2O with Aspen Plus® v12. (**a**) Comparison of the partial pressures of CO_2 and NH_3 calculated with ELECNRT property method and the experimental data available in Figures 5–12 and Figures 5–13 of Darde [27]. (**b**) Validation of the flash distillation by comparing the results of the simulation with the experimental data available in Figures 2 and 3 of Darde et al. [28]. (**c**) Validation of the solid formation by comparing the results of the simulation with the experimental data reported by Mullin [29], Green and Perry [34], and Haynes [29]. (**d**) Validation of the processing of manure digestate [19,22] with the addition of NaOH and HCl by comparison with the empirical titration of agrowaste digestate, food waste digestate [23], and animal slurry [32].

3.2. Optimization of the Conditions of the Flash Distillation to Produce NH_4HCO_3

In the initial development stage of the flash distillation process for stabilization of anaerobic digestate and manufacturing of NH_4HCO_3, the target processing capacity was set at 1 t per h (i.e., proposed organic slurry with a composition of 55 kmol H_2O/h, 0.1 kmol NH_3/h, and 0.1 kmol CO_2/h). The optimum temperatures for evaporation and condensation were investigated with this simplified composition of anaerobic digestate. Figure 3a shows the volatilization of NH_3, CO_2, and H_2O as a function of the temperature. In order to appreciate significant release of CO_2 from the anaerobic digestate, the temperature of the flash tank should be greater than 75 °C. There is not a big difference (i.e., around 0.2 °C) between the minimum temperature to obtain the maximum volatilization of NH_3 and that of H_2O (Figure 3a). However, it is this small difference that needs to be taken into account in the next stages of the design of the process to maximize the overall heat integration and profitability of the flash distillation process (Figure A2). Figure 3bpresents the concentrations of the different species in the distillate as a function of the condensation temperature. The condensation at a temperature below 70 °C maximizes the concentration of NH_4^+ and HCO_3^- (Figure 2b), which is necessary for the precipitation of NH_4HCO_3 when supersaturation is attained in the aqueous solution. In order to precipitate NH_4HCO_3

in the distillate cooled at 1 bar and 25 °C, the minimum temperature of the flash tank should be between 85 and 95 °C and the composition of the anaerobic digestate was found to be around 10 g NH_3/L and 13 g CO_2/L. This corresponds to a stream of anaerobic digestate fed to the flash distillation process at a rate of 55 kmol H_2O/h, 0.6 kmol NH_3/h, and 0.3 kmol CO_2/h. Under these conditions, the maximum production of NH_4HCO_3 at a rate of 35.4 mol/h or 2.8 kg/h was found when the temperature of the flash tank was 95 °C (Figure A2).

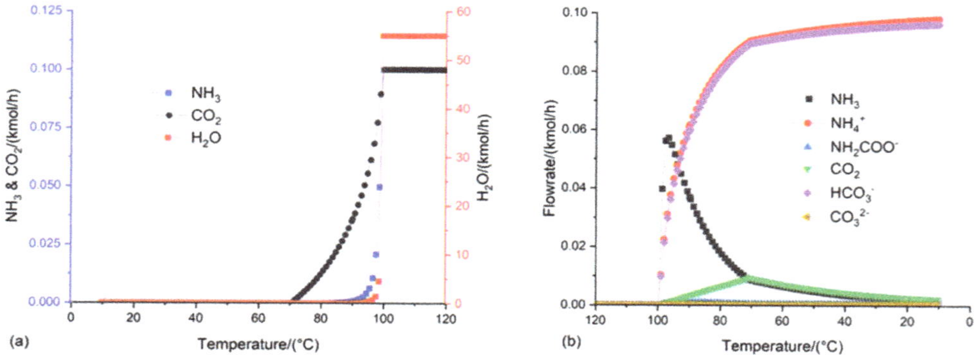

Figure 3. Optimization of the steps of evaporation and condensation that comprise the flash distillation process for stabilization of anaerobic digestate and isolation of NH_4HCO_3: (**a**) determination of the optimum flash-heating temperature to achieve a vapor stream with the greatest concentration of CO_2 and NH_3; (**b**) determination of the optimum condensation temperature to attain the maximum concentration of NH_4^+ and HCO_3^- in the distillate.

3.3. Anaerobic Fermentation of Amino Acids

As this minimum composition of digestate to attain stabilization via manufacturing of NH_4HCO_3 is not in agreement with the typical composition of anaerobic digestate [22], the next step in defining the specifications of the flash process was to find a source of anaerobic digestate that provides a concentration of 10 g NH_3/L and 13 g CO_2/L. Since amino acids are ultimately responsible for the content of NH_4^+-N in the anaerobic digestate, the composition of the feedstock of AD was investigated by starting with these organic molecules. Although this is not a realistic approach because AD is a technology for handling complex matrices of organic waste materials, this investigation with process engineering software could give an idea of the type of residues that are more suitable for stabilizing the resulting anaerobic digestate with NH_4HCO_3 extraction. The concentration profile of all chemical species was elucidated for the fermentation of 18 amino acids (Figure A3), considering the stoichiometry and kinetics defined by Rajendran et al. [19], which are shown in Equations (1)–(26). In all cases, the feed of the anaerobic digester was considered to have 90 wt.% moisture [35] and the remaining 10 wt.% of the feedstock corresponded to the mass of amino acids. The discontinuous fermenters were operated under thermophilic conditions (i.e., 55 °C and 1 bar) for 1000 h (i.e., 42 days) in order to be able to observe the constant profiles of all chemical species, with the exception of glutamic acid that required a longer residence time (Figure A3h and Figure S8). For the present theoretical work, the hydrolysis reactions that would break down proteins into amino acids were omitted and the model only included the acidogenic, acetogenic, and methanogenic reactions. As the fermentation kinetics proposed by Rajendran et al. [19] were implemented, it was not necessary to define a highly active inoculum for the calculation block of the bioreactor to function properly. Most amino acids required at least 500 h in the bioreactor, with the exception of cysteine (Figure A3n) that was rapidly converted in less than 100 h. This simplified model corresponds with the first steps in elucidating the extent to which the feasibility of the NH_4HCO_3 stabilization of anaerobic digestate is limited by the type of feedstock for AD. The concentrations of inorganic nitrogen and inorganic carbon that were found in the anaerobic digestates of amino acids (Figure 4) were similar to those reported by Akhiar et al. [22].

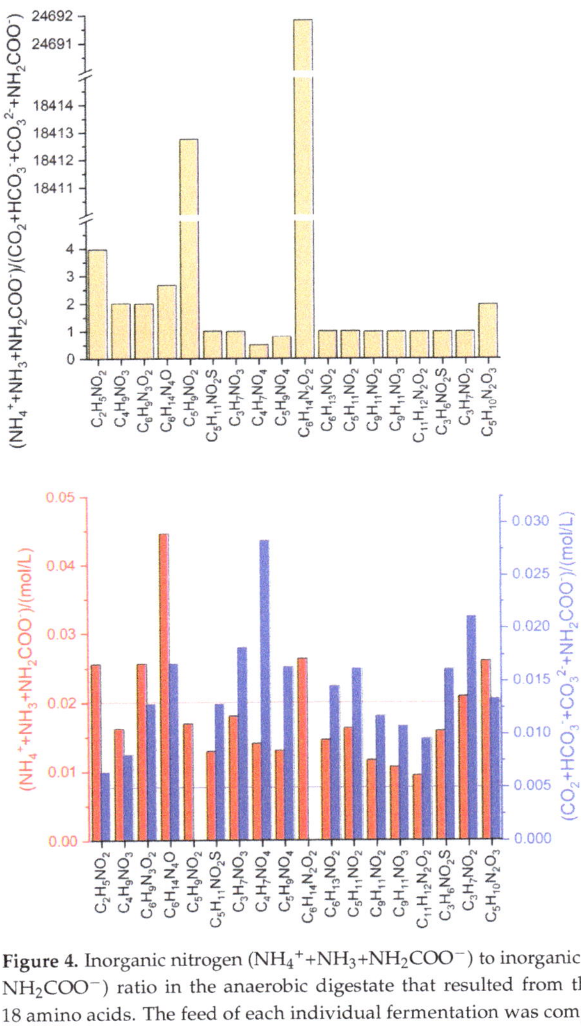

Figure 4. Inorganic nitrogen (NH_4^++NH_3+NH_2COO^-) to inorganic carbon (CO_2+HCO_3^-+CO_3^{2-}+ NH_2COO^-) ratio in the anaerobic digestate that resulted from the fermentation of each of the 18 amino acids. The feed of each individual fermentation was composed of 90 wt.% moisture and 10 wt.% amino acid.

3.4. Suitable Conditions for NH_4HCO_3-Stabilization of Anaerobic Digestate

As was expected according to the molecular formulas and stoichiometries of the amino acids' fermentations, arginine provided the greatest amount of mineralized nitrogen (Figure 4), but this corresponded to only 0.6 g/L. The arginine digestate was also considered in the study by Drapanauskaite et al. [13]. This means that it would be necessary to consider at least a feedstock for AD with 3.5 times less water in relation to arginine for the development of the process. In this way, the overall process of fermentation and stabilization via synthesis of NH_4HCO_3 (Figure 1) was found to be feasible for arginine ($C_6H_{14}N_4O$), glycine ($C_2H_5NO_2$), and histidine ($C_6H_9N_3O_2$) only when the feedstocks of these amino acids were prepared with a moisture content lower than 90 wt.%. For example, NH_4HCO_3 production rates were 15.16, 3.00, and 1.82 kg/h when processing 1 tonne/h of digestates resulting from the individual AD of arginine, glycine, and histidine with 50 wt.% moisture. The moisture content of the digestate was estimated considering the initial composition of the feedstock to be fermented. The results of the titrations showed that the extent of stabilization of the anaerobic digestates coming from these amino acids can be greatly improved via acidification (in the case of arginine digestate; Figure 5a) and alkalization (for glycine and histidine; Figure 5b). The rate of addition of HCl was not enough to deplete the partial and total alkalinities of arginine, glycine, and histidine digestates (Figure 5a), as the pH never dropped below a level of 7 (Figure 2d). Similarly, the addition of NaOH was not enough to overcome the OH alkalinity (Figure 2d), except for the alkalization of histidine digestate which reached a pH above 9.5 (Figure 5b). The greatest NH_4HCO_3 extraction rates were 22.71 kg/h from arginine digestate (50 wt.% moisture) with 20 kg HCl/h (Figure 5a), 22.08 kg/h from glycine digestate (50 wt.% moisture) with 31 kg NaOH/h, and 34.18 kg/h from histidine digestate (50 wt.% moisture) with 44 kg NaOH/h (Figure 5b). The detrimental effect that acidification of glycine digestate and histidine digestate have on the production rate of NH_4HCO_3 should be noted (Figure 5a). Similarly, any addition of NaOH to the arginine digestate reduces the amount of NH_4HCO_3 that can be recovered. This is important to consider when elaborating a more complex digestate (i.e., coming from the blend of several amino acids). The maximum tolerances observed for the anaerobic digestates with 50 wt.% moisture were 27 kg NaOH/h applied to arginine digestate yielding 327.96 g NH_4HCO_3/h, 4 kg HCl/h applied to glycine digestate yielding 245.54 g NH_4HCO_3/h, and 2 kg HCl/h applied to histidine digestate yielding 146.34 g NH_4HCO_3/h. However, these trends changed when anaerobic digestates with greater moisture content were considered for the NH_4HCO_3 stabilization. The highest moisture content of amino acid digestates that was suitable for production of NH_4HCO_3 was dependent on the dose of acid and base. NaOH was a more effective titrant than HCl in the case of arginine digestate (>50 wt.% moisture), and HCl was a more effective titrant than NaOH for cysteine and histidine digestates (>50 wt.% moisture). The models developed in Aspen Plus® v12 ELECNRTL show the feasibility of (a) extraction of 142.30 g NH_4HCO_3/h from arginine digestate (62 wt.% moisture) by adding a dose of 1 kg NaOH/h, (b) extraction of 23.16 g NH_4HCO_3/h from arginine digestate (65 wt.% moisture) by adding a dose of 14 kg HCl/h, (c) extraction of 21.94 g NH_4HCO_3/h from glycine digestate (82 wt.% moisture) by adding a dose of 26 kg NaOH/h, (d) extraction of 469.01 g NH_4HCO_3/h from glycine digestate (53 wt.% moisture) by adding a dose of 1 kg HCl/h, (e) extraction of 72.70 g NH_4HCO_3/h from histidine digestate (81 wt.% moisture) by adding a dose of 34 kg NaOH/h, and (f) extraction of 109.21 g NH_4HCO_3/h from histidine digestate (51 wt.% moisture) by adding a dose of 1 kg HCl/h.

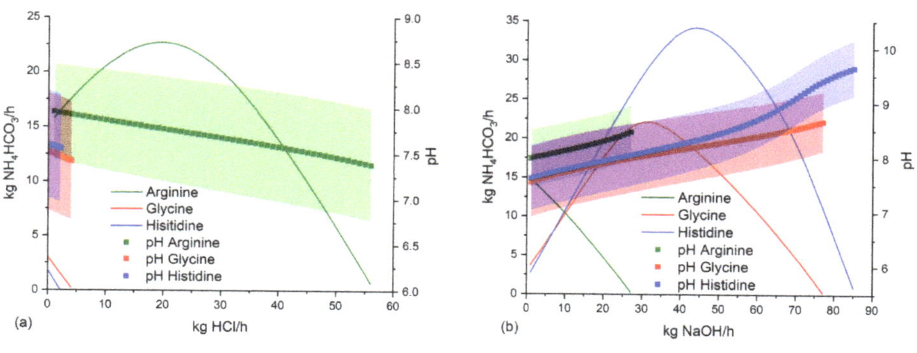

Figure 5. Optimization of the NH_4HCO_3 stabilization via addition of (**a**) HCl and (**b**) NaOH to enhance the flash distillation of amino acid digestates produced with a feedstock of 50 wt.% moisture.

The square symbols represent the pH of the stabilized digestates that leave the bottom of the flash tank (Figure 1).

4. Discussion

The present study questions the reliability of measurements of nitrogen in anaerobic digestate [22,36,37], since stoichiometric calculations show that the nitrogen content is expected to be lower than 0.63 g/L when moisture is 90 wt.% (Figure A3). Modeling of AD resulted in almost the complete consumption of amino acids by the end of the 42 days of residence time (Figure A3), and this could be regarded as a very favorable scenario for the production of white ammonia [38]. However, the content of organic nitrogen in anaerobic digestate is never less than 30% of the total nitrogen [22]. Still considering all the conversion of organic nitrogen to inorganic forms, the amino acid digestate with 90 wt.% moisture was found not to be enough for stabilization via synthesis of NH_4HCO_3. It was necessary to reduce the moisture content of the feedstock of AD to 50 wt.% to enable the synthesis of NH_4HCO_3 from arginine digestate, glycine digestate, and histidine digestate. The fermentation of alanine ($C_3H_7NO_2$; Figure A3o) and that of glutamine ($C_5H_{10}N_2O_3$; Figure A3r) also provided anaerobic digestates (50 wt.% moisture) with sufficient concentration of inorganic nitrogen and inorganic carbon (Figure 4), but the manufacturing of NH_4HCO_3 was not possible in these cases. The threshold values 0.02 mol N/L and 0.005 mol C/L, which were established based on the profile of glycine (Figure 4), explain why NH_4HCO_3 valorization was not attained with proline ($C_5H_9NO_2$) and lysine ($C_6H_{14}N_2O_2$), despite their anaerobic digestates having the greatest inorganic nitrogen to inorganic carbon ratio. The inorganic forms of nitrogen are more readily available for plants, since these compounds do not need to be metabolized by the microbes in the rhizosphere, unlike organic nitrogen. These inorganic forms are also more prone to be lost via leaching and volatilization, hence the digestate is less stable and does not behave as controlled-release fertilizer. Traditionally, acidification has been used to prevent ammonia emissions from organic manure and slurry during storage and land application, because it reduces the pH and the NH_3 can be kept in the aqueous solution as NH_4^+ [39]. However, in the approach of stabilizing the anaerobic digestate by NH_4HCO_3 extraction, the only purpose of using acids or alkalis is to promote the formation of a supersaturated distillate from where the crystals of the chemical grade fertilizer can be easily harvested. Hence, the doses of these endogenous agents to tune the pH of the anaerobic digestate need to be optimized accordingly, as long as NH_4HCO_3 is isolated.

A difference between the alanine digestate and glutamine digestate and the arginine digestate, glycine digestate, and histidine digestate is the presence of oxygen among the volatile compounds (Figure 6). Looking at the volatility of these gases, the use of the components of the residual biogas as stripping agents is appropriate because the volatility of H_2 ($129.87 \cdot 10^3$ $Pa \cdot m^3 \cdot mol^{-1}$) and O_2 ($76.92 \cdot 10^3$ $Pa \cdot m^3 \cdot mol^{-1}$) is greater than that of NH_3 (1.69 $Pa \cdot m^3 \cdot mol^{-1}$) and CO_2 ($3.03 \cdot 10^3$ $Pa \cdot m^3 \cdot mol^{-1}$). The effect of these endogenous stripping agents could be the reason that the manufacturing of NH_4HCO_3 was possible with arginine, glycine, and histidine digestates, even when these had lower concentrations than 0.5 mol NH_3/L and 0.3 mol CO_2/L (Figure A2). The most intuitive reason for the lack of formation of NH_4HCO_3 in the distillate of the alanine and glutamine digestates (50 wt.% moisture) could be the low flowrate of NH_3 in the stream of volatiles leaving the top of the flash tank at 95 °C (0.06 kmol NH_3/h; Figure 6a). From a broader perspective, there are resemblances among the H_2O and NH_3 trends of the streams of volatile elements leaving the top of the flash tank during the processing of the five untreated amino acid digestates with 50% moisture (Figure 6a). Together with the pH of the digestate, this could be an explanation for why the volatilization of NH_3 was the greatest in the arginine digestate (Figure 6a) and more NH_4^+ left the bottom of the flash tank when processing glycine, histidine, and glutamine digestates (Figure 6b). The pHs that were found in the stabilized digestates (50 wt.% moisture) leaving the bottom of the flash tank (Figure 1) were:

7.98 ± 0.52 (arginine); 7.53 ± 0.65 (glycine); 7.61 ± 0.59 (histidine); 7.12 ± 0.23 (alanine); and 7.14 ± 1.42 (glutamine).

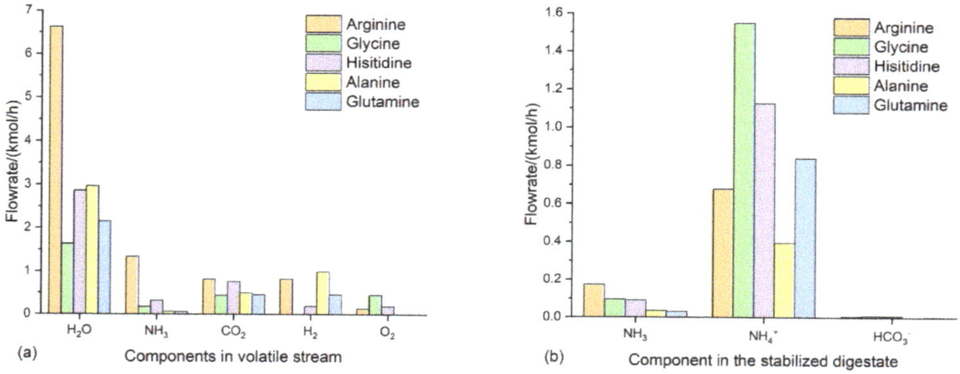

Figure 6. Flowrate of the most relevant chemical species for the manufacturing of NH_4HCO_3 leaving (**a**) the top and (**b**) the bottom of the flash distillation of five untreated amino acid digestates (50 wt.% moisture). The flowrate of water in the stabilized digestate was omitted to give a clear picture of the most relevant species of inorganic nitrogen and inorganic carbon. The flowrates of water in the stabilized digestates of arginine, glycine, histidine, alanine, and glutamine are 18.39, 24.78, 21.07, 23.79, and 24.21 kmol/h, respectively.

According to Ukwuani and Tao [40], at the high pH employed for NH_3 stripping, H_2S and VFAs are largely dissociated in the aqueous phase and render little volatilization. This could explain the fact that Ukwuani and Tao [40] did not find acetic acid in the sulfuric acid solution upon absorption of the stripped ammonia under vacuum pressure at different boiling-point temperatures. However, other volatile organic compounds (VOCs) can end up in the distillate stream depending on the operating conditions. For example, working at low pressures (i.e., under vacuum conditions) contributes more than operating at high temperatures to increasing emissions of VOCs [40]. Ukwuani and Tao [40] reported cyclohexene, 4-methylphenol, 4-ethylphenol, trichloromethane, and (p-hydroxyphenyl)-phosphoric acid as the most common VOCs in the acid NH_3 absorbent solution. The analysis of volatility that Ukwuani and Tao [40] carried out was based primarily on the concentration of these compounds in the absorbent solution at the different boiling points of the anaerobic digestate (ranging from 50 to 100 °C) under vacuum conditions, and secondly on the vapor pressure of the compound at 25 °C. Only for cyclohexene and chloroform are Henry's law volatility constants greater than that of ammonia. It is important to highlight that Henry's volatility constants are often reported as the ratio of vapor pressure and aqueous solubility [15]. Without considering the much greater solubility of NH_3 in the aqueous solution, since the vapor pressure of NH_3 (1002 kPa at 25 °C) is much greater than that of cyclohexane and chloroform (12 and 26 kPa, respectively), the VOC mass transfer is much slower. Cyclohexene ($3.45 \cdot 10^3$ Pa·m^3·mol^{-1}) has a volatility even greater than that of carbon dioxide but lower than that of methane. Ukwuani and Tao [40] explained the presence of (p-hydroxyphenyl)-phosphoric acid in the absorbent solution by the excessive foam formation in the anaerobic digestate at a pH 9 in such a way that the foam reached the sulfuric acid solution. This was the reason that Drapanauskaite et al. [13] used defoaming agents based on silicon for the operation of distillation and stripping processes. Hence, in addition to the titrant, the need for an antifoaming agent needs to be assessed experimentally [35,41–44]. Alternatively, the use of exogenous stripping agents, such as nitrogen ($156.25 \cdot 10^3$ Pa·m^3·mol^{-1}) of air, which has even greater volatility than oxygen, is interesting but makes more difficult the subsequent upgrading of the biogas [45], as this gaseous stream becomes diluted. Ideally, the synthesis of NH_4HCO_3 will be coupled with biogas upgrading [46], but the purity of the crystals according to the specifications of the chemical fertilizers needs to be confirmed [40]. The kinetic reactions detailed in Table 2, which were proposed by AspenTech [47] to develop a rate-based model of CO_2 capture process with NH_3, could also be implemented in the present model to enhance the predictions of speciation in the NH_3-CO_2-H_2O system.

Table 2. Kinetics of the NH_3-CO_2-H_2O system defined by AspenTech [47].

Kinetic Factor=$k \cdot e^{\frac{-E}{R \cdot T}}$	k/(d^{-1})	E/(cal/mol)
$CO_2 + OH^- \rightarrow HCO_3^-$	$4.32 \cdot 10^{13}$	13.249
$HCO_3^- \rightarrow CO_2 + OH^-$	$2.38 \cdot 10^{17}$	29.451
$NH_3 + CO_2 + H_2O \rightarrow NH_2COO^- + H_3O^+$	$1.35 \cdot 10^{11}$	11.585
$NH_2COO^- + H_3O^+ \rightarrow NH_3 + CO_2 + H_2O$	$2.14 \cdot 10^{21}$	17.203

5. Conclusions

The present study confirms the suitability of the conditions of the flash distillation (95 °C at 1 bar) for extracting NH_4HCO_3 from anaerobic digestates produced by the fermentation of arginine, glycine, and histidine. The use of HCl and NaOH was found necessary to maximize the stabilization of anaerobic digestates, although the rates of application of these titrants need to be further assessed from an economic point of view as well. This investigation elucidated an important role carried out by the most volatile compounds of biogas in the process of stabilization of anaerobic digestate. Therefore, it is recommended to carry out NH_4HCO_3 synthesis with anaerobic digestates saturated in biogas to maximize the endogenous stripping effect during the flash distillation. This process needs to be performed at the beginning of the storage of anerobic digestate to minimize the loss of the residual biogas.

Supplementary Materials: The following supporting information can be downloaded at: https://www.mdpi.com/article/10.3390/fermentation9080750/s1, Figure S1: Concentration profile of the species of inorganic nitrogen and inorganic carbon during the fermentation of glycine; Figure S2: Concentration profile of the species of inorganic nitrogen and inorganic carbon during the fermentation of threonine; Figure S3: Concentration profile of the species of inorganic nitrogen and inorganic carbon during the fermentation of arginine; Figure S4: Concentration profile of the species of inorganic nitrogen and inorganic carbon during the fermentation of proline; Figure S5: Concentration profile of the species of inorganic nitrogen and inorganic carbon during the fermentation of methionine; Figure S6: Concentration profile of the species of inorganic nitrogen and inorganic carbon during the fermentation of serine; Figure S7: Concentration profile of the species of inorganic nitrogen and inorganic carbon during the fermentation of aspartic acid; Figure S8: Concentration profile of the species of inorganic nitrogen and inorganic carbon during the fermentation of glutamic acid; Figure S9: Concentration profile of the species of inorganic nitrogen and inorganic carbon during the fermentation of lysine; Figure S10: Concentration profile of the species of inorganic nitrogen and inorganic carbon during the fermentation of leucine; Figure S11: Concentration profile of the species of inorganic nitrogen and inorganic carbon during the fermentation of valine; Figure S12: Concentration profile of the species of inorganic nitrogen and inorganic carbon during the fermentation of phenylalanine; Figure S13: Concentration profile of the species of inorganic nitrogen and inorganic carbon during the fermentation of tyrosine; Figure S14: Concentration profile of the species of inorganic nitrogen and inorganic carbon during the fermentation of cysteine; Figure S15: Concentration profile of the species of inorganic nitrogen and inorganic carbon during the fermentation of alanine; Figure S16: Concentration profile of the species of inorganic nitrogen and inorganic carbon during the fermentation of histidine; Figure S17: Concentration profile of the species of inorganic nitrogen and inorganic carbon during the fermentation of tryptophan; Figure S18: Concentration profile of the species of inorganic nitrogen and inorganic carbon during the fermentation of glutamine. The models of titration, anaerobic digestion, and subsequent flash distillation process for amino acids can be found at: https://zenodo.org/record/7738947 (accessed on 29 July 2023).

Author Contributions: Conceptualization, A.M.A.; methodology, A.M.A.; software, A.M.A. and F.A.; validation, A.M.A.; formal analysis, A.M.A.; investigation, A.M.A.; resources, A.M.A. and F.A.; data curation, A.M.A. and F.A.; writing—original draft preparation, A.M.A.; writing—review and editing, A.M.A. and F.A.; visualization, A.M.A.; supervision, G.A. and F.A.; project administration, A.M.A. and F.A.; funding acquisition, A.M.A. and F.A. All authors have read and agreed to the published version of the manuscript.

Funding: This research was funded by Doctoral Prize of the Engineering and Physical Sciences Research Council (EPSRC) of the United Kingdom, award reference number 1945857.

Institutional Review Board Statement: Not applicable.

Informed Consent Statement: Not applicable.

Data Availability Statement: Research data are publicly available as Aspen Plus® models hosted by ZENODO [48].

Conflicts of Interest: The authors declare no conflict of interest. The funders had no role in the design of the study, in the collection, analyses, or interpretation of data, in the writing of the manuscript, or in the decision to publish the results.

Abbreviations

ABC, ammonium bicarbonate process; AD, anaerobic digestion; ELECNRTL, electrolyte non-random two-liquid; NH4+-N, ammoniacal nitrogen; VOCs, volatile organic compounds.

Appendix A

Table A1. List of components [19,22] considered for the simulation of the titration of the digestate in Aspen Pus v12, following the experimental procedure of Moure Abelenda et al. [23].

Component Name	Alias	CAS Number	mol/L
WATER	H2O	7732-18-5	45.69365
CARBON-DIOXIDE	CO2	124-38-9	0.013835
AMMONIA	H3N	7664-41-7	0.100484
HYDROGEN-SULFIDE	H2S	7783-06-4	0.001921
ACETIC-ACID	C2H4O2-1	64-19-7	0.117655
GLYCEROL	C3H8O3	56-81-5	0.000788
OLEIC-ACID	C18H34O2	112-80-1	0.001519
DEXTROSE	C6H12O6	50-99-7	0.119788
PROPIONIC-ACID	C3H6O2-1	79-09-4	0.000969
ETHYL-CYANOACETATE	C5H7NO2	105-56-6	4.19E-05
DL-ALANINE	C3H7NO2-N9	302-72-7	0.000452
ARGININE	C6H14N4O2-N2	7004-12-8	0.000433
DL-ASPARTIC-ACID	C4H7NO4-N4	617-45-8	0.000462
L-CYSTEINE	C3H7NO2S-N1	52-90-4	0.000645
DL-GLUTAMIC-ACID	C5H9NO4-N5	617-65-2	0.000712
GLYCINE	C2H5NO2-D1	56-40-6	0.002407
L-ISOLEUCINE	C6H13NO2-N3	73-32-5	0.000448
L-LEUCINE	C6H13NO2-N2	61-90-5	0.000674
L-PHENYLALANINE	C9H11NO2	63-91-2	0.000347
DL-PROLINE	C5H9NO2-N9	609-36-9	0.001069
DL-SERINE	C3H7NO3-N5	302-84-1	0.001656
C4H9NO3-N5	C4H9NO3-N5	72-19-5	0.000452
C5H11NO2-N17	C5H11NO2-N17	516-06-3	0.000712
GLUTARIC-ACID	C5H8O4	110-94-1	0.037695
HYDROGEN	H2	1333-74-0	1.55E-06
METHANE	CH4	74-82-8	0.000238
MALTOSE	C12H22O11-N2	69-79-4	0.050679
TRIOLEIN	C57H104O6	122-32-7	5.62E-05

Table A1. *Cont.*

Component Name	Alias	CAS Number	mol/L
TRIPALMITIN	C51H98O6	555-44-2	6.18E-05
1-HEXADECANOL	C16H34O	36653-82-4	6.86E-08
2-OLEODIPALMITIN	C53H100O6-N1	2190-25-2	6.18E-05
TRILINOLENIN	C57H92O6	14465-68-0	8.4E-05
BETA-D-XYLOPYRANOSE	C5H10O5-D3	2460-44-8	0.009424
LINOLEIC-ACID	C18H32O2	60-33-3	0.000756
ETHANOL	C2H6O-2	64-17-5	0.13388
SODIUM-BICARBONATE	NAHCO3	144-55-8	0.044112
CALCIUM-CHLORIDE	CACL2	10043-52-4	0.002876
SODIUM-CHLORIDE	NACL	7647-14-5	0.003601
POTASSIUM-BICARBONATE	KHCO3	298-14-6	0.007789

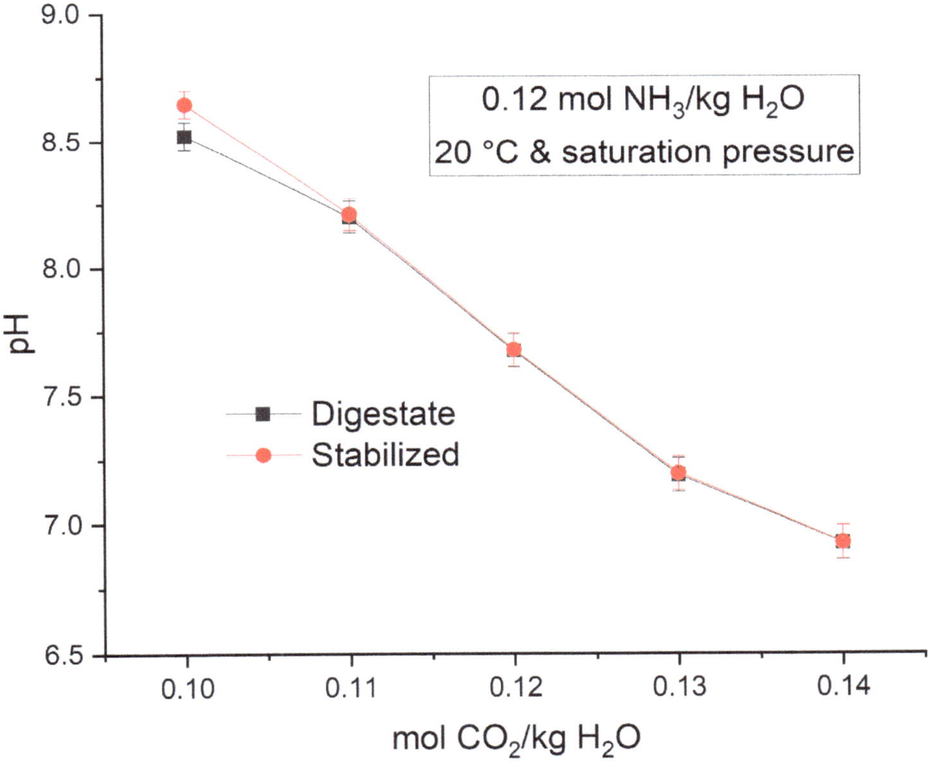

Figure A1. Dependence of the pH on the composition of the anaerobic digestates and the remaining stabilized digestates (Figure 1) obtained after the flash separations shown in Figure 2b.

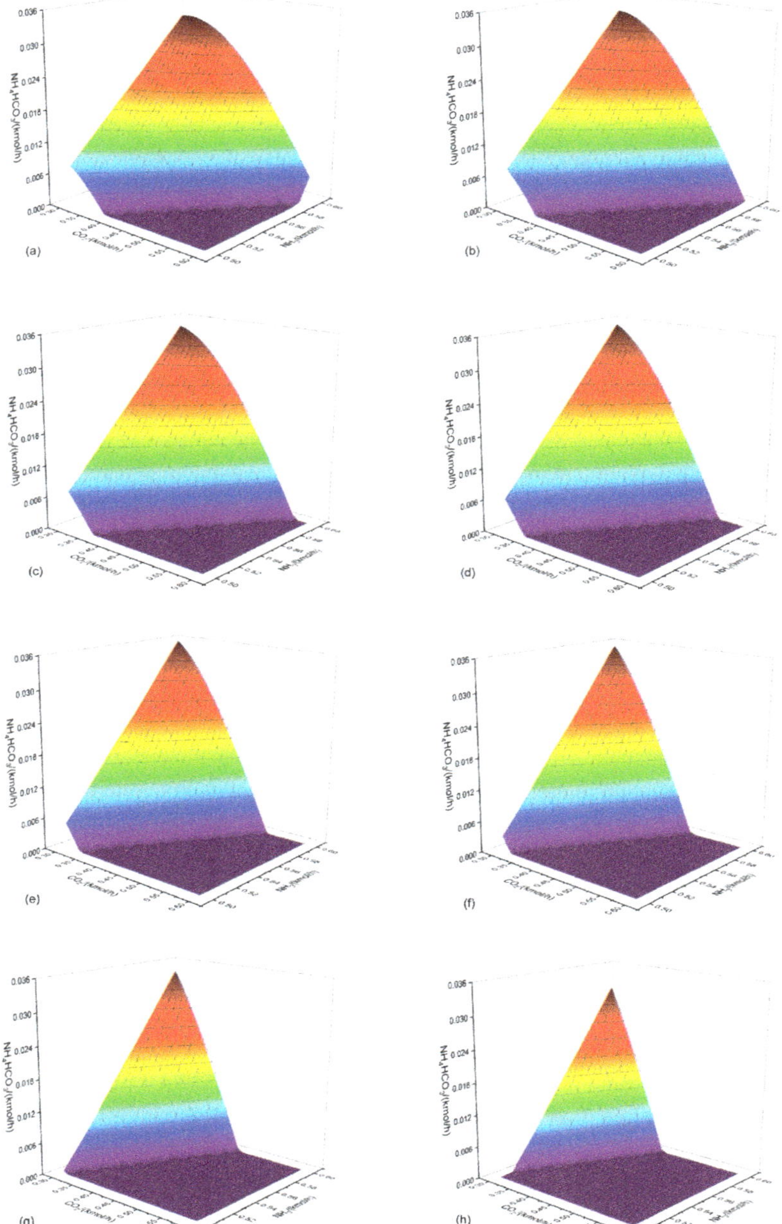

Figure A2. Optimization of the temperature of the flash tank at atmospheric pressure for the production of NH_4HCO_3: (**a**) 94.2 °C, (**b**) 94.4 °C, (**c**) 94.6 °C, (**d**) 94.8 °C, (**e**) 95.0 °C, (**f**) 95.2 °C, (**g**) 95.4 °C, and (**h**) 95.6 °C. The flowrates of NH_3 (0.5 to 0.6 kmol/h or 8 to 10 g/L) and CO_2 (0.3 to 0.6 kmol/h or 13 to 26 g/L) along with the 55 kmol/h of H_2O (or 990 kg/h), which is the main component of anaerobic digestate, were also considered as independent variables to maximize the production of NH_4HCO_3.

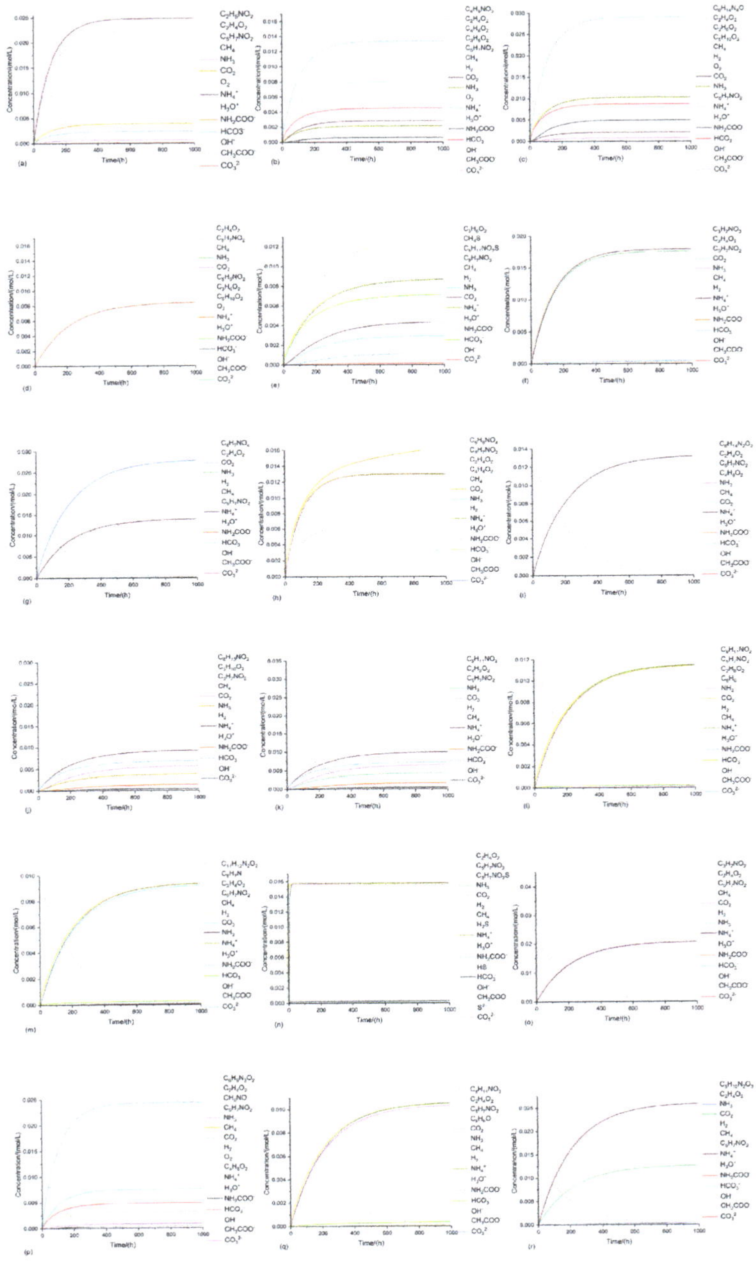

Figure A3. Concentration profile of the species of inorganic nitrogen and inorganic carbon during the fermentation of the amino acids: (**a**) glycine, (**b**) threonine, (**c**) arginine, (**d**) proline, (**e**) methionine, (**f**) serine, (**g**) aspartic acid, (**h**) glutamic acid, (**i**) lysine, (**j**) leucine, (**k**) valine, (**l**) phenylalanine, (**m**) tyrosine, (**n**) cysteine, (**o**) alanine, (**p**) histidine, (**q**) tryptophan, and (**r**) glutamine. The feedstocks that were considered to perform these anaerobic digestions consisted of 90% moisture. The Supplementary Material shows all these Figures separately (Figures S1–S18).

References

1. da Costa Gomez, C. Biogas as an energy option: An overview. In *The Biogas Handbook: Science, Production and Applications*, 1st ed.; Wellinger, A., Murphy, J.D., Baxter, D., Eds.; Woodhead Publishing Ltd.: Cambridge, UK, 2013; pp. 1–16.
2. Patel, S.K.S.; Das, D.; Kim, S.C.; Cho, B.-K.; Kalia, V.C.; Lee, J.-K. Integrating strategies for sustainable conversion of waste biomass into dark-fermentative hydrogen and value-added products. *Renew. Sustain. Energy Rev.* **2021**, *150*, 111491. [CrossRef]
3. Seadi, T.A.; Lukehurst, C. *Quality Management of Digestate from Biogas Plants Used as Fertiliser*; IEA Bioenergy, 2012; p. 40. Available online: http://www.iea-biogas.net/files/daten-redaktion/download/publi-task37/digestate_quality_web_new.pdf (accessed on 29 July 2023).
4. Lukehurst, C.; Frost, P.; Seadi, T.A. *Utilisation of Digestate from Biogas Plants as Biofertiliser*; IEA Bioenergy, 2010; Available online: https://task37.ieabioenergy.com/wp-content/uploads/sites/32/2022/02/Digestate_Brochure_Revised_12-2010.pdf (accessed on 29 July 2023).
5. Astals, S.; Martínez-Martorell, M.; Huete-Hernández, S.; Aguilar-Pozo, V.B.; Dosta, J.; Chimenos, J.M. Nitrogen recovery from pig slurry by struvite precipitation using a low-cost magnesium oxide. *Sci. Total Environ.* **2021**, *768*, 144284. [CrossRef] [PubMed]
6. Drosg, B.; Fuchs, W.; Al Seadi, T.; Madsen, M.; Linke, B. Nutrient Recovery by Biogas Digestate Processing. Available online: https://task37.ieabioenergy.com/wp-content/uploads/sites/32/2022/02/NUTRIENT_RECOVERY_RZ_web2.pdf (accessed on 14 December 2021).
7. Fuchs, W.; Drosg, B. Assessment of the state of the art of technologies for the processing of digestate residue from anaerobic digesters. *Water Sci. Technol.* **2013**, *67*, 1984–1993. [CrossRef]
8. Moller, K.; Muller, T. Effects of anaerobic digestion on digestate nutrient availability and crop growth: A review. *Eng. Life Sci.* **2015**, *12*, 242–257. [CrossRef]
9. Burke, D.A. Removal of Ammonia from Fermentation Effluent and Sequestration as Ammonium Bicarbonate and/or Carbonate. 12. 2010. Available online: https://patents.google.com/patent/US7811455B2/en (accessed on 29 July 2023).
10. Burke, D.A. Reclaiming Ammonia from Anaerobic Digestate As A Profitable Product. *Proc. Water Environ. Fed.* **2015**, *2015*, 1–12. [CrossRef]
11. Bassani, D.; Orentlicher, M.; Simon, M.M.; Pagano, S. Process to Recover Ammonium Bicarbonate from Wastewater. 2019. Available online: https://patents.google.com/patent/US10106447B2/en (accessed on 29 July 2023).
12. Jamaludin, Z.; Rollings-Scattergood, S.; Lutes, K.; Vaneeckhaute, C. Evaluation of sustainable scrubbing agents for ammonia recovery from anaerobic digestate. *Bioresour. Technol.* **2018**, *270*, 596–602. [CrossRef]
13. Drapanauskaite, D.; Handler, R.M.; Fox, N.; Baltrusaitis, J. Transformation of Liquid Digestate from the Solid-Separated Biogas Digestion Reactor Effluent into a Solid NH4HCO3 Fertilizer: Sustainable Process Engineering and Life Cycle Assessment. *ACS Sustain. Chem. Eng.* **2021**, *9*, 580–588. [CrossRef]
14. Budzianowski, W.M. Benefits of biogas upgrading to biomethane by high-pressure reactive solvent scrubbing. *Biofuels Bioprod. Biorefin.* **2012**, *6*, 12–20. [CrossRef]
15. Sander, R. Compilation of Henry's law constants (version 4.0) for water as solvent. *Atmos. Chem. Phys.* **2015**, *15*, 4399–4981. [CrossRef]
16. Walker, M.; Banks, C.; Heaven, S.; Frederickson, J. *Residual Biogas Potential Test for Digestates*; WRAP: Banbury, UK, 2010. Available online: https://www.ktbl.de/fileadmin/user_upload/Allgemeines/Download/Ringversuch-Biogas/Residual-Biogas-Potential.pdf (accessed on 29 July 2023).
17. Saveyn, H.; Eder, P. End-of-Waste Criteria for Biodegradable Waste Subjected to Biological Treatment (Compost & Digestate): Technical Proposals. Available online: https://publications.jrc.ec.europa.eu/repository/handle/JRC87124 (accessed on 14 December 2021).
18. WRAP. BSI PAS 110:2014 Specification for Whole Digestate, Separated Liquor and Separated Fibre Derived from the Anaerobic Digestion of Source-Segregated Biodegradable Materials. Available online: https://wrap.org.uk/sites/default/files/2021-03/PAS110_2014.pdf (accessed on 7 January 2022).
19. Rajendran, K.; Kankanala, H.R.; Lundin, M.; Taherzadeh, M.J. A novel process simulation model (PSM) for anaerobic digestion using Aspen Plus. *Bioresour. Technol.* **2014**, *168*, 7–13. [CrossRef] [PubMed]
20. Centorcelli, J.C.; Drapanauskaite, D.; Handler, R.M.; Baltrusaitis, J. Solar Steam Generation Integration into the Ammonium Bicarbonate Recovery from Liquid Biomass Digestate: Process Modeling and Life Cycle Assessment. *ACS Sustain. Chem. Eng.* **2021**, *9*, 15278–15286. [CrossRef]
21. Centorcelli, J.C.; Luyben, W.L.; Romero, C.E.; Baltrusaitis, J. Dynamic Control of Liquid Biomass Digestate Distillation Combined with an Integrated Solar Concentrator Cycle for Sustainable Nitrogen Fertilizer Production. *ACS Sustain. Chem. Eng.* **2022**, *10*, 7409–7417. [CrossRef]
22. Akhiar, A.; Battimelli, A.; Torrijos, M.; Carrere, H. Comprehensive characterization of the liquid fraction of digestates from full-scale anaerobic co-digestion. *Waste Manag.* **2017**, *59*, 118–128. [CrossRef] [PubMed]
23. Moure Abelenda, A.; Semple, K.T.; Lag-brotons, A.J.; Herbert, B.M.J.; Aggidis, G.; Aiouache, F. Strategies for the production of a stable blended fertilizer of anaerobic digestates and wood ashes. *Nat. Based Solut.* **2022**, *2*, 10014. [CrossRef]
24. Government, U.K. Annex 3: FETF 2023 Productivity and Slurry Eligible Items. 2023. Available online: https://www.gov.uk/government/publications/farming-equipment-and-technology-fund-fetf-2023/annex-3-fetf-2023-productivity-and-slurry-eligible-items (accessed on 29 July 2023).

25. García-Ochoa, F.; Santos, V.E.; Naval, L.; Guardiola, E.; López, B. Kinetic model for anaerobic digestion of livestock manure. *Enzym. Microb. Technol.* **1999**, *25*, 55–60. [CrossRef]
26. Canu, P.; Pagin, M. Biogas upgrading by 2-steps methanation of its CO_2—Thermodynamics analysis. *J. CO2 Util.* **2022**, *63*, 102123. [CrossRef]
27. Darde, V. CO2 Capture Using Aqueous Ammonia. 2011. Available online: https://www.cere.dtu.dk/-/media/centre/cere/publications/phd-thesis/2011/victor_darde_phd.pdf?la=da&hash=8A67D4D04AFF063B772D6A90CC6D88B9BEBAEABD (accessed on 29 July 2023).
28. Darde, V.; van Well, W.J.M.; Stenby, E.H.; Thomsen, K. Modeling of Carbon Dioxide Absorption by Aqueous Ammonia Solutions Using the Extended UNIQUAC Model. *Ind. Eng. Chem. Res.* **2010**, *49*, 12663–12674. [CrossRef]
29. Mullin, J.W. *Crystallization*, 4th ed.; Butterworth-Heinemann: Oxford, UK; Boston, MA, USA, 2001; Table A.4.
30. Haynes, W.M. *CRC Handbook of Chemistry and Physics: A Ready-Reference Book of Chemical and Physical Data*, 91st ed.; CRC Press: Boca Raton, FL, USA, 2010.
31. Brondi, M.; Eisa, M.; Bortoletto-Santos, R.; Drapanauskaite, D.; Reddington, T.; Williams, C.; Ribeiro, C.; Baltrusaitis, J. Recovering, Stabilizing, and Reusing Nitrogen and Carbon from Nutrient-Containing Liquid Waste as Ammonium Carbonate Fertilizer. *Agriculture* **2023**, *13*, 909. [CrossRef]
32. Vandré, R.; Clemens, J. Studies on the relationship between slurry pH, volatilization processes and the influence of acidifying additives. *Nutr. Cycl. Agroecosyst.* **1996**, *47*, 157–165. [CrossRef]
33. Mu, Z.X.; He, C.S.; Jiang, J.K.; Zhang, J.; Yang, H.Y.; Mu, Y. A modified two-point titration method for the determination of volatile fatty acids in anaerobic systems. *Chemosphere* **2018**, *204*, 251–256. [CrossRef]
34. Green, D.W.; Perry, R.H. *Perry's Chemical Engineers' Handbook*, 8th ed.; McGraw-Hill: New York, NY, USA; London, UK, 2008; pp. 140–145.
35. Limoli, A.; Langone, M.; Andreottola, G. Ammonia removal from raw manure digestate by means of a turbulent mixing stripping process. *J. Environ. Manag.* **2016**, *176*, 1–10. [CrossRef]
36. Astals, S.; Nolla-Ardèvol, V.; Mata-Alvarez, J. Anaerobic co-digestion of pig manure and crude glycerol at mesophilic conditions: Biogas and digestate. *Bioresour. Technol.* **2012**, *110*, 63–70. [CrossRef]
37. Astals, S.; Nolla-Ardèvol, V.; Mata-Alvarez, J. Thermophilic co-digestion of pig manure and crude glycerol: Process performance and digestate stability. *J. Biotechnol.* **2013**, *166*, 97–104. [CrossRef]
38. European Sustainable Phosphorus Platform. 1st White Ammonia Research Meeting (WARM)—7th June 2023. 2023. Available online: https://www.phosphorusplatform.eu/activities/conference/n-recovery (accessed on 29 July 2023).
39. Fangueiro, D.; Hjorth, M.; Gioelli, F. Acidification of animal slurry—A review. *J. Environ. Manag.* **2015**, *149*, 46–56. [CrossRef] [PubMed]
40. Ukwuani, A.T.; Tao, W. Developing a vacuum thermal stripping—Acid absorption process for ammonia recovery from anaerobic digester effluent. *Water Res.* **2016**, *106*, 108–115. [CrossRef]
41. Martín-Hernández, E.; Sampat, A.M.; Zavala, V.M.; Martín, M. Optimal integrated facility for waste processing. *Chem. Eng. Res. Des.* **2018**, *131*, 160–182. [CrossRef]
42. Morey, L.; Fernández, B.; Tey, L.; Biel, C.; Robles-Aguilar, A.; Meers, E.; Soler, J.; Porta, R.; Cots, M.; Riau, V. Acidification and solar drying of manure-based digestate to produce improved fertilizing products. *J. Environ. Manag.* **2023**, *336*, 117664. [CrossRef]
43. Moure Abelenda, A.; Semple, K.T.; Aggidis, G.; Aiouache, F. Dataset on the solid-liquid separation of anaerobic digestate by means of wood ash-based treatment. *Data Brief* **2022**, *44*, 108536. [CrossRef]
44. Meixner, K.; Fuchs, W.; Valkova, T.; Svardal, K.; Loderer, C.; Neureiter, M.; Bochmann, G.; Drosg, B. Effect of precipitating agents on centrifugation and ultrafiltration performance of thin stillage digestate. *Sep. Purif. Technol.* **2015**, *145*, 154–160. [CrossRef]
45. Baena-Moreno, F.M.; Saché, E.L.; Hurd Price, C.A.; Reina, T.R.; Navarrete, B. From biogas upgrading to CO_2 utilization and waste recycling: A novel circular economy approach. *J. CO2 Util.* **2021**, *47*, 101496. [CrossRef]
46. Moure Abelenda, A.; Ali, A.; Semple, K.T.; Aiouache, F. Aspen Plus® Process Simulation Model of the Biomass Ash-Based Treatment of Anaerobic Digestate for Production of Fertilizer and Upgradation of Biogas. *Energies* **2023**, *16*, 3039. [CrossRef]
47. Aspen Technology. *Rate-Based Model of the CO_2 Capture Process by NH3 Using Aspen Plus Aspen Plus*; Aspen Technology: Bedford, MA, USA, 2008.
48. Moure Abelenda, A. Aspen Plus Models of the Flash Distillation Process for Stabilization of Anaerobic Digestate and Synthesis of Ammonium Bicarbonate. 2023. Available online: https://www.zenodo.org/record/7738947 (accessed on 29 July 2023).

fermentation

MDPI

Article

Induced Autolysis of Engineered Yeast Residue as a Means to Simplify Downstream Processing for Valorization—A Case Study

Joana F. Fundo *, Teresa Deuchande, Daniela A. Rodrigues, Lígia L. Pimentel, Susana S. M. P. Vidigal, Luís M. Rodríguez-Alcalá, Manuela E. Pintado and Ana L. Amaro

CBQF—Centro de Biotecnologia e Química Fina—Laboratório Associado, Escola Superior de Biotecnologia, Universidade Católica Portuguesa, Rua Diogo Botelho 1327, 4169-005 Porto, Portugal; tdeuchande@ucp.pt (T.D.); darodrigues@ucp.pt (D.A.R.); lpimentel@ucp.pt (L.L.P.); svidigal@ucp.pt (S.S.M.P.V.); lalcala@ucp.pt (L.M.R.-A.); mpintado@ucp.pt (M.E.P.); aamaro@ucp.pt (A.L.A.)

* Correspondence: jfundo@ucp.pt; Tel.: +351-22-558-00-00

Abstract: The objective of this work was to study the efficiency of different autolysis processes, combining different temperatures and pH conditions, when applied to a genetically engineered yeast residue. The determination of the supernatants' dry weight showed that the autolysis time could be reduced to half, from 4 to 2 h, if the residue pH was increased from 5 to 8 at 50 °C (18.20% for 4 h and 18.70% for 2 h with a higher pH). This result allowed us to select a short autolysis time to proceed with the second part of the experiments. The application of this faster induced autolysis process enabled us to obtain supernatants with higher concentrations of relevant compounds, such as some amino acids and minerals. An increase in leucine (of around 7%), aspartic acid, valine, phenylalanine, isoleucine and serine (approximately 2%) was observed in the autolyzed samples, when compared to the untreated ones. Also, regarding minerals, the autolysis process allowed us to obtain significantly higher amounts of potassium in the treated samples' supernatants. This work allowed the selection of a fast and low-cost induced autolysis process for synthetic biotechnology-derived spent yeast residue to attain a product rich in high-value compounds, which can be used in commercial applications, for example, as an animal feed additive.

Keywords: synthetic biotechnology; spent yeast residue; nutrient source; autolysis; physicochemical composition; animal feed

Citation: Fundo, J.F.; Deuchande, T.; Rodrigues, D.A.; Pimentel, L.L.; Vidigal, S.S.M.P.; Rodríguez-Alcalá, L.M.; Pintado, M.E.; Amaro, A.L. Induced Autolysis of Engineered Yeast Residue as a Means to Simplify Downstream Processing for Valorization—A Case Study. *Fermentation* **2023**, *9*, 673. https://doi.org/10.3390/fermentation9070673

Academic Editors: Jose Luis García-Morales, Francisco Jesús Fernández Morales and Mohammad Taherzadeh

Received: 22 May 2023
Revised: 5 July 2023
Accepted: 7 July 2023
Published: 18 July 2023

1. Introduction

Spent yeast residue (SYR) is one of the major by-products generated by fermentation-based industries, such as brewing, baking or wine production. The brewing industry stands out with an annual spent brewer's yeast production of around 400 thousand tons [1]. For fermentation-based industries, the production of huge quantities of this residue represents a management challenge from both ecological and economical points of view. *Saccharomyces cerevisiae* is, by far, the most commonly used yeast in this type of industry, representing a powerful and versatile industrial tool for multiple purposes [2]. Its industrial use is mainly due to its fast growth, good ethanol-producing capacity and great tolerance against environmental stress, including high ethanol concentrations and low oxygen levels [3,4].

Currently, SYR also represents a challenging problem for the synthetic biotechnology industry. This industry conducts precision fermentations to produce complex and valuable biomolecules for applications in the cosmetics, pharmaceutical and agri-food industries. These fermentations are driven by genetically modified yeasts, mostly *S. cerevisiae* strains, which are used as cell factories to produce valuable molecules. This industry is under expansion, and an increasing number of synthetic biotechnology plants making use of innovative and sustainable processes to promote the conversion of plant sugars into stable, alternative and high-commercial-interest biomolecules is foreseen [5].

Spent yeast residue represents a cheap, attractive and easily available source of valuable molecules and bioactive compounds [6], which is suitable for different valorization strategies. The valorization of SYR can be achieved through the extraction of high-value components, such as proteins, polysaccharides, fibers, flavor compounds and phytochemicals, which can be reused as nutritionally, pharmacologically and cosmetically functional ingredients [7].

Regarding the applications of this waste by-product, the literature reports many possibilities of application in very distinct areas, unequivocally showing its versatility and valorization potential. Due to its high nutritional value and low cost, this residue can be used as a fermentation subtract and/or additive [7], as a biosorption element [8], and as a food ingredient or nutraceutical due to the presence of high levels of polyphenolic compounds [7]. Spent yeast extracts are also known to be rich sources of proteins, which can be incorporated as supplements in animal diets, namely for fish and ruminants [9–11]. Its incorporation in animal feed is currently one of the main strategies for reusing this by-product, and spent brewer' yeast, in particular, has long been incorporated in ruminant diets as a protein additive. It has been demonstrated that, under in vitro conditions, spent brewer's yeast from craft beer, which contains antimicrobial α- and β-acids, can prevent excessive rumen protein degradation by rumen hyper-ammonia-producing bacteria when used as a protein additive [7,11].

However, the use of the majority of spent yeast components requires yeast cell lysis, commonly known as autolysis. For valorization strategies, the development of an appropriate spent yeast residue cell disruption process is of the utmost importance for an efficient and cost-effective recovery of target compounds [6]. Although a naturally occurring event, autolysis can be induced and enhanced by exposing spent yeast residue to elevated temperatures, organic solvents or physical processes, such as ultrasounds or high pressure [12]. Depending on the final valorization objective, it is possible to choose the most appropriate autolysis process or even a combination of processes. The costs associated with this procedure are an important issue that needs to be considered; furthermore, due to the focus on waste recovery strategies, the target cost has to be as low as possible so that its application is worthwhile.

Several research studies have focused on induced autolysis to obtain additives/products with a relevant nutritional composition from SYR derived from beer fermentation. However, to the best of our knowledge, there have been no studies on the optimization of an efficient autolysis process for SYR derived from these platforms in which genetically modified yeasts strains are used. Based on previous studies on the characterization of such residues, it is possible to observe that SYRs derived from the production of different types of beer have significant compositional differences. Thus, it is important to assess if the autolysis methods currently applied to spent brewer's yeast are effective for the extraction of valuable compounds from these new waste streams, or if other methods should be developed for this purpose.

Therefore, as this study aimed to facilitate downstream valorization strategies for a synthetic-biology-derived SYR, a brief review of the methods used to autolyze yeast cells from spent brewer's yeast is given, emphasizing the efficiency, the cost and the process rate of these methods. An overview of the suitability of processed residue for application as a supplement for animal feeding is also discussed, considering the characteristics of existing animal feed supplements that are commercially available.

2. Yeast Cell Disruption Strategies—An Overview

The main objective of induced autolysis is to rupture yeast cell wall and release cell-soluble constituents, which are predominantly compounds of potential biological interest [13]. As mentioned above, this process is of the utmost importance to the success of a defined SYR valorization strategy and should be designed accordingly.

If the economic issues related to the induced autolysis process are not to be considered, the major difference between several cell disruption procedures is the size of the cell fragments generated and the method by which cells are sheared. Even though several mechanical methods (bead mills, high-pressure homogenization and ultrasonication), which are superior in terms of product recovery, are industrially preferred, their poor selectivity and complicated downstream processing are the major drawbacks. Non-mechanical methods are mild and result in large cell fragments, thus promoting downstream process operations. However, their limited recovery efficiency and expensiveness restrict their general applicability [6].

Nevertheless, if induced autolysis with temperature is combined with, for example, the addition of chemical compounds (such as sodium chloride and ethyl acetate), the disruption effectiveness is higher (around 98%) than some mechanical methods, namely bead mill and ultrasonication (80%), and leads to a higher cleavage of amino acids from cell protein (307, 155 and 115 mg g^{-1} of yeast when using autolysis with temperature/chemical agents, ultrasonication and bead mill, respectively) [7].

Particularly, if spent yeast is to be used as an additive for animal feed, the autolysis process plays an important role. In this case, disrupting yeast cell walls before feeding improves the availability of intracellular nutrients and facilitates their digestion and absorption [7,14]. In order to turn the incorporation of this additive in animal feeding profitable, the transformation process has to be as simple, fast, efficient and economical as possible.

Table 1 shows a brief summary of some of the published works that are relevant for an induced autolysis design. The different autolysis conditions and different methodologies used are described, and the main takeways are highlighted.

Table 1. Description of the different conditions and methodologies used in previous research, and the main takeaways for induced autolysis of wild-type yeasts.

Strain	Autolysis Conditions/Method	Main Takeways	Ref.
S. cerevisiae	-Pulse electric field (PEF): 5–20 kV/cm, 1–2000 pulses, and 15 µs pulse width. -Autolysis: 52 °C/72 h/ pH 5.5.	-PEF increased the final amino acid and total solid contents. -PEF was found to accelerate the progress of autolysis (up to 78%).	[15]
	-Autolysis: various pH values (4.0, 5.5, 7.0 and 8.5) and use of chemical autolysis promoters (ethyl acetate and chitosan).	-Best combination—pH 5.5 and ethyl acetate. -Good peptidase activities at these pH values. -Yeast extract with higher turbidity when produced at pH 7.0 and 8.5.	[16]
	-Influence of temperature, pH and ethanol concentration on PEF-induced autolysis.	-At the same incubation time, the amount of mannose released from PEF-treated cells ranged from 80 mg L^{-1}, when incubated with 25% ethanol, to 190 mg L^{-1}, when incubated at 43 °C.	[17]
	-Scale-up ultrasonic disruption of yeast (Barbell Horn Ultrasonic Technology—BHUT), a usual method for lab scale.	-BHUT can be successfully used on a large scale. -BHUT-based equipment allows efficient extraction of total protein and alkaline phosphatase from yeast cells.	[18]
	-Influence of ultrasound intensity, sonication time, temperature and yeast concentrations. -2 probe depths; ionic strengths at 0.05, 0.55 and 1.05 M; and levels of ethanol addition at 10, 50 and 100 mM.	-Release of polysaccharides and proteins was affected by most of the processing parameters. -The parameter, temperature, had the greatest influence on selectivity of released product.	[19]
	-Effect of temperature (45, 50, 55 and 60 °C) and reaction time (ranging from 8 to 72 h).	-Optimum temperature/time combination: 50 °C for 24 h, on the basis of α-amino nitrogen and the protein contents. - Also favorable for sensory analysis.	[20]
	-Effect of high pressure (HPH) (200 to 600 MPa) for 0 to 120 min. -Activity of the vacuolar proteases was monitored during the autolysis. The autolytic capacity of yeast was determined based on the physicochemical characteristics of the yeast extract.	-At 200 and 400 MPa, the proteolytic activity was enhanced up to 160% after 40 and 10 min, respectively. -Autolysis was significantly accelerated, in combination with cellular permeabilization, when achieved with HP treatment. -At 600 MPa, proteolytic enzymes were gradually inactivated, leading to the inhibition of autolysis.	[21]
	-Comparing conventional methods (autolysis and mechanical rupture) with enzymatic hydrolysis using proteolytic enzymes.	-The hydrolysate produced at pH of 5.5, 100% substrate, 10% enzyme/substrate ratio and 60 °C resulted in a maximized yield with enhanced antioxidant properties. -Enzymatic hydrolysis promoted more efficient release of solids, proteins and cell walls.	[22]

Table 1. *Cont.*

Strain	Autolysis Conditions/Method	Main Takeways	Ref.
S. pastorianus	-Mechanical disruption. -Separation of the β-glucan-rich fraction. -Extract rich in native proteins and enzymes.	-The best autolysis conditions were 36 °C/6 h	[23]
S. bayanus	-High-pressure homogenization (HPH) at 5, 100 and 150 MPa and comparison with thermolysis (121 °C/2 h).	-HPH seemed to be a promising technique (150 MPa was the best operation condition). -Thermolysis was more efficient.	[24]
S. cerevisiae/ *Sacch. uvarum*	Study on autolytic release of polysaccharides from cell walls, in a model medium, over a nine-month period of ageing over lees, and the effect of adding β-glucanase.	-The addition of enzyme promoted complete autolysis in less time (2–3 weeks instead 5 months) -Enzyme-assisted autolysis promoted the production of smaller-molecular-weight fragments. -The extension of autolysis was different for different strains.	[25]
Kluyveromyces fragilis	-NaCl-induced autolysis studied as a function of time (*t*), at different initial yeast concentrations (X_0) and reaction temperatures (*T*)	-Protein solubilization was temperature dependent. -Hydrolysis of total carbohydrates was found to be controlled firstly by yeast concentration and secondly by temperature.	[26]

This targeted literature review allows us to verify the use of *S. cerevisiae* as a key model strain for autolysis studies. As already stated, *S. cerevisiae* is undoubtedly the most known, studied and used yeast species. *S. cerevisiae*'s great adaptability and growth capacity makes it a model organism in eukaryotic biology and the first eukaryotic organism to have its genome sequenced [2].

As stated before, it is important to verify the efficacy of the autolysis process since synthetic biotechnology-derived SYRs may present a different chemical composition when compared to SYRs from native *S. cerevisiae.* Thus, a case study is herein presented to confirm if compositional differences in SYRs imply changes in the efficiency of the autolysis process and differences in the compositional profile of the generated products.

3. Case Study—Research Methodology and Main Results and Conclusions

Based on the brief literature review presented in the previous section, and considering the need to identify an effective, economical, fast and easy-to-carry-out autolysis process to be applied to a synthetic biotechnology-derived SYR, an induced autolysis process using different combinations of temperature, pH and time was studied. The use of different enzymes was also considered, as well as the use of some mechanical methods, namely ultrasounds and high-pressure homogenization. Even if these mechanical methods are, as reported in the literature, more efficient than autolysis based on temperature and pH, their incorporation into the process turns the industrial costs restrictive.

3.1. Samples

A SYR derived from a fermentation process driven by a *S. cerevisiae* strain and modified to produce β-farnesene, which was provided by Amyris, Inc. (Emeryville, CA, USA), was autolyzed under different conditions of pH, temperature and time.

Two different pH levels, 5.5 (the residue's pH at arrival) and 8, were studied. The higher pH value was attained using NaOH (Sigma, Aldrich, St. Louis, MO, USA) at 3 M. The autolysis process was implemented at two different controlled temperatures, 50 and 70 °C, using a water bath. The samples were collected for analysis at different treatment times (2, 4, 6, 8 and 24 h). The induced autolysis process was performed in triplicate in sterilized 500 mL glass jars using batches of 250 mL of whole spent yeast.

In the initial stage of the experiment, some tests were also carried out after the addition of different enzymes, namely alcalase (Sigma, Aldrich) at two different concentrations (1 and 3%) and a mixture of alcalase and celluclast (Sigma, Aldrich; both at 3%).

3.2. Methodology

3.2.1. Determination of the Best Combination of pH, Time and Temperature for Induced Autolysis

The first objective of this study was to select the most promising combination of temperature, pH and autolysis time, in order to achieve effective cellular breakdown while maintaining the process costs low. For this, the SYR supernatants' dry weight and protein content were selected as indicators of autolysis efficiency. The protein content is one of the main indicators of autolyzed spent yeast's suitability for animal feed supplementation [27]. To obtain the supernatants, the non-autolyzed and autolyzed spent yeast suspensions were centrifuged at 5000 rpm for 10 min, and the supernatants were collected for the dry weigh and protein content analysis. These determinations allow a prompt, easy and effective evaluation of the autolysis conditions. A brief comparison with the samples autolyzed with the addition of enzymes was also performed.

3.2.2. Comparison of Nutritional Composition of Non-Autolyzed and Autolyzed SYR

In the second stage of the experimental work, the SYR samples were autolyzed according to the conditions previously selected (pH and temperature), and the resulting product was compared with the non-autolyzed samples. In addition, the supernatants of the autolyzed and non-autolyzed SYR samples were also evaluated in terms of their nutritional characteristics [13]. These samples were subjected to different compositional analysis, namely dry weight, mineral content, total protein content, amino acid profile, and total sugar and total lipid contents and their respective profiles.

The samples' dry weight was determined using an oven at 105 °C for 24 h, according to the Association of Official Analytical Chemists (AOAC) [28]. For this analysis, 1 mL of the autolyzed and non-autolyzed supernatants and the whole spent yeast samples was used, and the analysis was carried out in triplicate.

The total protein content of SYR (whole sample and supernatants) was determined using the Dumas method. This method is based on the combustion of the whole sample in an oxygen-enriched atmosphere at a high temperature to ensure complete combustion [29]. A Dumatec^{TM} 8000 (Foss, Hilleroed, Denmark) was used.

The amino acid profile for the whole SYR samples (total amino acids) was determined through an acid hydrolysis of the samples according to [30]. Briefly, 10 mg of each sample was used and 3 mL of HCL (6 M) was added; then, the mixture was vortexed and flushed with nitrogen. The samples were then incubated for 20 h at 115 °C and diluted with 4 mL of water. The pH value was adjusted to 3.5 and deionized water was added until a final volume of 10 mL. The samples' supernatants were used directly, only diluted in 0.1 M of HCl. Both samples (whole and supernatants) were filtered using 0.45 µm filters to be analysed using high-performance liquid chromatography (HPLC). The amino acid profile was achieved using an HPLC equipped with a Chromolith® Performance RP18 (4.6 mm × 100 mm) column from Merck, a high-resolution fluorescence detector, and an "autosampler".

The quantification of total sugars and the determination of sugar profiles were performed through polysaccharide reduction and acetylation in alditol acetates. Before the derivatization reaction, polysaccharides were hydrolyzed and then separated and detected via GC-FID and quantified using 2-deoxyglucose as the internal standard [31,32]. Briefly, 2 mg of dried whole and supernatant samples was hydrolyzed with 200 µL of H_2SO_4 (58% *v*/*v*) for 3 h at room temperature and then with 1 M H_2SO_4 at 100 °C for 2.5 h. After this, reduction and acetylation of the hydrolyzed samples was carried out with the addition of the mixture of the internal standard at 2 mg mL^{-1} with 25% NH_3 and 15% $NaBH_4$ prepared in 3 M NH_3. The mixture was incubated at 30 °C for 1 h. Then, two washes with glacial acetic acid were performed, and 1-methylimidazole and acetic anhydride were added. After this, the tubes were kept at 30 °C for 30 min, and the organic phase was washed twice with 3.0 mL of distilled water, 2.5 mL of dichloromethane, and, then again, twice with 3 mL of distilled water. The organic phase was evaporated, and then anhydrous acetone was added and evaporated twice. Immediately before the GC-FID analysis, the dried sugars were suspended in 100 µL of anhydrous acetone.

For the extraction and quantification of total lipids, a mixture of solvents with different polarities was used since yeasts contain lipids with both polar and apolar properties. The extracted lipids were then purified via phase separation and gravimetrically quantified according to [33]. The lipid profiling was carried out using HPLC-ELSD as described by [34]. Briefly, the samples were weighted and dissolved in dichloromethane to a concentration of 3 mg mL^{-1}. Afterward, the samples were analyzed using an HPLC (model 1260 Infinity II; Agilent Technologies, Santa Clara, CA, USA) attached to an evaporative light scattering Detector (ELSD; 1290 Infinity II, Agilent Technologies, Santa Clara, CA, USA), using nitrogen as the nebulizing gas, coupled to a Zorbax RX-SIL column (2.1 mm × 150 mm, 5 µm; Agilent Technologies, Santa Clara, CA, USA). The analysis conditions were assayed as described by [35] with slight changes. The composition of the mobile phases was as follows: A, isooctane/ethyl acetate (99.8:0.2, *v*/*v*); B, acetone/ethyl acetate (2:1, *v*/*v*) containing 0.1% acetic acid (*v*/*v*); C, 2-propanol/water (85:15, *v*/*v*) containing 0.013% acetic acid (*v*/*v*) and 0.031% of TEA *v*/*v*; and D, EtAc. The flow rate was set at 0.275 mL min^{-1} with an injection volume of 20 µL. The detector was set as follows: evaporator and nebulizer temperature at 60 °C with nitrogen as the nebulizing gas at 1.20 SLM flow rate. For the determination of the elution order, the pure standards were injected, as well as available bibliography was used [36]. All samples were injected at least in triplicate.

The concentrations of minerals were determined using an optical emission spectrometer, Model Optima 7000 DV™ ICP-OES (Dual View, PerkinElmer Life and Analytical Sciences, Shelton, CT, USA), with a radial configuration, as described by [37]. Prior to the analysis, the autolyzed and non-autolyzed supernatants (2 mL) were mixed with 5 mL of HNO_3 (65%) and 1 mL of H_2O_2 (30%) in an appropriate vessel and digested in a microwave system (Speedwave MWS-3+, Berghof, Eningen, Germany), following an established program of times/temperatures. This analysis was carried out in triplicate and the result was expressed in mg L^{-1}.

Statistically significant differences between all samples, in relation to dry weigh, total protein, and total sugar and total lipid contents, were determined using an unpaired *t*-test for each parameter individually and assuming equal variances. In relation to amino acids, as well as sugar and lipid profiles, multiple *t*-tests corrected for multiple comparisons using the Holm–Sidak method for $p = 0.05$ were conducted, assuming consistent variances. These statistical analyses were performed using GraphPad Prism version 8.0.2 (San Diego, CA, USA). Hierarchical clustering analyses (HCA) were carried out using MetaboAnalyst 5.0 (McGill University, Canada). For the HCA of the whole samples before and after induced autolysis, a matrix including 38 variables (compositional data) and 6 observations (3 replicates of each condition) was constructed, whereas for the analysis of the supernatants of whole and autolyzed spent yeast, a data matrix consisting of 32 variables (compositional data) and 6 observations (3 replicates of each condition) was constructed. In both cases, the data were auto scaled before analysis, and the HCA was performed using the Euclidean distance measure and the ward's algorithm. The results were presented in the form of a heatmap.

4. Results and Discussion

4.1. Determination of the Best Combination of pH, Time and Temperature for Induced Autolysis

In Table 2, it is possible to compare the dry weight and protein yields (calculated using the day 0 values as a reference) in the supernatants of the SYR samples after being treated with different combinations of pH, temperature, enzyme addition and time.

Autolysis induced for 2 h at a lower temperature (50 °C) resulted in the highest dry weight yield (18.70%), even when compared with autolysis with the addition of enzymes (around 10%). According to the results reported by Tanguler et al. for spent brewer´s yeast [20], yields obtained at lower temperatures were higher than yields achieved when higher temperatures were used. In addition, Suphanthrika et al. [38] stated that autolysis between 45 and 50 °C led to a maximum yield with respect to the amount of solids released into the liquid yeast extract from baker's yeast.

Table 2. SYR supernatants' dry weight and protein contents after different induced autolysis conditions. The results are presented as the normalized values related to day 0 (D_0) of storage.

Treatment	Induced Autolysis Time (H)			
	Dry Weight (% D_0 Value)		Protein (% D_0 Value)	
	2	4	2	4
pH 5.5; T 70 °C	8.20	18.20	4.4	13.2
pH 8.0; T 50 °C	18.70	19.30	4.4	15.4
pH 8.0; T 70 °C	*	*	3.7	5.0
Alcalase 1%	10.20	34.35	2.2	*
Alcalase 3%	10.60	38.20	2.7	*
Enzyme Mix	6.90	15.00	3.8	13.1

* Value lower than the one obtained on day 0.

The increase in the samples' pH with the addition of NaOH also promoted this increase in autolysis efficacy when applied to spent yeast from beer manufacture [39].

When the autolysis time was extended to 4 h, the addition of alcalase was, without any doubt, the most efficient process with respect to dry weight yields (more than 30%). However, the addition of a mixture of enzymes (alcalase and celluclast) did not prove to be effective in increasing the efficiency of autolysis, since the values for dry weight, both at time 2 and 4, were lower when compared with the other treatments.

Protein yield varied with autolysis time. Although an increase of 2 h in the treatment resulted in an increase in protein yield by almost 10%, the 4 h treatment carried out at a lower temperature showed the best yield. These results are in accordance with Suphanthrika et al. [38] who reported the lowest protein yields at higher temperatures (55 and 60 °C) for the production of a baker's yeast extract.

Based on these results and with the objective of applying an autolysis process that was simultaneously fast, effective and economically viable, we decided to combine a time of 2 h and the addition of NaOH to increase the pH to 8 with a temperature of 50 °C.

4.2. *Compositional Comparison of Non-Autolyzed and Autolyzed Samples*

4.2.1. Dry Weight

The dry weight of the SYR samples was determined and compared, as shown in Table 3. As expected, no differences were attained for this parameter when the whole spent yeast samples were evaluated. However, with respect to the samples' supernatants, significant differences were observed between the untreated and the autolyzed samples.

Table 3. Dry weight of spent yeast residue samples. Results are expressed as mean ± standard deviation. Significant differences are represented by different letters in the same rows for $p < 0.05$. Lowercase letters refer to the whole spent yeast samples and uppercase letters refer to the supernatant samples.

Sample	Dry Weight (%)
Untreated whole spent yeast	17.48 ± 0.21^{a}
Autolyzed whole spent yeast	17.14 ± 0.03^{a}
Untreated supernatant	7.27 ± 0.28^{A}
Autolyzed supernatant	9.37 ± 0.00^{B}

The supernatants of the autolyzed samples showed higher values of dry weight, which suggests higher amounts of solid components [40] resulting from the release of intracellular constituents. This could be due to the amounts of free amino acids, peptides, sugars and nucleotides that became available through induced autolysis [41,42].

4.2.2. General Nutritional Composition: Proteins, Sugars and Lipid Contents

The nutritional composition of the SYR samples, regarding proteins, sugars, and lipid contents, was determined, and the results are presented in Table 4. It was possible to verify some differences when comparing the general nutritional composition of the SYR used in this study with the native spent brewer's yeast. Generally, and using values reported in the literature, it is possible to conclude that the residue used in this study had slightly lower levels of proteins (in some cases, less than 23.5%) but higher percentages of total sugars and lipids (more than 1.62% and 3.30%, respectively) [7]. These results allowed us to evaluate the overall efficacy of the autolysis process.

Table 4. Total proteins, sugars, and lipid content of SYR. Results are expressed as mean ± standard deviation. Significant differences are represented by different letters in the same rows for $p < 0.05$. Lowercase letters refer to the whole spent yeast samples and uppercase letters refer to supernatant samples.

Sample	Protein (% $_{DW}$)	Total Sugars (% $_{DW}$)	Total Lipids (% $_{DW}$)
Untreated whole spent yeast	40.6 ± 0.01 ^{a}	14.52 ± 0.36 ^{a}	4.60 ± 0.22 ^{a}
Autolyzed whole spent yeast	37.9 ± 0.04 ^{b}	16.27 ± 1.01 ^{a}	4.68 ± 0.18 ^{a}
Untreated supernatant	41.9 ± 0.08 ^{A}	5.92 ± 0.68 ^{A}	ND
Autolyzed supernatant	45.2 ± 0.23 ^{B}	6.30 ± 0.95 ^{A}	ND

ND—not determined; DW—dry weight.

In the case of whole SYR, no significant differences were expected between the non-autolyzed and autolyzed samples. However, this was not verified for the protein content parameter, where the amount of protein in the untreated whole yeast sample was found to be higher than that in the autolyzed yeast sample. This difference, despite being lower than 2.7% DW, might be related to the heterogeneity of the yeast samples.

Protein is, without any doubt, the main component in yeast extracts, constituting approximately half of their composition (more than 40% on a dry weight basis for most of the cases [20,43]). The highest protein content was found in the autolyzed samples' supernatants (more than 45% DW), when compared with the non-autolyzed ones (around 42% DW), revealing the release of these components from the intracellular region and indicating an effective cell wall breakdown [41,42], thus confirming the success of the smooth autolysis process. The same results were reported in the literature for extracts from the native spent brewer's yeast [20].

No significant differences were observed with respect to total sugars for both the non-autolyzed and autolyzed samples' supernatants. The total lipid content was only determined in the whole yeast samples since lipids are found on spent yeasts' cellular walls and not in supernatants. The total lipid content of the autolyzed and non-autolyzed SYR samples was around 4.60% DW for both. Regarding the spent yeast supernatant samples and, similarly, the whole yeast samples, autolysis had no significant effect on total sugar content.

4.2.3. Amino Acid Profile

The SYR samples, in general, have a high nutritional value mainly due to their high contents in some essential amino acids. Table 5 shows the amino acid profiles obtained for both the whole yeast and supernatant samples, whether untreated and autolyzed.

The high contents of essential amino acids and the abundance of some of them stand out, thus characterizing this SYR as an excellent material to complement animal diets [27,44]. For instance, lysine and threonine are excellent amino acids to complement an animal cereal diet. Animal diets based on cereals are composed of proteins and are typically deficient in these amino acids [27,44].

Table 5. Amino acid profiles of SYR samples. Results are expressed as mean ± standard deviation. Significant differences are represented by different letters in the same rows for $p < 0.05$. Lowercase letters refer to the whole spent yeast samples and uppercase letters refer to the supernatant samples.

Amino Acid Profile	Samples			
	Untreated Whole (% $_{DW}$)	Autolyzed Whole (% $_{DW}$)	Untreated Supernatant (% $_{FW}$)	Autolyzed Supernatant (% $_{FW}$)
Aspartic acid	10.4 ± 0.3 ^{a}	10.7 ± 0.4 ^{a}	5.1 ± 0.3 ^{A}	8.5 ± 0.8 ^{B}
Glutamic acid	17.8 ± 0.2 ^{a}	18.4 ± 0.4 ^{a}	30.5 ± 1.4 ^{A}	6.4 ± 0.7 ^{B}
Cysteine	1.8 ± 0.1 ^{a}	1.8 ± 0.2 ^{a}	1.0 ± 0.1 ^{A}	0.8 ± 0.1 ^{A}
Asparagine	ND	ND	2.8 ± 0.1 ^{A}	4.2 ± 0.1 ^{A}
Serine	5.9 ± 0.0 ^{a}	5.5 ± 0.1 ^{a}	1.9 ± 0.1 ^{A}	4.0 ± 0.1 ^{B}
Histidine	ND	ND	2.8 ± 0.0 ^{A}	2.5 ± 1.0 ^{A}
Glutamine	ND	ND	12.3 ± 0.7 ^{A}	8.6 ± 0.8 ^{B}
Glycine	4.8 ± 0.1 ^{a}	5.0 ± 0.2 ^{a}	1.8 ± 0.1 ^{A}	2.8 ± 0.7 ^{A}
Threonine	6.4 ± 0.0 ^{a}	6.9 ± 0.1 ^{a}	5.7 ± 0.3 ^{A}	3.6 ± 0.6 ^{B}
Arginine	6.8 ± 0.2 ^{a}	3.7 ± 1.0 ^{b}	5.1 ± 0.3 ^{A}	2.8 ± 0.2 ^{B}
Alanine	10.7 ± 0.1 ^{a}	9.3 ± 1.1 ^{b}	14.9 ± 0.7 ^{A}	12.9 ±0.6 ^{B}
Tyrosine	3.9 ± 0.0 ^{a}	3.6 ± 0.2 ^{a}	1.8 ± 0.0 ^{A}	3.4 ± 0.2 ^{B}
Valine	5.5 ± 0.1 ^{a}	6.8 ± 0.1 ^{b}	5.4 ± 0.2 ^{A}	7.7 ± 1.0 ^{B}
Methionine	2.0 ± 0.0 ^{a}	2.1 ± 0.0 ^{a}	0.8 ± 0.1 ^{A}	1.4 ± 0.1 ^{A}
Tryptophan	ND	ND	0.3 ± 0.1 ^{A}	1.3 ± 0.0 ^{A}
Phenylalanine	4.0 ± 0.1 ^{a}	4.3 ± 0.1 ^{a}	1.9 ± 0.0 ^{A}	4.3 ± 0.1 ^{B}
Isoleucine	3.9 ± 0.2 ^{a}	4.9 ± 0.0 ^{a}	2.6 ± 0.0 ^{A}	5.3 ± 0.2 ^{B}
Leucine	6.8 ± 0.1 ^{a}	7.2 ± 0.0 ^{a}	2.6 ± 0.2 ^{A}	9.4 ± 0.6 ^{B}
Lysine	9.3 ± 0.4 ^{a}	9.7 ± 0.2 ^{a}	9.5 ± 0.1 ^{A}	10.2 ± 1.0 ^{A}

DW—dry weight; FW—fresh weight; ND—not detectable.

As expected, the differences in the amino acid profiles of the whole yeast samples for both untreated and autolyzed ones were not as evident as the ones obtained for the supernatants (free amino acids). Regarding free amino acids, further significant differences could be observed. The glutamic acid and glutamine contents significantly decreased after the induced autolysis process. These are not essential amino acids, and they are closely related in a chemical sense [45].

In contrast, regarding the spent yeast supernatants, a general and significant increase in amino acid concentrations was observed after induced autolysis. These results are in accordance with the ones reported by Podpora et al. [46] in relation to spent brewer's yeast, which stated that along with autolysis time, an increase in free amino acid content occurred. Also, the obtained result is in agreement with the ones reported by [20,38]. Tanguler et al. (2008) [20] and Suphantharika et al. (1997) [38] studied native spent brewer's yeast. These authors asserted that there was a higher breakdown of proteins and peptides at 50 °C when compared to other temperatures. Yeast proteases were inactive at 55 and 60 °C but active at 45 and 50 °C.

Leucine, for example, increased around 7%, while aspartic acid, valine, phenylalanine, isoleucine and serine increased more than 2% after the induced autolysis process. Some animal studies showed that a supplementation with leucine and/or phenylalanine could effectively improve the intestinal starch digestion of ruminants [47]. Another study stated that serine supplementation provided to laying hens fed on low-crude protein diets enhanced humoral and ileal mucosal immunity and attenuated the ileal inflammation of layers [48].

4.2.4. Neutral Sugar and Lipid Profiles

The neutral sugar profile (Table 6) was also determined and analyzed for both whole yeast and supernatant samples. The induced autolysis process carried out using selected combinations of temperature, pH and time conditions promoted a slight but non-significant increase in glucose levels in the autolyzed samples of spent yeast residue.

Table 6. Neutral sugar content in the whole spent yeast samples and corresponding supernatants. Results are expressed as mean ± standard deviation. Significant differences are represented by different letters in the same rows for $p < 0.05$. Lowercase letters refer to the whole spent yeast samples.

Sample	Mannose (% $_{DW}$)	Glucose (% $_{DW}$)
Untreated whole spent yeast	6.10 ± 0.24 ^{a}	8.42 ± 0.12 ^{a}
Autolyzed whole spent yeast	6.67 ± 0.50 ^{a}	9.59 ± 0.51 ^{a}
Untreated supernatant	2.36 ± 0.31 ^{a}	3.56 ± 0.37 ^{a}
Autolyzed supernatant	2.75 ± 0.47 ^{a}	3.55 ± 0.49 ^{a}

The lipid profile of the non-autolyzed and autolyzed whole yeast samples is shown in Figure 1. Phospholipids represented the main moiety. The phospholipid contents were 28.62 ± 1.88 g/100 g and 26.95 ± 0.12 g/100 g for the non-autolyzed and autolyzed whole spent yeast samples, respectively. Other relevant compounds were hydrocarbons (25.15 ± 1.20 g/100 g and 25.05 ± 0.19 g/100 g for the non-autolyzed and autolyzed whole spent yeast samples, respectively). These results agree with the fact that the analyzed fraction is related to lipids recovered from the membrane of *S. cerevisiae.* where they exert structural functions. On the other hand, this yeast is a synthetic biotechnology-derived organism, which was engineered to produce terpenes.

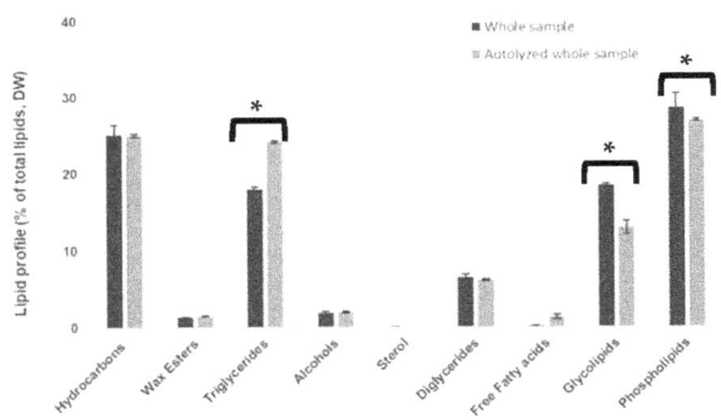

Figure 1. Lipid profile of non-autolyzed and autolyzed whole spent yeast residue samples. Results are expressed as mean ± standard deviation. Significant differences are represented by * for $p < 0.05$.

After the induced autolysis process, a decrease in the concentration of glycolipids (18.40 ± 0.21 g/100 g for non-autolyzed vs. 12.88 ± 0.81 g/100 g for autolyzed samples, $p < 0.05$) was observed, followed by increments in the concentrations of triglycerides (17.91 ± 0.31 g/100 g for non-autolyzed vs. 24.15 ± 0.17 g/100 g for autolyzed samples, $p < 0.05$) and free fatty acids (0.17 ± 0.02 g/100 g for non-autolyzed vs. 1.36 ± 0.31 g/100 g for autolyzed whole yeast samples). It is known that alkali conditions can lead to the hydrolysis of glycolipids, releasing sugars and fatty acids. This is supported by the fact that a decrease in glycolipids is accompanied by an increment in free fatty acids as well as in glucose, as shown in Table 6.

4.2.5. Macro- and Micro-Mineral Determination

The macro-mineral and micro-mineral composition of the SYR samples is presented in Figure 2. The mineral profile was determined for the whole yeast samples and their respective supernatants. The amount of minerals found in the whole non-autolyzed and autolyzed samples should be identical. Autolysis should not change the concentration of these compounds when dealing with the entire spent yeast sample. No significant differences were attained between the autolyzed and non-autolyzed whole spent yeast samples for the majority of minerals analyzed. The only significant difference was found in the aluminum content, which was 1.52 ± 0.03 and 0.51 ± 0.06 mg L^{-1} for the non-autolyzed and autolyzed supernatants, respectively. This result could be related to the heterogeneity of the samples. As we added NaOH to some samples to raise their pH, sodium was excluded from this analysis.

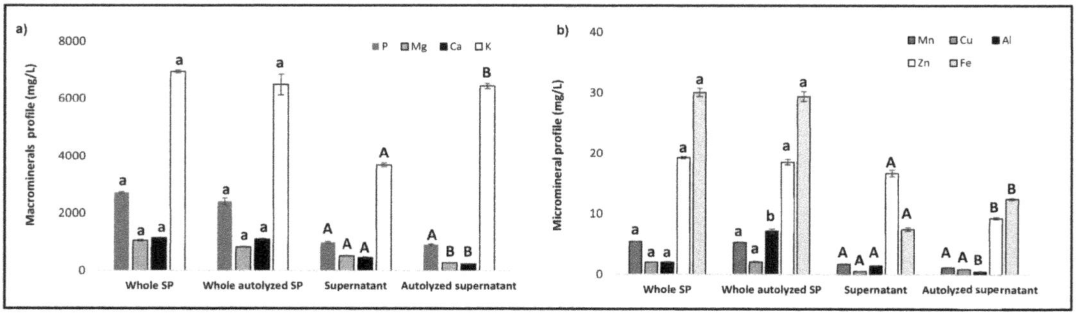

Figure 2. Macro- (**a**) and micro-minerals (**b**) contents of non-autolyzed and autolyzed whole spent yeast residue and respective supernatants. Results are expressed as mean ± standard deviation. Significant differences are represented by different letters above the bars for $p < 0.05$. Lowercase letters refer to the whole spent yeast samples and uppercase letters to the supernatant samples.

Similar to native brewer's spent yeast, potassium is one of the macro-minerals (Figure 2a) which is present in greater amounts [49]. This mineral, together with sodium, plays an important role in the regulation of the cell acid–base balance and water retention, and is essential for ribosomal protein synthesis [49]. Our results showed that the autolysis process induced a significant increase in this mineral in the samples' supernatants.

Regarding micro-minerals (Figure 2b), the results were also similar to what happens for native brewer's spent yeast, with iron and zinc being the prevalent elements. When comparing the untreated and autolyzed samples' supernatants, it is possible to observe that the induced autolysis process decreased the zinc concentration while it slightly increased the iron concentration (Figure 2b) [50]. In other studies, the mineral zinc was found at 30% in yeast cell wall's mannoprotein fractions, at 56% in the vacuole, at 5% in the cytosol, and the rest in other organelles [43]. This fact suggests that the conditions applied during autolysis were not sufficient to allow the release of this mineral to the yeast extract. If the objective was to produce a zinc-enriched yeast extract, the method for the breakdown of cell wall should be directed toward this goal and carried out with the use of cell wall-cleaving β-1,3-glucanase [50].

5. Overall Effect of Induced Autolysis Process

As shown in Figure 3, it is possible to observe the overall impact of the induced autolysis process on both the whole yeast and supernatant samples (Figure 3a,b, respectively). In accordance with what has been discussed in the above sections, the overall difference observed after the induced autolysis process is that the whole SYR samples are not as clustered as the supernatants. This means that, although differences between the autolyzed and non-autolyzed samples exist, they are more evident in the case of the supernatants, and more tenuous and less standardized in the case of the whole yeast samples (Figure 3).

For these samples, the important features identified by the analysis are related to the lipid profile, particularly with the free fatty acid content. These parameters appear with more intensity in the whole SYR samples, when compared with the autolyzed ones.

With respect to the sample supernatant analysis (Figure 3b), it is possible to identify a color pattern that unambiguously separates the untreated and autolyzed samples. In these samples, the highest incidence of the majority of the parameters analyzed is found in the autolyzed samples. This means that these parameters are present in higher levels in the autolyzed supernatants, when compared with the untreated ones, with the exception of minerals and glutamic acid contents.

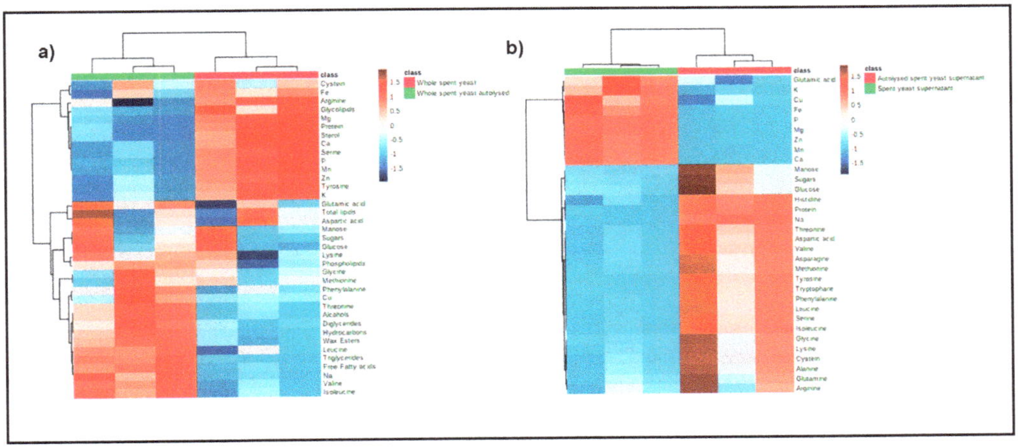

Figure 3. Heatmap of the hierarchical clustering analysis generated from SYR characterization data, reflecting the effect of the induced autolysis process on the whole yeast residue samples (**a**) and respective supernatants (**b**).

6. General Conclusions

This case study shows the possibility of selecting an autolysis process as a fast and low-cost technology to attain products from a synthetic biotechnology-derived SYR, which are rich in high-value nutritional compounds; this process can be targeted to commercial applications.

The production of complex biomolecules through precision fermentation constitutes a sustainable and innovative practice. This technology allows the production of a myriad of biomolecules using renewable resources such as sugarcane, while preventing the depletion of natural resources. Most of the molecules produced through this technology were previously and are still extracted or captured from nature due to their unique properties. However, to close the sustainability loop and effectively contribute to the establishment of a circular economy, efforts should be made to valorize the increasing amounts of residues produced during these fermentation processes.

The work presented herein clearly shows that the studied SYR derived from synthetic biotechnology is a rich source of amino acids, minerals and other components, with potential for valorization. The determination of the supernatants' dry weight revealed that the induced autolysis time, when at 50 °C, can be reduced from four to two hours if pH is raised from 5 to 8 (18.20% dry weight for 4 h and 18.70% dry weight for 2 h at a higher pH).

The impact of the selected induced autolysis process on the composition of this SYR, in both its whole yeast and respective supernatants, was characterized. From the overall analysis of the autolyzed supernatants, it is possible to attain an improved residue, with a generally higher availability of the compounds of interest. The highest positive impact of the selected autolysis process was observed for the sample's supernatants in terms of proteins, free amino acids and mineral contents.

An increase in leucine (around 7%), aspartic acid, valine, phenylalanine, isoleucine, and serine (approximately 2%) was observed in the autolyzed samples, when compared with the untreated ones. Also, regarding minerals, the autolysis process allowed us to obtain significantly higher amounts of potassium in the treated supernatants.

These parameters are of utmost importance in the animal feed supplementation field. SYRs derived from precision fermentation platforms are rich sources of valuable components, and induced autolysis contributes to their bioavailability, increasing their potential for the development of different applications. Also, the autolyzed whole biomass may be bulked for further processing in biorefineries, aiming at producing new bioproducts that are capable of replacing some of the currently used oil-based products. For this purpose, knowledge must be established for such innovative waste streams since it will allow the development of the most suitable valorization strategies for these new residues.

Author Contributions: Conceptualization, J.F.F. and A.L.A.; methodology, J.F.F. and D.A.R.; software, T.D.; validation, M.E.P., L.M.R.-A. and A.L.A.; investigation, J.F.F., D.A.R., T.D., L.L.P. and S.S.M.P.V.; writing—original draft preparation, J.F.F.; writing—review and editing, J.F.F., T.D., M.E.P. and A.L.A.; supervision, M.E.P. and A.L.A. All authors have read and agreed to the published version of the manuscript.

Funding: This work was supported by Amyris Bio Products Portugal Unipessoal Lda and Escola Superior de Biotecnologia—Universidade Católica Portuguesa through the Alchemy project—Capturing high value from industrial fermentation bioproducts (POCI-01-0247-FEDER-027578).

Institutional Review Board Statement: Not applicable.

Informed Consent Statement: Not applicable.

Data Availability Statement: The data presented in this study are available on request from the corresponding author. The data are not publicly available due to confidentiality agreements.

Conflicts of Interest: The authors declare no conflict of interest.

References

1. Marson, G.V.; de Castro, R.J.S.; Machado, M.T.C.; Zandonadi, F.S.; Barros, H.D.F.Q.; Júnior, M.R.M.; Sussulini, A.; Hubinger, M.D. Proteolytic enzymes positively modulated the physicochemical and antioxidant properties of spent yeast protein hydrolysates. *Process. Biochem.* **2020**, *91*, 34–45. [CrossRef]
2. Perruchon, O.; Schmitz-Afonso, I.; Afonso, C.; Abdelhakim Elomri, A. State-of-the-art in analytical methods for metabolic profiling of *Saccharomyces cerevisiae*. *Microchem. J.* **2021**, *170*, 106704. [CrossRef]
3. Bastos, R.; Oliveira, P.G.; Gaspar, V.M.; Mano, J.F.; Coimbra, M.A.; Coelho, E. Brewer's yeast polysaccharides—A review of their exquisite structural features and biomedical applications. *Carbohydr. Polym.* **2022**, *277*, 118826. [CrossRef] [PubMed]
4. Ferreira, I.M.P.L.V.O.; Pinho, O.; Vieira, E.; Tavarela, J.G. Brewer's Saccharomyces yeast biomass: Characteristics and potential applications. *Trends Food Sci. Technol.* **2010**, *21*, 77–84. [CrossRef]
5. Patras, P.; Das, M.; Kundu, P.; Ghosh, A. Recent advances in systems and synthetic biology approaches for developing novel cell-factories in non-conventional yeasts. *Biotechnol. Adv.* **2021**, *47*, 107695. [CrossRef] [PubMed]
6. Liu, D.; Ding, L.; Sun, J.; Boussetta, N.; Vorobiev, E. Yeast cell disruption strategies for recovery of intracellular bio-active compounds—A review. *Innov. Food Sci. Emerg. Technol.* **2016**, *36*, 181–192. [CrossRef]
7. Puligundla, P.; Mok, C.; Park, S. Advances in the valorization of spent brewer's yeast. *Innov. Food Sci. Emerg. Technol.* **2020**, *62*, 102350. [CrossRef]
8. Soh, E.Y.S.; Lim, S.S.; Chew, K.W.; Phuang, X.W.; Ho, V.M.V.; Chu, K.Y.H.; Wong, R.R.; Lee, L.Y.; Tiong, T.J. Valorization of spent brewery yeast biosorbent with sonication-assisted adsorption for dye removal in wastewater treatment. *Environ. Res.* **2022**, *204*, 112385. [CrossRef]
9. Estévez, A.; Padrell, L.; Iñarra, B.; Orive, M.; San Martin, D. Brewery by-products (yeast and spent grain) as protein sources in gilthead seabream (*Sparus aurata*) feeds. *Aquaculture* **2021**, *543*, 736921. [CrossRef]
10. Castro, C.; Pérez-Jiménez, A.; Coutinho, F.; Pousão-Ferreira, P.; Brandão, T.M.; Oliva-Teles, A.; Peres, H. Digestive enzymes of meagre (*Argyrosomus regius*) and white seabream (*Diplodus sargus*). Effects of dietary brewer's spent yeast supplementation. *Aquaculture* **2013**, *416*, 322–327. [CrossRef]
11. Harlow, B.E.; Bryant, R.W.; Cohen, S.D.; O'Connell, S.P.; Flythe, M.D. Degradation of spent craft brewer's yeast by caprine rumen hyper ammonia-producing bacteria. *Lett. Appl. Microbiol.* **2016**, *63*, 307–312. [CrossRef] [PubMed]
12. Alexandre, H. Autolysis of Yeasts. In *Comprehensive Biotechnology*, 2nd ed.; Murray, M.-Y., Ed.; Academic Press: Cambridge, MA, USA, 2011; pp. 641–649.
13. Amorim, M.; Pinheiro, H.; Pintado, M. Valorization of spent brewer's yeast: Optimization of hydrolysis process towards the generation of stable ACE-inhibitory peptides. *LWT-Food Sci. Technol.* **2019**, *111*, 77–84. [CrossRef]
14. Liu, Q.; Kang, J.; Zhang, Z.; Zhou, D.; Zhang, Y.; Zhuang, S. Comparative study on the nutrient digestibility of diets containing brewer's yeast products processed by different techniques fed to T-cannulated growing pigs. *Anim. Feed Sci. Technol.* **2021**, *278*, 114981. [CrossRef]

15. Dimopoulos, G.; Stefanou, N.; Andreou, V.; Taoukis, P. Effect of pulsed electric fields on the production of yeast extract by autolysis. *Innov. Food Sci. Emerg. Technol.* **2018**, *48*, 287–295. [CrossRef]
16. Champagne, C.P.; Barrette, J.; Gouleta, J. Interaction between pH, autolysis promoters and bacterial contamination on the production of yeast extracts. *Int. Food Res. J.* **2000**, *32*, 575–583. [CrossRef]
17. Martinez, J.M.; Delso, C.; Aguilar, D.; Cebrián, G.; Álvarez, I.; Raso, J. Factors influencing autolysis of *Saccharomyces cerevisiae* cells induced by pulsed electric fields. *Food Microbiol.* **2018**, *73*, 67–72. [CrossRef]
18. Bystryak, S.; Santockyte, R.; Peshkovsky, A.S. Cell disruption of *S. cerevisiae* by scalable high-intensity ultrasound. *Biochem. Eng. J.* **2015**, *99*, 99–106. [CrossRef]
19. Zhang, L.; Jin, Y.; Xie, Y.; Wu, X.; Wu, T. Releasing polysaccharide and protein from yeast cells by ultrasound: Selectivity and effects of processing parameters. *Ultrason. Sonochem.* **2014**, *21*, 576–581. [CrossRef]
20. Tanguler, H.; Erten, H. Utilisation of spent brewer's yeast for yeast extract production by autolysis: The effect of temperature. *Food Bioprod. Process.* **2008**, *86*, 317–321. [CrossRef]
21. Dimopoulos, G.; Limnaios, A.; Aerakis, R.; Andreou, V.; Taoukis, P. Effect of high pressure on the proteolytic activity and autolysis of yeast *Saccharomyces cerevisiae*. *Innov. Food Sci. Emerg. Technol.* **2021**, *74*, 102865. [CrossRef]
22. Marson, G.V. Sequential hydrolysis of spent brewer's yeast improved its physico-chemical characteristics and antioxidant properties: A strategy to transform waste into added-value biomolecules. *Process Biochem.* **2019**, *84*, 91–102. [CrossRef]
23. Ferreira, M.P.L.V.O. Autolysis of intracellular content of Brewer's spent yeast to maximize ACE-inhibitory and antioxidant activities. *LWT-Food Sci. Technol.* **2017**, *82*, 255–259.
24. Comuzzo, P.; Calligaris, S.; Lacumin, L.; Ginaldi, F.; Paz, A.E.P.; Zironi, R. Potential of high-pressure homogenization to induce autolysis of wine yeasts. *Food Chem.* **2015**, *185*, 340–348. [CrossRef] [PubMed]
25. Palomero, F.; Morata, A.; Benito, S.; González, M.C.; Suárez-Lepe, J.A. Conventional and enzyme-assisted autolysis during ageing over lees in red wines: Influence on the release of polysaccharides from yeast cell walls and on wine monomeric anthocyanin content. *Food Chem.* **2007**, *105*, 838–846. [CrossRef]
26. Orban, E.; QuaglialCasini, G.B.; Moresi, M. Effect of temperature and yeast concentration on the autolysis of *Kluyverommyces fragilis* grown on lactose-based media. *J. Food Eng.* **1994**, *21*, 245–261. [CrossRef]
27. Schlabitz, C.; Lehn, D.N.; Volken de Souza, C.F. A review of *Saccharomyces cerevisiae* and the applications of its byproducts in dairy cattle feed: Trends in the use of residual brewer's yeast. *J. Clean. Prod.* **2022**, *332*, 130059. [CrossRef]
28. Horwitz, W.L.; George, W., Jr. *Official Methods of Analysis of the Association of Official Analytical Chemists*; AOAC: Rockville, MD, USA, 2010.
29. Serrano, S.; Rincón, F.; García-Olmo, J. Cereal protein analysis via Dumas method: Standardization of a micro-method using the EuroVector Elemental Analyser. *J. Cereal Sci.* **2013**, *58*, 31–36. [CrossRef]
30. Wang, L.; Yang, J.; Wang, J.; Zhang, J.; Gao, Y.; Yuan, J.; Su, A.; Ju, X. Study on Antioxidant Activity and Amino Acid Analysis of Rapeseed Protein Hydrolysates. *Int. J. Food Prop.* **2016**, *19*, 1899–1911. [CrossRef]
31. Coimbra, M.A.; Delgadillo, I.; Waldron, K.W.; Selvendran, R.R. *Modern Methods of Plant Analysis*; Springer: Berlin/Heidelberg, Germany, 1996; Volume 17, pp. 19–44.
32. Bastos, R.; Coelho, E.; Coimbra, M.A. Modifications of *Saccharomyces pastorianus* cell wall polysaccharides with brewing process. *Carbohydr. Polym.* **2015**, *124*, 322–330. [CrossRef]
33. Bligh, E.G.; Dyer, J. A rapid method of total lipid extraction and purification. *Can. J. Biochem. Physiol.* **1959**, *37*, 911–917. [CrossRef]
34. Teixeira, F.S.; Vidigal, S.M.P.; Pimentel, L.L.; Costa, P.T.; Valente, D.T.; Silva, J.A.; Pintado, M.E.; Fernandes, J.C.; Rodríguez-Alcalá, L.M. Phytosterols and Novel Triterpenes Recovered from Industrial Fermentation Coproducts Exert In Vitro Anti-Inflammatory Activity in Macrophages. *Pharmaceuticals* **2021**, *14*, 583. [CrossRef] [PubMed]
35. Abreu, S.; Solgadi, A.; Chaminade, P. Optimization of normal phase chromatographic conditions for lipid analysis and comparison of associated detection techniques. *J. Chrom. A* **2017**, *1514*, 54–71. [CrossRef]
36. Homan, R.; Anderson, M.K. Rapid separation and quantification of combined neutral and polar lipids classes by high-performance liquid chromatography and evaporation light-scattering mass detection. *J. Chromatogr. B Biomed. Sci. Appl.* **1998**, *708*, 21–26. [CrossRef] [PubMed]
37. Chatelain, E.G.; Pintado, M.E.; Vasconcelos, M.W. Evaluation of chitooligosaccharide application on mineral accumulation and plant growth in Phaseolus vulgaris. *Plant Sci.* **2016**, *215*, 134–140. [CrossRef] [PubMed]
38. Suphantharika, M.; Varavinit, S.; Shobsngob, S. Determination of optimum conditions for autolyzed Yeast extract production. *ASEAN J. Sci. Technol. Dev.* **1997**, *14*, 21–28.
39. Jacob, F.F.; Striegel, L.; Rychlik, M.; Hutzler, M.; Methner, F.-J. Yeast extract production using spent yeast from beer manufacture: Influence of industrially applicable disruption methods on selected substance groups with biotechnological relevance. *Eur. Food Res. Technol.* **2019**, *245*, 1169–1182. [CrossRef]
40. Chae, H.J.; Joo, H.; In, M.-J. Utilization of brewer's yeast cells for the production of food-grade yeast extract. Part 1: Effects of different enzymatic treatments on solid and protein recovery and flavor characteristics. *Biores. Technol.* **2001**, *76*, 253–258. [CrossRef]
41. Stam, H.; Hoogland, M.; Laane, C. Food flavours from yeast. In *Microbiology of Fermented Foods*; Wood, B.J.B., Ed.; Springer: Boston, MA, USA, 1998.

42. Erten, H.H.; Tanguler. The Production Methods of Yeast Extract. Feed Info News Service. 2006. Available online: http://www.feedinfo.com (accessed on 12 January 2023).
43. Jacob, F.F.; Hutzler, M.; Methner, F.J. Comparison of various industrially applicable disruption methods to produce yeast extract using spent yeast from top-fermenting beer production: Influence on amino acid and protein content. *Eur. Food Res. Technol.* **2019**, *245*, 95–109. [CrossRef]
44. Barbosa, E.N.R.; Rabello, C.D.-V.; Lopes, C.C.; Silva, E.P.; Freitas, E.R. Amino acid composition, and determination and prediction of the protein digestibility of different sugarcane yeasts in broilers. *Rev. Ciênc. Agron.* **2018**, *49*, 334–342. [CrossRef]
45. Amorim, M.; Pereira, J.O.; Gomes, D.; Pereira, C.D.; Pinheiro, H.; Pintado, M. Nutritional ingredients from spent brewer's yeast obtained by hydrolysis and selective membrane filtration integrated in a pilot process. *J. Food Eng.* **2016**, *185*, 42–47. [CrossRef]
46. Podpora, B.; Świderski, F.; Sadowska, A.; Piotrowska, A.; Rakowska, R. Spent Brewer's Yeast Autolysates as a New and Valuable Component of Functional Food and Dietary Supplements. *J. Food Process. Technol.* **2015**, *6*, 1000526.
47. Cao, Y.C.; Yang, X.J.; Guo, L.; Zheng, C.; Wang, D.D.; Cai, C.J.; Yao, J.H. Regulation of pancreas development and enzymatic gene expression by duodenal infusion of leucine and phenylalanine in dairy goats. *Livest. Sci.* **2018**, *216*, 9–15. [CrossRef]
48. Zhou, J.-M.; Qiu, K.; Wang, J.; Zhang, H.-J.; Qi, G.-H.; Wu, S.-G. Effect of dietary serine supplementation on performance, egg quality, serum indices, and ileal mucosal immunity in laying hens fed a low crude protein diet. *Poult. Sci.* **2012**, *100*, 101465. [CrossRef]
49. Vieira, E.F.; Carvalho, J.; Pinto, E.; Cunha, S.; Almeida, A.A.; Ferreira, I.M.P.L.V.O. Nutritive value, antioxidant activity and phenolic compounds profile of brewer's spent yeast extract. *J. Food Compos. Anal.* **2016**, *52*, 44–51. [CrossRef]
50. Gaudreau, H.; Conway, J.; Champagne, C.P. Production of zinc-enriched yeast extracts. *J. Food Sci. Technol. Mysore* **2001**, *38*, 348–351.

fermentation

MDPI

Article

Sustainability Assessment of Food Waste Biorefineries as the Base of the Entrepreneurship in Rural Zones of Colombia

Carlos Ariel Cardona [1,*], Mariana Ortiz-Sanchez [1], Natalia Salgado [1], Juan Camilo Solarte-Toro [1], Carlos Eduardo Orrego [2], Alexander Perez [3], Carlos Daniel Acosta [4], Eva Ledezma [5], Haminton Salas [5], Javier Gonzaga [6] and Steven Delgado [7]

1 Instituto de Biotecnología y Agroindustria, Departamento de Ingeniería Química, Universidad Nacional de Colombia Sede Manizales, Manizales 170001, Colombia; mortizs@unal.edu.co (M.O.-S.); nsalgadoa@unal.edu.co (N.S.); jcsolartet@unal.edu.co (J.C.S.-T.)
2 Departamento de Física y Química, Facultad de Ciencias Exactas y Naturales, Universidad Nacional de Colombia Sede Manizales, Manizales 170001, Colombia; ceorregoa@unal.edu.co
3 Departamento de Agricultura y Zootecnia, Facultad de Ciencias Agronómicas, Universidad de Sucre, Sincelejo 700001, Colombia; alexander.perez@unisucre.edu.co
4 Departamento de Matemáticas y Estadística, Facultad de Ciencias Exactas y Naturales, Universidad Nacional de Colombia Sede Manizales, Manizales 170001, Colombia; cdacostam@unal.edu.co
5 Facultad de Ciencias Naturales, Universidad Tecnológica del Choco, Quibdó 270001, Colombia; d-eva.ledezma@utch.edu.co (E.L.); hasamo49@gmail.com (H.S.)
6 Facultad de Derecho, Universidad de Caldas, Manizales 170001, Colombia; directorcientifico.posconflicto@ucaldas.edu.co
7 Consultorio Administrativo, Facultad de Administración, Universidad Nacional de Colombia, Bogotá 110111, Colombia; stdelgadoco@unal.edu.co
* Correspondence: ccardonaal@unal.edu.co

Citation: Cardona, C.A.; Ortiz-Sanchez, M.; Salgado, N.; Solarte-Toro, J.C.; Orrego, C.E.; Perez, A.; Acosta, C.D.; Ledezma, E.; Salas, H.; Gonzaga, J.; et al. Sustainability Assessment of Food Waste Biorefineries as the Base of the Entrepreneurship in Rural Zones of Colombia. *Fermentation* **2023**, *9*, 609. https://doi.org/10.3390/fermentation9070609

Academic Editors: Jose Luis García-Morales and Francisco Jesús Fernández Morales

Received: 18 May 2023
Revised: 26 June 2023
Accepted: 26 June 2023
Published: 28 June 2023

Abstract: The sustainability of food value chains is affected by the large amounts of waste produced with a high environmental impact. Food waste valorization applying the biorefinery concept has emerged as an alternative to reduce the generation of greenhouse gases and to promote the socio-economic development of value chains at local, regional, and national levels. This paper analyzes the sustainability of food waste biorefineries designed for boosting rural economic development in Colombia. These biorefineries were designed following a strategy based on a portfolio of bio-processes involving fractions based on the composition of the raw materials. The valorization of six food residues produced in three representative rural areas of Colombia (i.e., Chocó, Caldas, and Sucre) was analyzed. Acai, annatto, sugarcane bagasse, rejected plantain and avocado, and organic kitchen food waste (OKFW) were selected as food wastes for upgrading. The biorefinery design strategy comprised five steps for filtering the most promising bioprocesses to be implemented. The OKFW was analyzed in detail, applying the design strategy to provide a step-by-step guide involving a portfolio of bioproducts, the technological maturity index, and the socio-economic context. This strategy implementation for OKFW valorization resulted in a scenario where biorefineries with levulinic acid production were the most feasible and sustainable, with high techno-economic performances and low environmental impacts. For the valorization of the other food residues, the processes with the greatest feasibility of being implemented in rural areas were bioactive compounds, oil, flour, and biogas production.

Keywords: food waste; biorefineries; sustainability assessment; design strategy; entrepreneurship

1. Introduction

Food residues are one of the most important issues in the world due to the high per capita residues generated [1]. Developed and developing countries are making efforts to mitigate food residue generation by implementing strategies and policies [2]. A food supply chain (FSC) generates food losses (FLs) and food waste (FW) [3]. FLs obtained during the agricultural and farming production, post-harvest, handling, slaughter, and storage, and process distribution and transformation stages can be grouped as agronomic losses (ALs) [4]. These wastes are generated at a single FSC location. For example, the FSC for fruit production has, in the first three stages, the same location as the crop or is in neighboring areas. In the fruit agricultural production stage, stems, leaves, roots, flowers, and fruits are generated with low-quality standards (overripe) [5]. On the other hand, the post-harvest, storage, process distribution, and transportation stages produce low-quality standard fruits. FLs generated during the processing, packaging, and distribution stages can be grouped as agroindustry losses (AgLs) [6]. In the case of the FSC for fruit, residues such as peel, seeds, liquids, and solid residues are generated during the above-mentioned stages. Finally, the FW obtained in the last stages of an FSC can be classified as manufacturing and domestic food waste [7]. This type of waste contains a mixture of agronomic and agroindustrial products such as vegetables, fruits, farming products, or processed products [8]. The main characteristic of this waste is its non-standard composition [9]. Regarding the above information, two characterizations can be approached according to the FW composition. Agronomic and agroindustrial residues are classified as standard food waste. On the other hand, manufacturing and domestic food waste are considered to be non-standard food waste [10].

Sustainability has been defined as the perfect balance between the economic, environmental, and social aspects of a system, product, or process [11]. This concept has been applied to describe the performances of different food residue upgrading alternatives to obtain value-added products and energy vectors at the laboratory, pilot, bench, or industrial scale [12]. Food residue valorization is the base for closing the loop in several value chains since the residues produced in one link (e.g., food losses) can be used to produce marketable products with commercial value and to move forward to carbon neutrality [13]. Moreover, FW valorization is in line with the sustainable development goals (SDGs) proposed by the UN since actions to reduce and upgrade FW are being researched and implemented at different scales [14]. SDG 12, "Sustainable production and consumption", can be accomplished through waste upgrading since unsustainable patterns (e.g., excessive reliance on natural resources and high per capita food residue production) can be reduced without affecting the dynamics of any value chain. Then, the integral FW upgrading plays a key role in developing a more sustainable production–consumption dynamics since reducing and valorizing waste streams results in more income (i.e., fewer disposal expenses) and lower environmental impacts. In this way, FWs should be upgraded by applying the biorefinery concept as a strategy to increase the product portfolio of an FSC [15].

Biorefineries are complex systems where a biomass is processed to obtain a portfolio of value-added products and energy vectors after integral processing that applies biotechnological, thermochemical, and physic-chemical processes [16]. FLs and FWs have been studied as raw materials to be valorized in conceptually designed biorefineries [17]. There are several reports in the literature of techno-economic (TEA) and environmental analyses. Nevertheless, most studies have not involved other crucial factors for designing more reliable and feasible processes. Factors such as (i) context (i.e., specific territory knowledge), (ii) processing scale, (iii) logistics and location, (iv) technological readiness level (TRL), (v) local and regional market needs, and (vi) national and international policies must be involved to propose more accurate processes for the reality of the situation [18]. These factors are important when designing biorefineries since the portfolio of products and biorefinery configuration can change depending on the biomass fractions and context.

Developed countries (e.g., Germany, Italy, and the United States of America) have great potential for establishing bioeconomies through the implementation of large-scale biorefineries to produce value-added products such as biosurfactants, organic molecules, and pharmaceuticals [19]. Large-scale processes require adequate infrastructure and a high industrialization level [20]. These processes are favored by the economy–scale concept. Nevertheless, their most important disadvantages are their raw material acquisition and logistics [21]. Developing countries (e.g., Latin American countries) have a great potential to develop a rural bioeconomy based on implementing small-scale biorefineries since these processes do not require a high industrialization level. Small-scale biorefineries must be addressed to produce local products and energy vectors [22]. The starting point to develop these processes are rural areas in developing countries since a large amount of FL is produced [23]. Several efforts to involve rural zones as the bases for establishing bioeconomies have been reported in the open literature. For instance, Solarte-Toro et al. [24] reported different small-scale configurations to upgrade avocado (*Persea americana var. americana*) residues into local marketable products such as avocado oil and guacamole. Moreover, Serna-Loaiza et al. [25] published small-scale processes addressed to upgrade cocoyam (*Xanthosoma sagittifolium*) into local products such as animal feed and starch. These efforts have demonstrated the great potential of FLs as raw materials to contribute to the socio-economic growth of a region.

Regarding the potential of small-scale biorefineries to improve the socio-economic conditions of a region, these facilities can be considered as entrepreneurship since small-scale processes can generate new job positions and contribute to decreasing the number of informal jobs. Moreover, implementing entrepreneurships based on biomasses in rural zones can establish rural bioeconomies. Thus, FW upgrading in small-scale biorefineries is the first step towards the sustainable development of a region. The objective of this work was to evaluate the potential for upgrading different FWs produced in representative rural zones of Colombia for a series of marketable products and energy vectors. A design methodology based on selecting the bioproducts portfolio reported by Ortiz-Sanchez and Cardona Alzate [26] was applied. The studied FWs in this manuscript come from avocado, plantain, acai, brown sugarcane, annatto, and OKFW.

2. Methodology

In Colombia, there are different rural areas dedicated to agricultural activities with problems such as armed conflicts, low production yields, and high waste generation. In this work, the sustainability analysis of food waste biorefineries was conducted considering three rural zones of Colombia. The rural zones analyzed are located on the north coast (close to Caribbean Sea), the Montes de María in the Sucre department; the west coast (close to Pacific Ocean), the Unión Panamericana, Quibdó, and Bojayá in the Chocó department; and Samaná, in the Caldas department. The most representative food crops of the analyzed zones are avocado, plantain, açai, annatto, and sugarcane. Figure 1 shows the three zones selected and the raw materials analyzed in this paper. The raw materials were classified into FLs and FWs. The three rural regions analyzed in this manuscript represent three different ecosystems and thermal floors (i.e., different types of soil, crops, agricultural practices, productivities, and yields) that allowed for the analyses of various raw materials with diverse chemical compositions. In this sense, raw materials such as achiote and acai allow for valorizing extractive fractions to obtain bioactive compounds (e.g., colorants). Plantain and avocado allow for the analysis of valorization routes for producing foods such as flour and avocado oil. Finally, OKFW, due to its content of fats, pectin, starch, fiber, and extractives, requires more complex recovery routes to be proposed. In countries located in the tropics, this type of analysis demonstrates how FW can be valued in different ecosystems. The sustainability of the biorefineries was analyzed considering the methodology reported by Ortiz-Sanchez and Cardona Alzate [26].

In Sucre (zone 1), the valorization of rejected avocado (*Persea americana* sp.) and plantain (*Musa paradisiaca* sp.) was analyzed. Raw material flows of 150 kg/h of rejected avocado and 2145 kg/h of rejected plantain were considered. The flows were equivalent to 100% of the rejected avocado and plantain generated in the rural zone of Montes de María. In Chocó (zone 2), the use of non-marketed açai (*Euterpe oleracea*) and waste food additives generated from the extraction of annatto dye (*Bixa orellana* L.) were evaluated.

The raw material flows for the analysis of the biorefineries were 13.5 kg/h of açai and 51.8 kg/h of annatto. The flows were selected considering 50% of the açai and annatto production in Unión Panamericana, Quibdó, and Bojayá from Chocó. Finally, sugarcane bagasse (*Saccharum officinarum*) and OKFW were analyzed in Caldas (zone 3). The sugarcane milling generated the bagasse corresponding to 44% w/w of the raw material. This work analyzed the valorization of 50% of the sugarcane bagasse generated in zone 3 (80.4 kg/h). OKFW was considered an optional source of raw material in zone 3 due to its current use of sugarcane bagasse. Given the impossible standardization of the OKFW, a model based on Colombian food consumption was used. In this case, a use of 40% of the OKFW generated in zone 3 (93.2 ton/h) was analyzed. The valorization of FL analyzed in this work was carried out considering small-scale biorefineries. On the other hand, the valorization of FW was analyzed considering high-scale biorefineries.

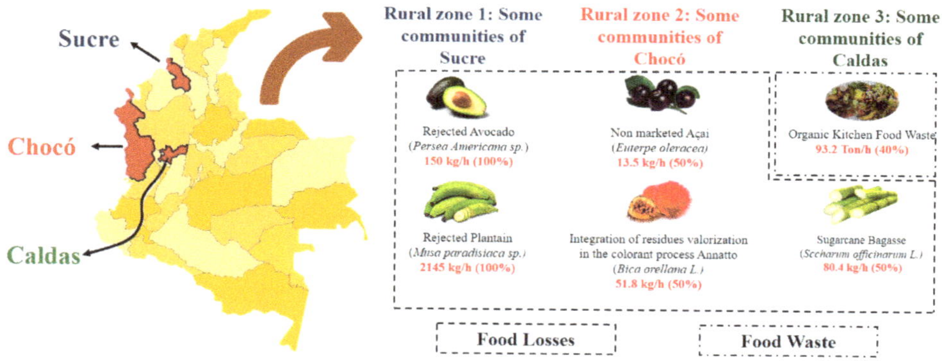

Figure 1. Zones and raw materials selected for the food waste biorefineries analysis.

2.1. Sustainability Analysis of Biorefineries—Design Strategy of the Biomass Valorization

The sustainability analysis of the food waste biorefineries was carried out considering the strategy reported by Ortiz-Sanchez and Cardona Alzate [26]. This work defined a design and evaluation strategy considering different biomass processing routes based on chemical composition. The strategy comprises five steps where filtration of the bioprocesses is developed as a function of each fraction of the raw material (i.e., cellulose, hemicellulose, lignin, starch, pectin, extracts, and fats). The steps of the design strategy are presented in Figure 2.

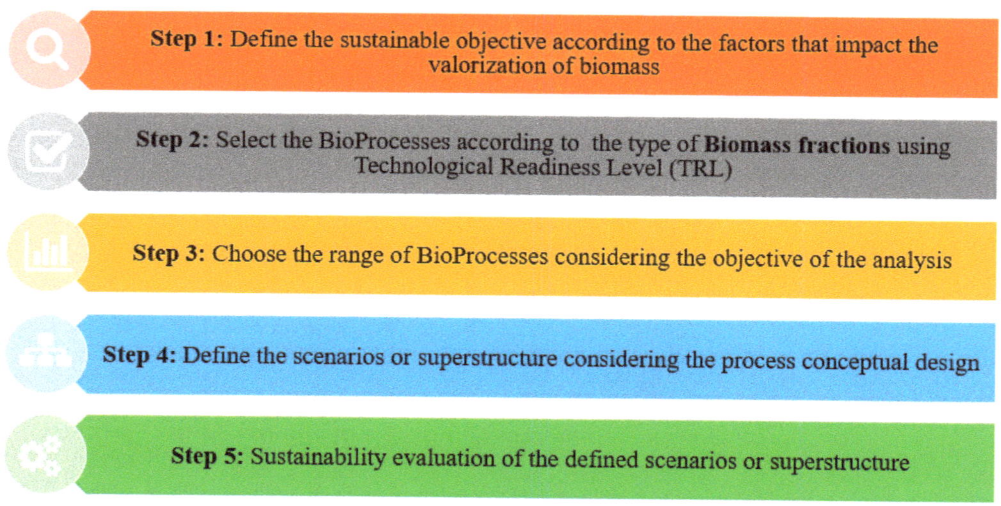

Figure 2. The steps of the design strategy for analyzing the sustainability of the biorefineries.

The first step in the design strategy is to define the sustainability objective considering limiting factors such as production chain, scale, technological context, and product type to be obtained. In this step, it is necessary to be clear about the biomass uses in the specific context. The second step is to select the bioprocesses according to the sustainability objective. In this step, the TRL of the bioprocesses must be analyzed. Therefore, the selection of bioprocesses is considered as the first filter. The third step is the second bioprocess filter. This second filter is based on the sustainability objective (technical, economic, or environmental). If the sustainability objective is to seek the economic and environmental viability of biomass uses, the bioprocesses must be defined with favorable economic and environmental indicators. Step four defines the scenarios or superstructure according to the selected bioprocesses. Scenarios must be considered using the conceptual design methodology. Finally, step five evaluates the scenarios or superstructure considering the technical, economic, or environmental indicators.

2.1.1. Step 1: The Sustainable Objective

The sustainability objective was to define the best route for FL and FW valorization in economic and environmental terms as the basis of entrepreneurship. The main limiting factors for the valorization of FL and FW under the biorefinery concept are the low technological level (zones) and low raw material flows (low-scale biorefineries). These considerations limit the type of bioprocess that can be implemented in the study zones. For this reason, processes with high TRLs and easy-to-market products should be considered.

2.1.2. Step 2: First Filter of the Bioprocesses According to the TRL

The second step of the biorefinery design strategy was carried out considering the portfolio reported by Ortiz-Sanchez and Cardona Alzate [26]. Table 1 shows the bioprocesses considered in the portfolio.

Table 1. Bioprocesses portfolio considered to upgrade raw material fractions. Based on [26].

Raw Material Fraction	Bioprocesses	Bioproducts	Technology
Extractives	2	Bioactive compounds	Agitated solvent extraction Supercritical fluid extraction with carbon dioxide
Fats	4	Essential oil and oil Biodiesel	Steam distillation and hydrodistillation Extrussion Trasesterification
Cellulose	9	Glucose platform Ethanol and ABE * Lactic acid PHB ** Itaconic acid Polylactic acid	Catalytic and enzymatic hydrolysis glucosa production Fermentation—*Saccharomyces cerevisiae* *Clostridium acetobutylicym*, *Lactobacillus casei*, *Bacillus megaterium*, and *Aspegillus terreus* Catalytic upgrading
Hemicellulose	4	Xylose platform Furfural Xylitol Pentane	Acid hydrolysis Catalytic upgrading Fermentation—*Candida giillermondii*
Lignin	4	Soda lignin Organosolv lignin Kraft lignin Vainillin and vanilic acid	Alcaline pretreatment Organosolv pretreatment Kraft process Catatytic upgrading
Pectin	4	Pectin Mucic acid Galacturonic acid and sugars platform	Acid hydrolysis Fermentation Enzymatic hydrolysis

Table 1. *Cont.*

Raw Material Fraction	Bioprocesses	Bioproducts	Technology
Starch	2	Glucose plarform Flour	Enzymatic hydrolysis Extraction
All fractions	6	Biogas Biomethane Syngas Hydrogen Heat and power	Anaerobic digestion Pressure swing absorption Chemical absorption Gasification Water gas shift Combustion and cogeneration

* Acetone, Butanol and Ethanol, ** Polyhydroxybutyrate.

The selection of the bioprocesses was carried out considering a TRL implementation level. This was completed based on the technological context of the zone. The selection of the bioprocesses was developed considering the raw material fractions shown in Table 1 (in the Figure 3 of the reference (Ortiz-Sanchez and Cardona Alzate [26]), the TRL for these technologies is described). Thus, the chemical characterizations of the raw materials were taken from available literature reports. Table 2 shows the chemical compositions of the raw materials used in the analyzed zones.

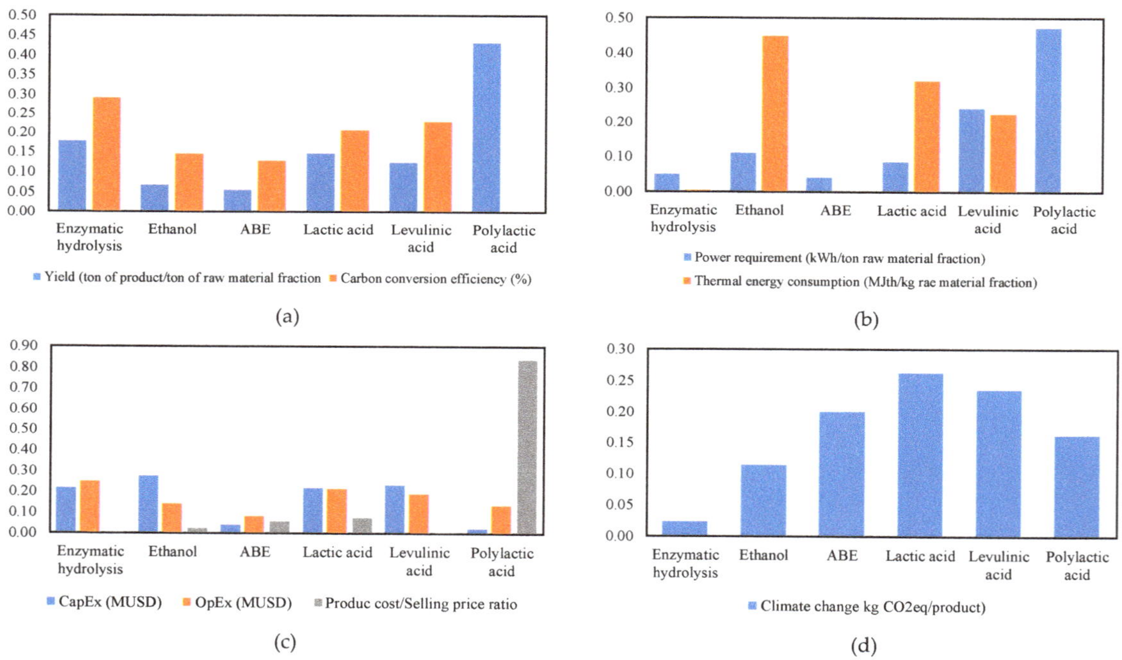

Figure 3. Indicators for cellulose fraction valorization: (**a**) mass indicators; (**b**) energy indicators; (**c**) economic indicators; and (**d**) environmental indicator. Based on [26].

The technical, economic, and environmental indicators of the bioprocesses presented in Table 1 were calculated considering the fermentable sugars that could be obtained from the cellulose fraction (for the methodology details, please see Figure 4 in the study by Ortiz-Sanchez and Cardona Alzate [26]). For the hemicellulose fraction, the bioprocesses for obtaining furfural, xylitol, and pentane were obtained from xylose. Additionally, the production of galacturonic acid and mucic acid was completed based on the pectin fraction.

Table 2. Chemical characterization of the raw materials used in the zones in Colombia.

Item	Food Losses								Food Waste	
	Zone 1 [27,28]					Zone 2 [29]			Zone 3	
	Avocado		Plantain			Annatto	Açai		Sugarcane Bagasse [30]	OKFW [18]
	Peel	Seed	Peel	Peel and Pulp	Pseudosteam	Seed	Seed	Pulp		
Share of fruit (% *w*/*w*)	13.03	15.33	28.65	100	N.A.		90	10	N.A.	N.A.
Chemical composition (% *w*/*w*, dry basis)										
Moisture *	13.17	11.09	87.16	71	82.74	40.01	31.26	89.63	21.83	79.13
Extractives	28.09	32.01	31.58	42.41	46.80	27.00	21.36	N.R.	11.36	21.13
Cellulose	14.21	22.50	11.04	11.96	18.78	17.85	12.49		43.42	19.91
Hemicellulose	9.88	15.64	9.66	18.95	16.12	10.76	40.85	16.81	20.20	5.17
Lignin	8.26	10.35	7.42	14.32	4.01	13.21	15.23		22.61	13.83
Pectin	N.R.	N.R.	N.R.	N.R.	N.R.	15.18	N.R.	N.R.	N.R.	5.28
Protein	N.R.	N.R.	N.R.	N.R.	N.R.	8.26	N.R.	6.01	N.R.	N.R.
Starch	26.10	1.66	29.17	N.R.	N.R.	N.R.	N.R.	N.R.	N.R.	26.03
Fats	10.42	14.44	N.R.	N.R.	N.R.	2.62	2.85	73.99	N.R.	5.39
Ash	3.04	3.40	11.13	12.36	14.28	5.11	7.22	3.20	2.41	3.26
TOTAL	100	100	100	100	100	100	100	100	100	100
Total and volatile solids ** (% *w*/*w*)										
Total solids	89.92	43.99	90.95	89.43	88.05	94.5	N.R.	25.4	91.6	27.98
Volatile solids	87.91	42.40	71.13	75.31	76.99	90.3	N.R.	24.3	88.2	25.61

N.A., not applicable; N.R., not reported; *, raw moisture content; **, total and volatile solids measured based on raw materials as received.

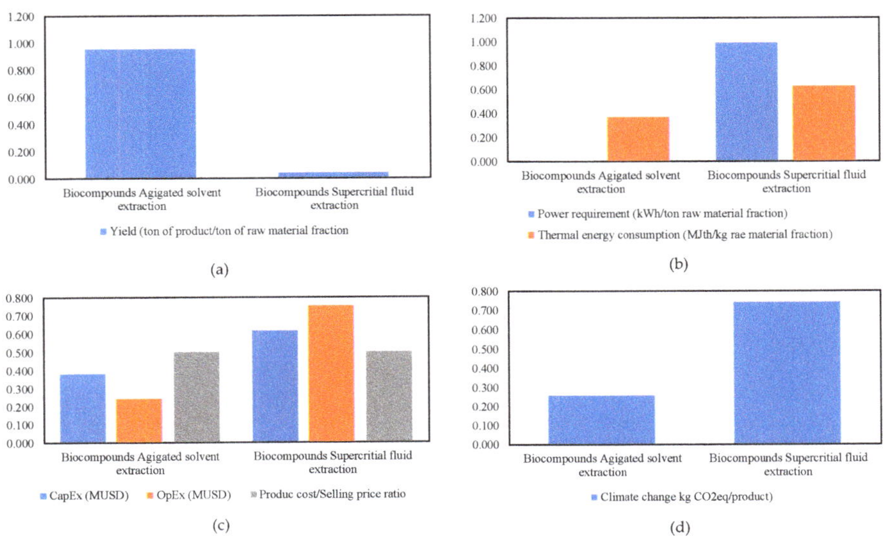

Figure 4. Indicators for the extractives fraction valorization: (**a**) mass indicators; (**b**) energy indicators; (**c**) economic indicators; and (**d**) environmental indicator. Based on [26].

2.1.3. Step 3: Second Filter Based on the Technical, Economic, and Environmental Indicators

Once the bioprocesses were selected according to the TRL, the second filtration step was carried out considering the sustainability objective. The bioprocess portfolio reported by Ortiz-Sanchez and Cardona Alzate [26] presented technical, economic, and environmental indicators for the bioproducts presented in Table 1. The technical indicators considered in the portfolio were yield (ton of product/ton of raw material fraction), carbon conversion efficiency (%), power requirements (kWh/ton raw material fraction), and thermal energy consumption (MJth/ton raw material fraction). The thermal energy consumption for the bioprocesses presented in Table 1 indicate the distribution of utilities (i.e., cooling water, low-pressure steam, medium-pressure steam, and high-pressure steam). The economic indicators were capital expenditures and operational expenditures. Finally, the environmental indicators referred to climate change (kg CO_2 eq/product). The second bioprocess filter was carried out according to the sustainability objective for food waste biorefineries considering the lowest values of capital costs and operational costs and the lowest environmental impact.

2.1.4. Step 4: Biorefineries Scenarios

The biorefinery scenarios were proposed considering the bioprocesses selected up to the previous step. In this step, the conceptual design methodology reported by Cardona et al. [16] was considered. The conceptual design methodology encompassed the use of hierarchy and process sequencing. The hierarchy concept implied the hierarchical decomposition of the fractions of the raw materials. On the other hand, sequencing defined the logical synthesis of the bioprocesses.

2.1.5. Step 5: Biorefinery Analysis

Based on the biorefinery scenarios, an economic and environmental evaluation was completed. In economic terms, the analysis of the biorefineries was carried out considering the net present value (NPV) of the process. The methodology described by Towler and Sinnott [31] was considered. Operational expenditures and capital expenditures were obtained from the bioprocess portfolio reported by Ortiz-Sanchez and Cardona Alzate [26]. In addition, the economic assessment of the biorefinery was completed considering the straight line as the depreciation method. Moreover, a continuous operation was assumed (i.e., 8000 h per year). Then, three (3) shifts were required. The project lifetime was presumed to be 20 years.

3. Results and Discussion

The methodological steps of the biorefinery design strategy are presented in detail using OKFW as example. The results obtained for the other raw materials are presented and avoid a deeper explanation following the steps described in the methodology section. Nevertheless, the economic analyses of the small-scale biorefineries are described and analyzed.

3.1. Results for OKFW Valorization Applying the Design Strategy

3.1.1. Step 2: Results of First Filter of Bioprocesses

The TRL was selected as the starting point to specify a preliminary list of bioprocesses for upgrading each fraction of the raw material (i.e., OKFW). In this case, the biorefineries should be proposed to involve bioprocesses with a TRL value of between seven and nine (i.e., system prototype to system proven in an operational environment) since the process objective was addressed to establish reliable and feasible entrepreneurships. Then, those bioprocesses with TRLs higher than seven were selected. Table 3 presents the options available for upgrading the raw material into a series of value-added products and energy vectors.

The analysis of the raw material chemical composition serves as the basis for selecting the most relevant bioprocesses to be involved in the process configuration. For instance,

OKFW has a low content of pectin and fats (<10% w/w). Therefore, an upgrading of these fractions is not suitable since low yields would be obtained and higher capital costs would be required. Therefore, the bioprocesses addressed to upgrade these fractions were not considered. In addition, the physical characteristics of the raw materials played a key role when selecting the valorization route for all fractions together. Indeed, high moisture content, as in the case of the OKFW, makes such a raw material unsuitable for thermochemical processing (i.e., gasification and combustion). Thus, these bioprocesses should not be considered since high energy must be supplied to reduce the moisture content, affecting the global energy balance of the process. Once this specification related to the raw material composition was obtained, a list of 15 bioprocesses was established for upgrading the cellulose, hemicellulose, lignin, and extractives fractions. Products such as levulinic acid, butanol, polylactic acid (PLA), lignin, xylose, and biogas constituted options for upgrading OKFW. Nevertheless, a second filter needed to be applied to define the most promising alternatives to be implemented based on technical, economic, and environmental indicators.

Table 3. List of selected bioprocesses for raw materials upgrading (e.g., for OKFW).

Fraction	Bioprocess	Bioproduct	TRL *
Cellulose	Enzymatic hydrolysis	Glucose platform	9
	Fermentation	Ethanol	9
	Fermentation	Butanol	8
	Fermentation	Lactic acid	9
	Catalytic upgrading	Levulinic acid	9
	Catalytic upgrading	Polylactic acid	9
Hemicellulose	Acid hydrolysis	Xylose platform	9
Lignin	Alkaline pretreatment	Soda lignin	9
	Organosolv pretreatment	Organosolv lignin	8
	Kraft/pulping	Kraft lignin	9
Extractives	Agitated solvent extraction	Bioactive compounds	9
	Supercritical fluids extraction	Bioactive compounds	8
Fats	Steam distillation	Essential oil	9
	Hydrodistillation	Essential oil	9
	Extrusion	Oil	9
	Transesterification	Biodiesel	9
Pectin	Enzymatic hydrolysis	Galacturonic acid	9
	Enzymatic hydrolysis	Glucose platform	9
	Starch production	Starch	9
All fractions	Anaerobic digestion (AD)	Biogas	9
	AD plus pressure swing absorption	Biomethane	9
	AD plus chemical absorption	Biomethane	9
	Gasification	Synthesis gas	9
	Cogeneration	Heat and Power	9

*, based on Figure 4 in the study by Ortiz-Sanchez and Cardona Alzate [26].

3.1.2. Step 3: Results of Second Filter According to the Technical, Economic, and Environmental Indicators

The second filter applied to the selected bioprocesses in Figure 2 was completed considering technical, economic, and environmental indicators. The indicators values for the fractions defined in step 2 are presented in Table 4. The economic indicators were calculated considering the OKFW flow.

Table 4. Technical, economic, and environmental indicators for each raw material fraction [26].

Fraction	Bioprocesses	Technical Indicators				Economic Indicators			Environmental Indicators
		Mass Indicators		Energy Indicators					
		Yield (Ton of Product/Ton of Raw Material Fraction	Carbon Conversion Efficiency (%)	Power Requirement (kWh/Ton Raw Material Fraction)	Thermal Energy Consumption (MJth/kg Raw Material Fraction)	CapEx (MUSD)	OpEx (MUSD)	Product Cost/Selling Price Ratio	Climate Change kg CO_2 (eq/product)
Cellulose	Enzymatic hydrolysis	0.8	92.6	1.8	0.73	132.87	43.92	0.21	0.28
	Ethanol	0.3	46.2	4	61.9	170.29	24.87	1	1.38
	ABE	0.25	40.51	1.5	0.007	24.82	14.32	2.41	2.41
	Lactic acid	0.66	65.75	3.1	44.3	134.74	37.70	3.11	3.16
	Levulinic acid	0.56	72.75	8.8	31	143.17	33.42	0.2	2.84
	Polylactic acid	1.95	N.A.	17.17	0.1	15.27	23.87	35.5	1.96
Extractives	Biocompounds agitated solvent extraction	0.2–0.8	N.A.	0.42–1.58	106.52–402.41	543.58	14.31	0.33–1.25	0.16
	Biocompounds supercritical fluid extraction	0.009–0.33	N.A.	54.23–208.87	180.32–681.21	878.3	43.94	0.33–1.25	0.46
All fractions	Biogas	190–750	86.94	33.4–132	0.56	1.12	1.02	1.53	0.75
	Biomethane pressure swing absorption	190–750	86.95	42.5–144.2	2.75	1.24	0.61	1.64	1.3
	Biomethane chemical absorption	190–750	86.94	42.5–144.2	2.98	1.27	0.72	1.83	1.42

The enzymatic hydrolysis of cellulose to obtain glucose was defined as a stage before the bioprocesses presented in Figure 2. Figure 3a,b shows the normalized results for the technical indicators (i.e., mass and energy indicators). The bioprocesses with the best yields were PLA, levulinic acid, lactic acid, ethanol, and ABE production. The carbon conversion efficiency for these processes had a similar behavior. PLA did not present this indicator due to the polymerization process and the increase in molecular weight that took place in the process. The power requirement indicator had low consumption levels for the ABE, lactic acid, ethanol, levulinic acid, and PLA bioprocesses. Regarding thermal energy consumption, the bioprocesses with the lowest consumption were ABE, PLA, levulinic acid, lactic acid, and ethanol. In this sense, the bioprocesses with the best technical behaviors were ABE, levulinic acid, lactic acid, ethanol, and PLA.

Figure 3c presents the normalized economic indicators for the bioprocesses proposed for the cellulose fraction. The bioprocesses with the lowest CapEx were PLA, ABE, lactic acid, levulinic acid, and ethanol. Regarding OpEx, the bioprocesses in ascending order were ABE, PLA, ethanol, levulinic acid, and lactic acid. Finally, the relation between the production cost and the sale price presented better values for the bioprocesses of levulinic acid, ethanol, ABE, lactic acid, and PLA. Based on these indicators, it was determined that the bioprocesses with the highest economic pre-feasibility were ABE, levulinic acid, ethanol, and lactic acid.

Figure 3d shows the climate change related to the bioprocesses for the cellulose fraction. Ethanol, PLA, ABE, levulinic acid, and lactic acid were the bioprocesses, in ascending order, regarding greenhouse gas emissions.

Based on the comprehensive analysis of the technical, economic, and environmental indicators for the cellulose fraction, ABE, PLA, and levulinic acid were the analyzed bioprocesses with the greatest potential to be implemented.

The normalized technical, economic, and environmental indicators for the extractive fraction bioprocesses are presented in Figure 4. The technical indicators in terms of yield and energy consumption (i.e., thermal and electrical) showed that bioactive compounds extraction with stirred solvent presented higher prefeasibility than supercritical fluid extraction (see Figure 4a,c). In economic and environmental terms, parameters such as CapEx, OpEx, and climate change presented the same behaviors described for extracting bioactive compounds with stirred solvents. Therefore, this bioprocess was selected as the best alternative to valorize OKFW.

Finally, the technical, economic, and environmental indicators of the OKFW valorization considering all the fractions (i.e., cellulose, hemicellulose, lignin, starch, pectin, etc.) are presented in Figure 5. For the bioprocesses, the anaerobic digestion was considered as the base for the raw material fractions. For the biomethane production, two purification technologies were considered: pressure swing absorption and chemical adsorption by using amines. Biogas production was the process that presented the best indicators due to its high yield, low capital investment, and environmental impact compared to biomethane production.

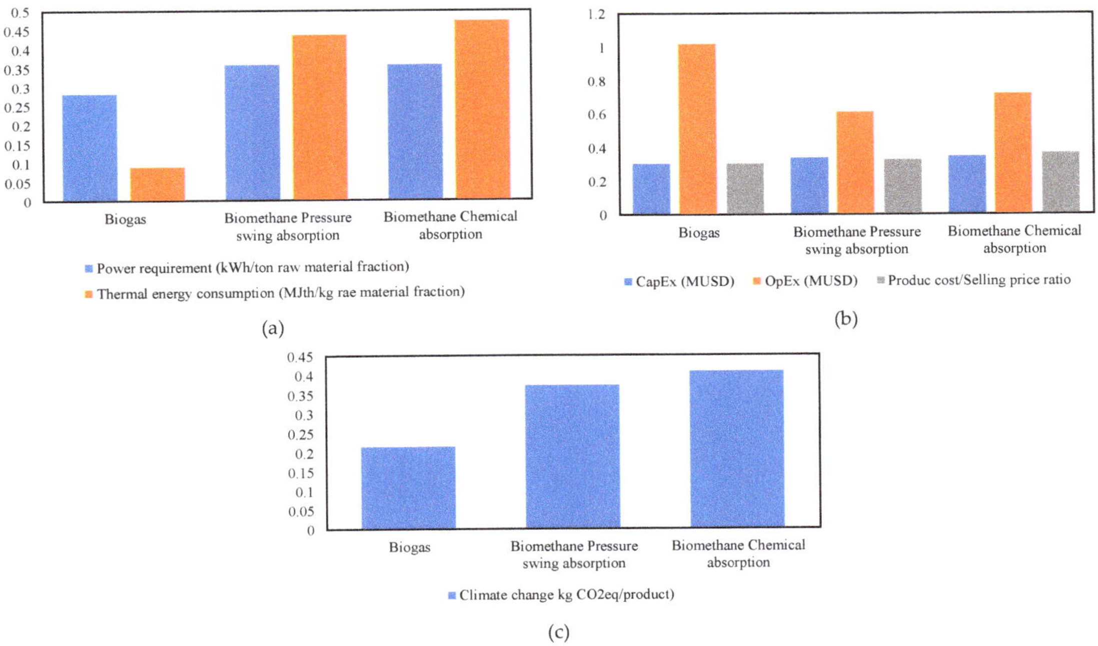

Figure 5. Indicators for the fractions valorization: (**a**) energy indicators; (**b**) economic indicators; and (**c**) environmental indicator [26].

3.1.3. Step 4: Biorefinery Scenarios

The bioprocesses with the best technical, economic, and environmental indicators to be evaluated were ABE, PLA, levulinic acid, stirred solvent extraction, and biogas production. From the conceptual design of the biorefineries, three biorefinery scenarios were generated. The proposed scenarios are presented in Figure 6.

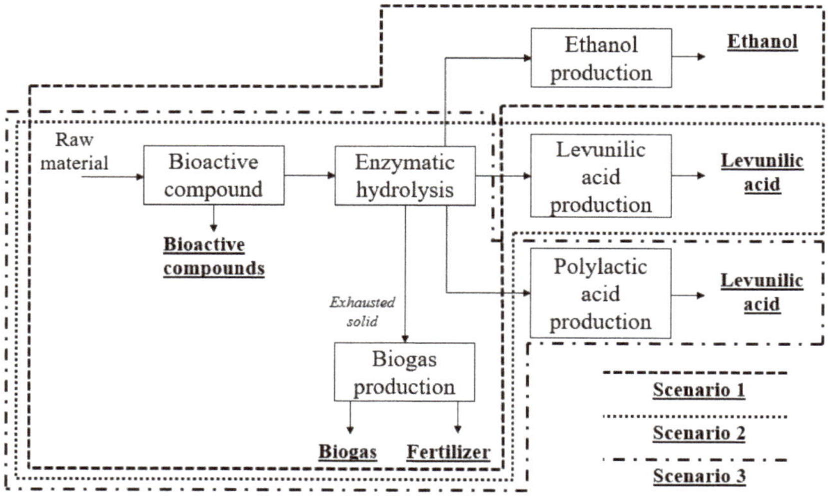

Figure 6. Scenarios for OKFW valorization after incorporating the design strategy for the biorefinery application.

3.1.4. Step 5: Economic Prefeasibility

The levulinic acid production scenario was selected for the economic pre-feasibility analysis. Figure 7 shows the NVP of the levulinic acid production biorefinery for the following three OKFW scales:

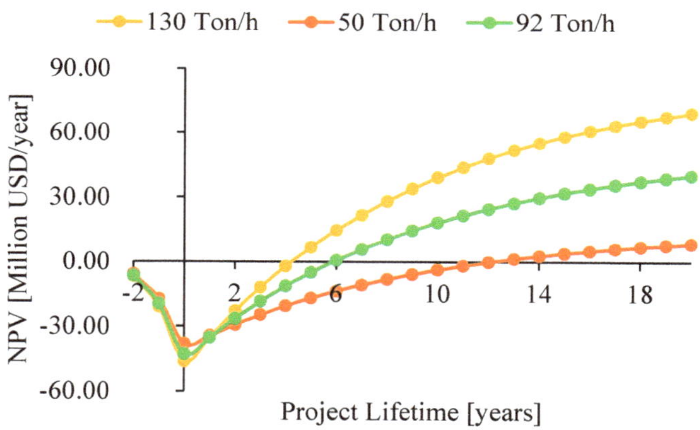

Figure 7. Economic prefeasibility of OKFW valorization.

For all the proposed scales, the biorefinery had viability, and the return periods for investment were located between 4 and 12 years. The CapEx for the biorefinery was between 34 and 48 MUSD. One factor contributing to the biorefinery's economic viability was the high commercial value of levulinic acid compared to other products such as ethanol, butanol, and lactic acid.

3.2. Results for the Other Raw Materials

All the process configurations for the other raw materials are presented in Figure 8. In the case of the small-scale biorefineries (i.e., biorefineries addressed to upgrade FLs), the scenarios introduced low technological complexity processes while the food waste upgrading introduced high technological complexity processes (e.g., levulinic acid). The proposed scenarios for FL and FW valorization are described per zone.

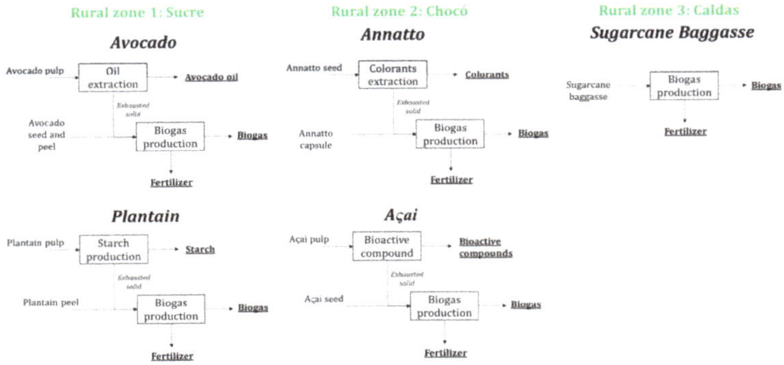

Figure 8. Selected scenarios for food residue biorefineries based on the design strategy.

In Zone 1, rejected avocados are upgraded to produce biogas and avocado oil. The avocado pulp is used to produce avocado oil through cold pressing to guarantee product quality. The process temperature does not exceed 40 °C. The exhausted pulp, peels and seeds are used as raw material to produce biogas. This energy vector is produced by using sludge as inoculum from the local wastewater treatment plant. Rejected plantain is converted into starch and biogas. Plantain pulp is used as a raw material in the starch production line. Additives such as citric acid and sodium hydroxide are used to extract the starch. Biogas is produced using plantain peels as a raw material since this fraction has a considerable biogas production yield. The products obtained by implementing the proposed biorefinery configuration can be commercialized at the local level.

For Zone 2 (Choco), annatto seeds are used as a raw material. First, colorants are extracted by using a green solvent such as ethanol. Afterward, the exhausted solids are used to produce biogas as an energy vector and possible source of electricity. Instead, acai pulp is also used to extract bioactive compounds by using green solvents. The exhausted pulp is co-digested with acai seeds to produce biogas. Finally, in Zone 3 (Samana), sugarcane bagasse is upgraded to produce biogas which can be used to improve the thermal efficiency of the brown-sugar production. Regarding the OKFW, first, the fat content is extracted by using a pressing machine. Then, the solid is used to extract bioactive compounds by using green solvents. Afterward, the extracted solid is subjected to an enzymatic hydrolysis process to produce fermentable sugars as platforms for obtaining added-value products. The liquor of the saccharification process is used to produce levulinic acid when implementing the Biofine process. The exhausted solid after enzymatic hydrolysis is used to produce biogas as thermal energy source and power.

The economic assessment of the proposed biorefineries to upgrade FL and FW is presented in Figure 9. The economic analysis demonstrated the potential of using rejected avocado in Zone 1, acai and annatto as raw materials in Zone 2, and OKFW in Zone 3. Rejected avocado was more feasible than plantain since the avocado oil production process has a lower capital investment than the starch processing line. Moreover, starch has a lower commercial value (USD 0.83 per kg) than avocado oil (USD 8.15 per kg). Then, the economic feasibility of the proposed scenarios was determined by the selected products for upgrading. Annatto and acai are potential raw materials for producing colorants and bioactive compounds. Both products have a high market cost since the food, cosmetic, and pharmaceutical sectors have well-defined uses.

The processing scale was not an issue since a low processing scale has a good economic performance. Sugarcane bagasse used only as a biogas source is not feasible at the economic level since biogas has a low commercial value. Even if the biogas is converted to energy (electricity), the process is unfeasible since Samana has hydro-energy as a renewable energy source. Therefore, biogas production should be considered as a complement for other process.

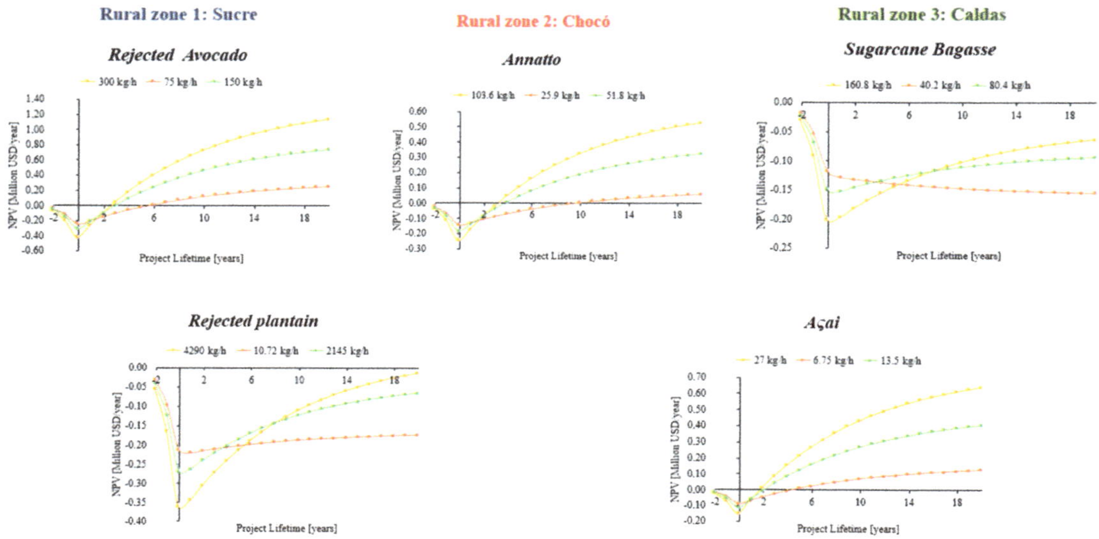

Figure 9. Economic prefeasibility of the food residue biorefineries.

3.3. Sustainability Analysis

The sustainability analysis of the proposed biorefineries for upgrading FLs and FWs must involve the triple bottom line (i.e., economic, environmental, and social benefits). The economic and environmental performances of the biorefineries were ensured by the bioprocess screening conducted in the previous steps. Then, social aspects must be involved to understand the complete impacts of the biorefineries. Indicators related to job creation and access to material resources should be included. Despite the numbers of these indicators, the implementation of new processes addressed to upgrade FLs and FWs can promote the development of more sustainable communities at the local and regional levels. In addition, the sustainability analysis of the proposed biorefineries ensures the possibility of implementing these processes in real life. The results obtained for each zone reflect the potential of the development of bio-based products for boosting rural bioeconomies.

In the rural zones of Sucre, the rejected avocado valorization presented a better economic feasibility than upgrading the rejected plantain. The same behavior was found for the environmental perspective due to the low carbon dioxide emissions of rejected avocado being upgraded to avocado oil and biogas. In the rural zones of Choco, the acai and annatto valorization were feasible from the economic perspective due to the production and commercialization of added-value products (i.e., colorants and bioactive compounds). The acai valorization presented the best economic performance because of the high selling price of its bioactive compounds. From the environmental perspective, both scenarios in Choco presented similar environmental impacts.

Finally, in the rural zones of Caldas, the sugarcane bagasse valorization was not feasible for producing biogas. Moreover, the OKFW had a good economic performance. Nevertheless, the technological context for the biorefinery implementation could not be a rural zone. This scenario was presented as a future alternative to be implemented in more developed regions with better logistics and technological development. This study demonstrated the possible development of a rural bioeconomy under the biorefinery concept.

3.4. Entrepreneurship Alternatives in Rural Zones

The results obtained for the FL and FW valuation scenarios can make it possible to define sustainability before generating ventures. In addition, technical and environmental indicators can be differential factors that promote the positioning of products through seals that denote the extensive use of resources. For example, calculations of air emissions generated determine the carbon footprint of products that can be used on a label to increase marketing potential. The valuation schemes proposed for the FLs and FWs generated in the three areas analyzed in this paper serve as a fundamental basis for the development of enterprises. Furthermore, this perspective defines the viability of the schemes considering limiting factors such as waste generation flows, socio-economic contexts, technological contexts, and bioprocess TRLs. Based on these results, the probability of success in formulating projects and creating ventures can be increased.

The upgrading alternatives for the agricultural products proposed in this research paper can be applied to other agricultural products obtained in other rural regions of Colombia. For instance, new alternatives for valorizing cocoa residues can be proposed based on cocoa's high production rate in South Colombia (e.g., in Nariño and Putumayo). These alternatives can involve biorefineries addressed to produce cellulose fibers, food additives, and bioenergy. On the other hand, the methodology applied for upgrading agricultural products and residues can be extrapolated to other crops such as cassava, corn, palm oil, rice, mango, and coffee. Thus, the methodology and results reported in this research paper can be considered as the basis for boosting new alternatives for sustainably upgrading biomasses.

4. Conclusions

The sustainability analysis of a biorefinery is delimited through the selection of a bioprocess portfolio based on the food residues' chemical compositions (fractions). For the case studies, technical and economic prefeasibility was demonstrated using the bioprocess portfolio. For the OKFW, the filtration processes of the biorefinery strategy resulted in the scenarios with the greatest potential for evaluation being the production of levulinic acid, PLA, and ABE from the cellulose fraction. In addition, the production of biogas was determined to be the best process for the integral use of the raw material considering the exhausted solid generated after the extraction process and enzymatic hydrolysis. On the other hand, the design strategy allowed for the identification of biorefinery schemes for the valorization of the FLs generated in rural areas (i.e., Caldas, Chocó, and Sucre) with commercialization potential at the local level. The foregoing highlights the role of conceptual design in the proposition of ventures in different areas and contexts. Finally, the implantation of the design strategy for food waste biorefineries allowed for elucidating the best scenarios to be implemented as entrepreneurship initiatives in rural zones in Colombia.

Author Contributions: Conceptualization, C.A.C., M.O.-S., N.S., J.C.S.-T., S.D. and C.E.O.; formal analysis, C.A.C., M.O.-S., N.S., J.C.S.-T. and C.E.O.; investigation, M.O.-S., N.S. and J.C.S.-T.; writing—original draft preparation, C.A.C., M.O.-S., N.S., J.C.S.-T. and C.E.O.; writing—review and editing, M.O.-S., N.S. and J.C.S.-T.; supervision, C.A.C.; funding acquisition, C.A.C., C.E.O., A.P., C.D.A., E.L., H.S. and J.G. All authors have read and agreed to the published version of the manuscript.

Funding: This paper is the result of the research developed through the following project: PROGRAMA DE INVESTIGACIÓN RECONSTRUCCIÓN DEL TEJIDO SOCIAL EN ZONAS DE POSCONFLICTO EN COLOMBIA SIGP, coded: 57579, within the research project "COMPETENCIAS EMPRESARIALES Y DE INNOVACIÓN PARA EL DESARROLLO ECONÓMICO Y LA INCLUSIÓN PRODUCTIVA DE LAS REGIONES AFECTADAS POR EL CONFLICTO COLOMBIANO" SIGP, coded: 58907. This research was funded within the framework of Colombia Científica, contract no. FP44842-213-2018.

Institutional Review Board Statement: Not applicable.

Informed Consent Statement: Not applicable.

Data Availability Statement: Not applicable.

Conflicts of Interest: The authors declare no conflict of interest. The funders had no role in the design of the study; in the collection, analyses, or interpretation of data; in the writing of the manuscript; or in the decision to publish the results.

References

1. Withanage, S.V.; Dias, G.M.; Habib, K. Review of Household Food Waste Quantification Methods: Focus on Composition Analysis. *J. Clean. Prod.* **2021**, *279*, 123722. [CrossRef]
2. Pawlak, K.; Kołodziejczak, M. The Role of Agriculture in Ensuring Food Security in Developing Countries: Considerations in the Context of the Problem of Sustainable Food Production. *Sustainability* **2020**, *12*, 5488. [CrossRef]
3. Girotto, F.; Alibardi, L.; Cossu, R. Food Waste Generation and Industrial Uses: A Review. *Waste Manag.* **2015**, *45*, 32–41. [CrossRef] [PubMed]
4. O'Connor, J.; Hoang, S.A.; Bradney, L.; Dutta, S.; Xiong, X.; Tsang, D.C.W.; Ramadass, K.; Vinu, A.; Kirkham, M.B.; Bolan, N.S. A Review on the Valorisation of Food Waste as a Nutrient Source and Soil Amendment. *Environ. Pollut.* **2021**, *272*, 115985. [CrossRef]
5. Jia, L.; Zhang, J.; Qiao, G. Scale and Environmental Impacts of Food Loss and Waste in China—A Material Flow Analysis. *Int. J. Environ. Res. Public Health* **2022**, *20*, 460. [CrossRef]
6. Hoehn, D.; Vázquez-Rowe, I.; Kahhat, R.; Margallo, M.; Laso, J.; Fernández-Ríos, A.; Ruiz-Salmón, I.; Aldaco, R. A Critical Review on Food Loss and Waste Quantification Approaches: Is There a Need to Develop Alternatives beyond the Currently Widespread Pathways? *Resour. Conserv. Recycl.* **2023**, *188*, 106671. [CrossRef]
7. Kavitha, S.; Kannah, R.Y.; Kumar, G.; Gunasekaran, M.; Banu, J.R. Introduction: Sources and Characterization of Food Waste and Food Industry Wastes. *Food Waste Valuab. Resour. Appl. Manag.* **2020**, 1–13. [CrossRef]
8. Cristóbal, J.; Caldeira, C.; Corrado, S.; Sala, S. Techno-Economic and Profitability Analysis of Food Waste Biorefineries at European Level. *Bioresour. Technol.* **2018**, *259*, 244–252. [CrossRef]
9. Slopiecka, K.; Liberti, F.; Massoli, S.; Bartocci, P.; Fantozzi, F. Chemical and Physical Characterization of Food Waste to Improve Its Use in Anaerobic Digestion Plants. *Energy Nexus* **2022**, *5*, 100049. [CrossRef]
10. Tiwari, A.; Khawas, R.; Tiwari, A.; Khawas, R. Food Waste and Agro By-Products: A Step towards Food Sustainability. In *Innovation in the Food Sector through the Valorization of Food and Agro-Food By-Products*; IntechOpen: London, UK, 2021. [CrossRef]
11. Palmeros Parada, M.; Osseweijer, P.; Posada Duque, J.A. Sustainable Biorefineries, an Analysis of Practices for Incorporating Sustainability in Biorefinery Design. *Ind. Crops Prod.* **2017**, *106*, 105–123. [CrossRef]
12. Caldeira, C.; Vlysidis, A.; Fiore, G.; De Laurentiis, V.; Vignali, G.; Sala, S. Sustainability of Food Waste Biorefinery: A Review on Valorisation Pathways, Techno-Economic Constraints, and Environmental Assessment. *Bioresour. Technol.* **2020**, *312*, 123575. [CrossRef]
13. D'Amato, A.; Lerma-García, M.J.; Fasoli, E.; Simó-Alfonso, E.F.; Righetti, P.G. Orange Proteomic Fingerprinting: From Fruit to Commercial Juices. *Food Chem.* **2015**, *196*, 739–749. [CrossRef]
14. Granato, D.; Carocho, M.; Barros, L.; Zabetakis, I.; Mocan, A.; Tsoupras, A.; Cruz, A.G.; Pimentel, T.C. Implementation of Sustainable Development Goals in the Dairy Sector: Perspectives on the Use of Agro-Industrial Side-Streams to Design Functional Foods. *Trends Food Sci. Technol.* **2022**, *124*, 128–139. [CrossRef]
15. Gubeladze, D.; Pavliashvili, S. Linear Economy and Circular Economy. Current State Assessment and Future Vision. *Int. J. Innov. Technol. Econ.* **2020**, *32*, 1–6. [CrossRef]
16. Cardona, C.; Moncada, J.; Aristizabal, V. Design Strategies for Sustainable Biorefineries. *Biochem. Eng. J.* **2016**, *116*, 122–134. [CrossRef]
17. Chaturvedi, T.; Torres, A.I.; Stephanopoulos, G.; Thomsen, M.H.; Schmidt, J.E. Developing Process Designs for Biorefineries—Definitions, Categories, and Unit Operations. *Energies* **2020**, *13*, 1493. [CrossRef]
18. Cardona-Alzate, C.A.; Ortiz-Sanchez, M.; Solarte-Toro, J.-C. Design Strategy of Food Residues Biorefineries Based on Multifeed-stocks Analysis for Increasing Sustainability of Value Chains. *Biochem. Eng. J.* **2023**, *194*, 108857. [CrossRef]
19. Barragán-Ocaña, A.; Merritt, H.; Sánchez-Estrada, O.E.; Méndez-Becerril, J.L.; del Pilar Longar-Blanco, M. Biorefinery and Sustainability for the Production of Biofuels and Value-Added Products: A Trends Analysis Based on Network and Patent Analysis. *PLoS ONE* **2023**, *18*, e0279659. [CrossRef]
20. Zetterholm, J.; Bryngemark, E.; Ahlström, J.; Söderholm, P.; Harvey, S.; Wetterlund, E. Economic Evaluation of Large-Scale Biorefinery Deployment: A Framework Integrating Dynamic Biomass Market and Techno-Economic Models. *Sustainability* **2020**, *12*, 7126. [CrossRef]
21. Sur, S.; Dave, V.; Prakesh, A.; Sharma, P. Expansion and Scale up of Technology for Ethanol Production Based on the Concept of Biorefinery. *J. Food Process. Eng.* **2021**, *44*, e13582. [CrossRef]
22. Orejuela-Escobar, L.M.; Landázuri, A.C.; Goodell, B. Second Generation Biorefining in Ecuador: Circular Bioeconomy, Zero Waste Technology, Environment and Sustainable Development: The Nexus. *J. Bioresour. Bioprod.* **2021**, *6*, 83–107. [CrossRef]
23. Lopez Barrera, E.; Hertel, T. Global Food Waste across the Income Spectrum: Implications for Food Prices, Production and Resource Use. *Food Policy* **2021**, *98*, 101874. [CrossRef]

24. Solarte-Toro, J.C.; Cardona Alzate, C.A. Perspectives of the Sustainability Assessment of Biorefineries. *Chem. Eng. Trans.* **2021**, *83*, 307–312. [CrossRef]
25. Serna-Loaiza, S.; Carmona-Garcia, E.; Cardona, C.A. Potential Raw Materials for Biorefineries to Ensure Food Security: The Cocoyam Case. *Ind. Crops Prod.* **2018**, *126*, 92–102. [CrossRef]
26. Ortiz-Sanchez, M.; Cardona Alzate, C.A. Analysis of the Routes for Biomass Processing towards Sustainable Development in the Conceptual Design Step: Strategy Based on the Compendium of Bioprocesses Portfolio. *Bioresour. Technol.* **2022**, *350*, 126852. [CrossRef]
27. Alonso-Gómez, L.A.; Solarte-Toro, J.C.; Bello-Pérez, L.A.; Cardona-Alzate, C.A. Performance Evaluation and Economic Analysis of the Bioethanol and Flour Production Using Rejected Unripe Plantain Fruits (*Musa paradisiaca* L.) as Raw Material. *Food Bioprod. Process.* **2020**, *121*, 29–42. [CrossRef]
28. Solarte-Toro, J.C.; Ortiz-Sanchez, M.; Cardona Alzate, C.A. Environmental Life Cycle Assessment (E-LCA) and Social Impact Assessment (SIA) of Small-Scale Biorefineries Implemented in Rural Zones: The Avocado (*Persea americana var. americana*) Case in Colombia. *Environ. Sci. Pollut. Res.* **2022**, *30*, 8790–8808. [CrossRef]
29. Poveda-Giraldo, J.A.; Piedrahita-Rodríguez, S.; Salgado-Aristizábal, N.; Cardona-Alzate, C.A. Prefeasibility Analysis of Low-Scale Biorefineries: Annatto and Açai Case. *Res. Sq.* 2022; *preprint*. [CrossRef]
30. Quintero, J.A.; Moncada, J.; Cardona, C.A. Techno-Economic Analysis of Bioethanol Production from Lignocellulosic Residues in Colombia: A Process Simulation Approach. *Bioresour. Technol.* **2013**, *139*, 300–307. [CrossRef]
31. Towler, G.P.; Sinnott, R.K. *Chemical Engineering Design. Principles, Ptractice and Economics of Plant and Process Design*, 1st ed.; Elsevier: Amsterdam, The Netherlands, 2013.

fermentation

Article

The Effect of pH on the Production and Composition of Short- and Medium-Chain Fatty Acids from Food Waste in a Leachate Bed Reactor at Room Temperature

Pooja Radadiya [1,†], Ashika Latika [2,†], Xunchang Fei [2,*], Jangho Lee [3], Saurabh Mishra [4] and Abid Hussain [1,*]

1 Department of Civil and Environmental Engineering, Carleton University, Ottawa, ON K1S 5B6, Canada
2 Department of Civil and Environmental Engineering, Nanyang Technological University, Singapore 639798, Singapore
3 Department of Civil and Environmental Engineering, Cornell University, Ithaca, NY 14853, USA
4 College of Civil and Transportation Engineering, Hohai University, Nanjing 210024, China
* Correspondence: xcfei@ntu.edu.sg (X.F.); abid.hussain@carleton.ca (A.H.)
† These authors contributed equally to this work.

Abstract: This study evaluated the hydrolysis and acidogenesis of food waste at different operating pHs (uncontrolled, 5.5, 6.5, 7.5, 8.5) in a leachate bed reactor (LBR) at room temperature. LBR operation at pH 6.5–8.5 resulted in a hydrolysis yield of 718–729 g SCOD/kg VS_{added}, which was statistically ($p \leq 0.05$) higher than that obtained at pH 5.5 (577 g SCOD/kg VS_{added}) and the uncontrolled pH (462 g SCOD/kg VS_{added}). The hydrolysis rate at pH 6.5 was the highest amongst all the pH values. Stabilization at pH at 6.5 also resulted in a high fatty acid (FA) yield of 643 g COD_{FA}/kg VS_{added}. Butyrate was the main FA at the pH of 5.5–6.5, while acetate was the main FA at the pH of 7.5–8.5. At the uncontrolled pH, lactate production was the highest, indicating a shift in the microbial community from fatty-acid-producing bacteria to lactate-producing bacteria. The compositions of medium-chain fatty acids, such as caproate, were the highest at pH of 5.5.

Keywords: food waste fermentation; leachate bed reactor; pH; short-chain fatty acids; medium-chain fatty acids

Citation: Radadiya, P.; Latika, A.; Fei, X.; Lee, J.; Mishra, S.; Hussain, A. The Effect of pH on the Production and Composition of Short- and Medium-Chain Fatty Acids from Food Waste in a Leachate Bed Reactor at Room Temperature. *Fermentation* **2023**, *9*, 518. https://doi.org/10.3390/fermentation9060518

Academic Editors: Jose Luis García-Morales and Francisco Jesús Fernández Morales

Received: 16 April 2023
Revised: 24 May 2023
Accepted: 25 May 2023
Published: 27 May 2023

1. Introduction

Over 2 billion tons of municipal solid waste (MSW) was generated globally in 2016. At the current rate of generation, MSW is anticipated to further increase by 69% to reach 3.40 billion tons by 2050 [1,2]. A large portion of MSW is food waste, constituting up to 45% of MSW [1,2]. The main components of food waste are fruits and vegetables, which are disposed in large quantities by local markets and grocery stores. The bulk of the food waste in many developing and developed countries is disposed of through landfilling, resulting in adverse health and environmental effects including greenhouse gas (GHG) emissions, the contamination of subsurface environments and loss of habitats [3,4]. Therefore, sustainable approaches for managing food waste are being intensively researched. The conversion of food waste to fatty acids via the acidogenic fermentation process is an emerging biotechnology that combines the sustainable management of food waste with resource recovery.

Acidogenic fermentation is carried out by a consortium of bacteria (mixed microbial culture) under anaerobic conditions to produce different fatty acids (FAs) from heterogenous waste such as food waste through multi-step concurrent biochemical reactions. Fatty acids with 2–5 carbon atoms (e.g., acetate, propionate, butyrate, valerate) are categorized as short-chain fatty acids (SCFAs), and those with 6–8 carbon atoms (e.g., caproate, hexanoate, etc.) are called medium-chain fatty acids (MCFAs) [5–7]. FAs (SCFAs and MCFAs) are industrially important chemicals that are currently derived from petrochemicals causing substantial GHG emissions [1,8–10]. These are widely used in pharmaceutical, chemical, food processing, cosmetic, textile, paint and other industries. Additionally, these FAs can be used as a substrate in the microbial production of bioplastics and biofuels [11–15]. Comparatively, MCFAs have higher economical value than SCFAs because of their higher carbon to oxygen ratio (C:O) and energy potential [16].

Dry fermenters such as leachate bed reactors (LBRs) are being widely studied as an energy-efficient and cost-effective bioreactor platform for the production of FAs. Unlike the commonly used continuously stirred tank reactors (CSTRs), LBRs can handle higher solid contents (30–40% of total solid). Consequently, no dilution of food waste is required (no process water). Furthermore, no mechanical stirring is required in LBRs, which significantly reduces energy consumption. An added advantage of the LBR design is the separation of the degraded food waste from FAs containing broth, which eliminates or reduces the downstream costs associated with solid–liquid separation [17,18].

The operating parameters of LBRs significantly impact FA production (yield) and the percentage composition of individual FAs (e.g., the composition of acetate, butyrate and propionate) during acidogenic fermentation. Amongst the different operating parameters, pH is one of the most crucial parameters affecting FA production and composition, since pH has an impact on the microbial community's composition and metabolic activity of the bacteria [12,18,19]. Many studies have investigated the impacts of pH on the production and composition of FAs from food waste in LBRs [18,20–24]. However, FA compositional analysis in these studies has been limited to primarily three SCFAs, namely, acetate, propionate and butyrate. The production of other SCFAs, such as iso-butyrate, n-valerate and iso-valerate, as well as MCFAs (i.e., n-caproate, iso-caproate, heptanoate), has not been extensively analyzed. In addition to limited compositional analysis, most of the studies on LBRs have tested food waste fermentation at temperatures above 35 °C. Maintaining LBRs at such temperatures (>35 °C) requires external heating, which can significantly impact the net energy gain as well as the reduction in GHG emissions. Therefore, the characterization of SCFAs and MCFAs under room conditions (i.e., without external heating) is of interest. However, there is limited information in the literature in this regard. This study was performed to fill these research gaps.

This study evaluated the impacts of five different pHs (uncontrolled, 5.5, 6.5, 7.5 and 8.5) on the full range of SCFAs (acetate, propionate, butyrate, iso-butyrate, n-valerate, iso-valerate) and MCFAs (n-caproate, iso-caproate, heptanoate) produced from food waste in an LBR at the room temperature of 22 °C. Firstly, the hydrolysis yields and rates were compared at different pHs. Secondly, the production and composition of SCFAs and MCFAs were analyzed to elucidate the impacts of pH on the range of acidogenic products obtained from food waste, and finally, the microbial community composition was analyzed to elucidate the hydrolytic and fermentative bacteria at different pHs.

2. Material and Methods

2.1. Characteristics of Food Waste and Inoculum

Simulated food waste was used in this study because of the closure of restaurants, cafeterias and commercial centers due to the COVID-19 pandemic. The simulated food waste consisted of (weight basis) 13% apples, 7% bananas, 17% capsicums, 14% tomatoes, 26% potatoes and 22% guavas with a total solids (TS) content of 13.23 ± 0.20% and volatile solids (VS) content of 9.86 ± 0.25%. To prevent degradation, each component of the food waste was stored at −10 °C until the time of use for the experiments. The required quantity of an individual component of food waste was defrosted at room temperature (22 °C) for two hours before use for the experiments. The food waste was then immediately shredded to an average particle size of 5–10 mm.

Anaerobic digestion sludge (AD-sludge) was used as the inoculum in this study. The AD sludge was collected from an anaerobic wastewater treatment plant (Ulu Pandan Water Reclamation Plant, Singapore). The AD sludge was filtered to remove solid particles with a size of over 10 mm so as to prevent the clogging of the LBR and then stored at 4 °C until use.

To kill the methanogens, the AD sludge was heated at 75 °C for 15 min before addition to the LBR. The AD sludge contained a TS content of 1.68 ± 0.04% and VS content of 1.08 ± 0.02%.

2.2. LBR Design

A cylindrically shaped LBR was used in this study, which was fabricated using acrylic materials. The LBR was cylindrical with a height of 620 mm and a diameter of 130 mm; thus, the total volume of the LBR was 8 L (volume of a cylinder). The LBR comprised three sections (Figure 1A): (1) a top section with a headspace of 2.5 L, (2) a middle section with a food-waste-holding basket of 1.5 L and (3) a bottom section with leachate-holding bed of 4 L. The headspace was equipped with a detachable cover with a customized sprinkler nozzle and a gas collection port. The gas produced during acidogenic fermentation was collected in the gas collection bag (Tedlar Multilayer Gas Sampling Bags, 10 L) connected to the gas collection port. The food-waste-holding basket was made of stainless steel with a height of 185 mm and a diameter of 100 mm. The side wall and base of the food-waste-holding basket were perforated with a pore size of 4 mm to enable the percolation of the leachate into the leachate-holding bed while preventing food waste particles from entering the leachate-holding bed (Figure 1B). The leachate-holding bed had side sampling ports to collect the leachate samples for different analyses. The pH probe was installed on the retaining wall of the leachate-holding bed to monitor the pH using a pH controller (MODEL MC122, Milwaukee, WI, USA). The pH controller was connected to a dosing pump to adjust the pH of the leachate to the desired level by injecting 1 M NaOH. The leachate in the leachate-holding bed was gently mixed using a peristaltic pump (Masterflex Standard Digital Drive, Model 77200-62, Cole Parmer, Vernon Hills, IL, USA). Another peristaltic pump (Masterflex Digital Economy Drive, Model 77800-62, Cole Parmer, Vernon Hills, IL, USA) was used to recirculate and spray the leachate from the bottom of the leachate-holding bed to the food-waste-holding basket. The recirculation and mixing of the leachate with the peristaltic pumps were controlled using a timer (33 Multifunction timer relay, RS pro, Singapore). To ensure anaerobic conditions inside the LBR, the joints of the LBR had O-rings and rubber gaskets.

2.3. LBR Experimental Procedure

All LBR experiments were performed using the same procedure unless otherwise specified. The LBR was operated in batch mode at room temperature (22 °C) for a fixed period of 14 days. For each run, 1.5 kg of food waste was loaded into the food-waste-holding basket along with 0.6 L of heat-treated AD sludge in the leachate-holding bed, with an inoculum to substrate ratio (ISR) of 4%. This loading of food waste (VS of 98.63 g/kg) and AD sludge (VS of 10.81 g/kg) provided a volumetric organic loading of 19 g $VS/L_{reactor}$ for each LBR run. Nitrogen gas was purged from the LBR to ensure anaerobic conditions inside the LBR. The impacts of pH on hydrolysis and acidification (SCFA and MCFA production) were evaluated by operating the LBR at different pHs: uncontrolled pH (designated as LBR-UC), 5.5 ± 0.5 (designated as LBR-5.5), 6.5 ± 0.5 (designated as LBR-6.5), 7.5 ± 0.5 (designated as LBR-7.5) and 8.5 ± 0.5 (designated as LBR-8.5). The pH for LBR-UC was measured to be in the range of 3.5–4. During the experiments, FA (SCFA and MCFA) generation causes the leachate to be acidic, which affects microbial activity. Therefore, it is crucial to use alkaline solution to maintain the pH. To maintain the pH at the required level in the LBR, 1 M solution of caustic soda (NaOH) was used during the experiments. The leachate was recirculated from the leachate-holding bed to the food-waste-holding basket at a leachate recirculation rate of 3 L/h in all the LBRs.

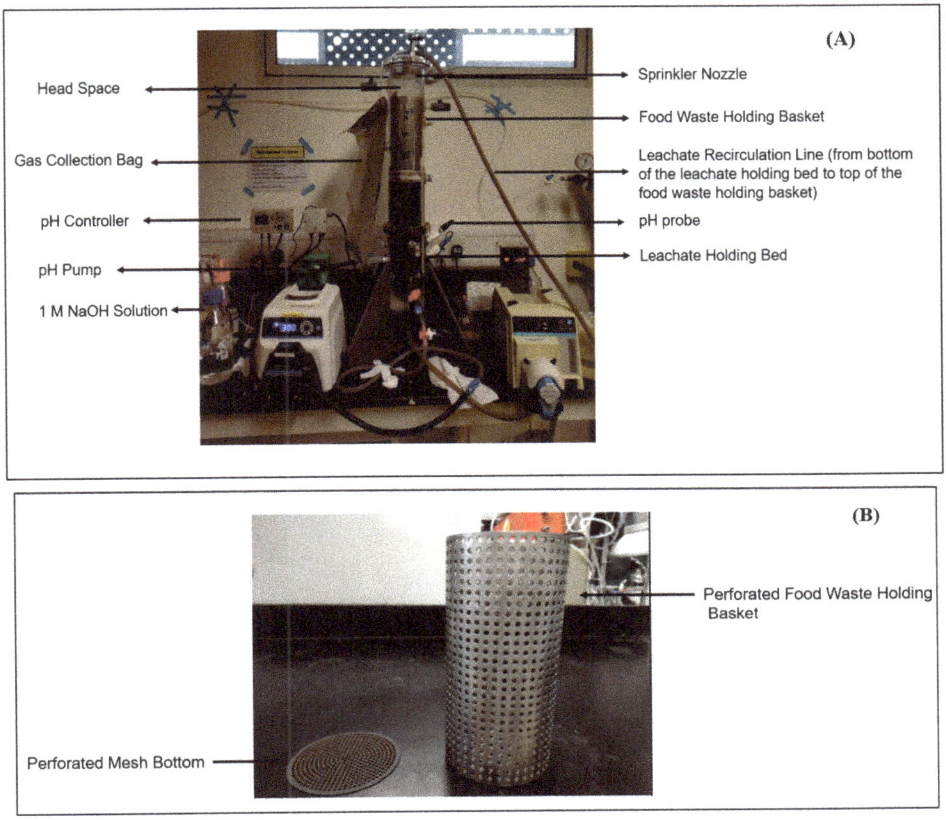

Figure 1. Image of (**A**) LBR and (**B**) Food-Waste-Holding Basket.

2.4. Sampling and Analytical Methods

About 30 mL of leachate was sampled every second day for all the LBRs to analyze the soluble chemical oxygen demand (SCOD) and VFAs. TS and VS were analyzed at the beginning and the end of each experimental run. For the SCOD and VFA analyses, the leachate sample was centrifuged at 6000 rpm for 10 min to obtain the supernatant, and then the supernatant was filtered with a syringe filter with a 0.45 μm pore size filter membrane. Subsequently, the filtered sample was used to analyze the SCOD using a COD reagent tube (Hatch, Los Angeles, CA, USA). Fatty acids (FAs) in the leachate were analyzed by injecting the filtered sample into a gas chromatograph (GC 7890A, Agilent, Santa Clara, CA, USA) equipped with a flame ionization detector (FID) and a DB-FFAP fused-silica capillary column. The injector and detector were both set to a temperature of 260 °C. The column temperature was initially adjusted to 80 °C for 1 min and then raised to 120 °C at a rate of 20 °C/min, and after that, it was increased to 205 °C at a rate of 10 °C/min and maintained at this temperature for 2 min.

In this study, the total FA (TFA) content was estimated as the sum of SCFAs (acetate, propionate, n-butyrate, iso-butyrate, iso-valerate and n-valerate) and MCFAs (iso-caproate, n-caproate and heptanoate). The lactate in the leachate was quantified with a high-performance liquid chromatograph (HPLC, CA) equipped with a refractive index detector (RID) and an ion exclusion column (300 × 7.8 mm diameter, 9 μm particle size, Aminex HPX-87H, Biorad, CA, USA). The temperature of the column was maintained at 65 °C with a sample injection flow rate of 0.6 mL/min. The concentration of total FAs and lactate was expressed as a chemical oxygen demand (COD) equivalent using half reactions for FAs and lactate with O_2 [25]. All analyses were performed in triplicate unless otherwise specified.

2.5. Calculations

The impacts of pH on food waste degradation and acidification were evaluated based on the hydrolysis yield, acidification yield, FA yield, lactate yield and the ratios of FAs and lactate to SCOD [3,13,17,18,26]. The hydrolysis yield was calculated as the ratio of the mass of cumulative SCOD produced in the leachate to the initial amount vs. that added to the LBR (Equation (1)) [3]:

$$\text{Hydrolysis yield } (\text{g SCOD}/\text{kg VS}_{\text{added}}) = \frac{\text{cumulative SCOD produced (g SCOD)}}{\text{VS}_{\text{add edinitially}}(\text{kg})} \tag{1}$$

where

Cumulative SCOD produced (g SCOD) = Final SCOD of leachate(g SCOD) − Initial SCOD of the inoculum (g SCOD).

$VS_{\text{added initially}}$ (kg) = VS of food waste (kg) + vs. of inoculum (kg).

The FA yield was computed as the cumulative TFA (sum of SCFAs and MCFAs) produced to the initial, amount vs. that added to the LBR (Equation (2)) [17]:

$$\text{TFA yield } (\text{g COD}_{\text{FA}}/\text{kg VS}_{\text{added}}) = \frac{\text{cumulative TFA produced (g COD}_{\text{FA}})}{\text{VS}_{\text{added initially}}(\text{kg})} \tag{2}$$

where:

Cumulative TFA produced (g COD_{FA}) = Final total TFA of leachate (g COD_{FA}) − Initial total TFA of inoculum (g COD_{FA}).

$VS_{\text{added initially}}$ (kg) = VSof food waste (kg) + VSof inoculum (kg).

The TFA/SCOD ratio (%) was calculated as the ratio of the TFA yield to the hydrolysis yield.

The lactate yield was calculated based on the ratio of cumulative lactate produced (g $COD_{Lactate}$) to the initial amount vs. that added to the LBR (Equation (3)) [17]. The lactate/SCOD ratio (%) was calculated as the ratio of the lactate yield to the hydrolysis yield:

$$\text{Lactate yield } (\text{g COD}_{\text{Lactate}}/\text{kg VS}_{\text{added}}) = \frac{\text{cumulative lactate produced (g COD}_{\text{Lactate}})}{\text{VS}_{\text{added initially}}(\text{kg})} \tag{3}$$

where:

Cumulative lactate produced (g $COD_{Lactate}$) = Final total lacate of leachate (g $COD_{Lactate}$) − Initial total lactate of inoculum (g $COD_{Lactate}$).

$VS_{\text{added initially}}$ (kg) = VSof food waste (kg) + VSof inoculum (kg).

The acidification yield was calculated as the sum of the TFA yield and lactate yield (Equation (4)) [17]:

$$\text{Acidification yield } (\text{g COD}/\text{kg VS}_{\text{added}}) = \frac{\text{cumulative TFA produced (g COD}_{\text{FA}}) + \text{cumulative lactate produced (g COD}_{\text{Lactate}})}{\text{VS}_{\text{added initially}}(\text{kg})} \tag{4}$$

where:

Cumulative TFA produced (gCOD_{VFA}) = Final total TFA of leachate (g COD_{VFA}) − Initial total TFA of inoculum (g COD_{VFA}).

Cumulative lactate produced (g$COD_{Lactate}$) = Final total lacate of leachate (g $COD_{Lactate}$) 25 May 2023 Initial total lactate of inoculum (g $COD_{Lactate}$).

$VS_{\text{added initially}}$ (kg) = VSof food waste (kg) + VSof inoculum (kg).

2.6. Microbial Community and Statistical Analysis

The residual food waste (in the food waste basket) and centrifuged biomass from the leachate in different LBRs were collected at the end of the batch cycle for microbial community analysis. As described by Xiong [18], genomic DNA (gDNA) was extracted from the food waste and the biomass samples using the Sox DNA Isolation Kit (Genewiz, Singapore) according to the protocol provided by the supplier. Polymerase chain reaction (PCR) for 16S rRNA genes was performed for each sample of gDNA (25 μL each) in triplicate, containing 0.5 μL of 10 mM dNTP, 2.5 μL of PCR buffer, 5.0 μL of 1 μM forward primer, 5.0 μL of 1 μM reverse primer, 0.25 μL of BSA (20 mg/mL), 5.0 μL DNA, 0.2 μL of Taq DNA polymerase (5u/μL) and 6.55 μL of PCR water. The forward and reverse primers were used to target 16S rRNA genes in both bacteria and archaea: Pro341F: CCTACGGGNBGCASCAG, Pro805R: GACTACNVGGGTATCTAATCC [27]. The PCR cycle included: (1) initial DNA denaturation at 95 °C for 5 min, (2) 35 cycles of DNA denaturation at 95 °C for 30 s, primer annealing at 30 °C for 30 s, primer extension at 72 °C for 50 s and then (3) a final extension at 72 °C for 10 min [27].

An equal amount of PCR amplicons were pooled and quantified using the NanoDrop 1000 (Thermo Fisher Scientific Inc.). The DNA sequences were produced in FASTQ files with a MiSeq Reagent Kit v2 (2 × 250 cycles) using an Illumina MiSeq sequencer (Illumina Inc, San Diego, CA, USA). The demultiplexing sequences, including the truncation of forward and reverse reads to 245 nucleotides, primer removal and the merging of paired reads, were processed using the DADA2 v1.6 tool [28] in QIIME 2 v.2018.2 [29].

After the chimera-containing sequences' removal, clustering was performed at 97% identity, and then taxonomy was assigned to representative sequences from each cluster using a naive Bayesian classifier implemented in QIIME 2 based on SILVA release 132.

Single-factor analysis of variance (ANOVA) analysis was used to verify the impacts of different pHs on food waste degradation and acidification ($p \leq 0.05$) using Microsoft Excel software version 2019.

3. Results and Discussion

3.1. Hydrolysis of Food Waste at Different pHs

The hydrolysis of food waste was assessed based on the cumulative SCOD production. Figure 2 illustrates the impact of pH on cumulative SCOD (g SCOD) production in the LBR throughout the fermentation time of 14 days. The cumulative SCOD (g SCOD) production differed depending on the pH range. A nearly neutral to alkaline pH (i.e., 6.5–8.5 pH) resulted in statistically higher cumulative SCOD production than acidic pH ranges (uncontrolled pH–5.5). The highest cumulative SCOD production of 112.5 ± 2.6 g SCOD was obtained in LBR-7.5, followed by 111.7 ± 4.1 g SCOD in LBR-8.5, 110.8 ± 1.7 g SCOD in LBR-6.5, 89.0 ± 3.4 g SCOD in LBR-5.5 and 71.34 ± 0.9 g SCOD in LBR-UC. Notably, no statistical difference ($p \geq 0.05$) was found for the cumulative SCOD (g SCOD) production from pH 6.5 to 8.5 (i.e., LBR-6.5, LBR-7.5 and LBR-8.5) after 14 days of fermentation time; however, these values were statistically ($p \leq 0.05$) higher than the cumulative SCOD (g SCOD) obtained at pH 5.5 (LBR-5.5) and the uncontrolled pH (LBR-UC). A similar trend was also observed for the hydrolysis yields.

Table 1 summarizes the hydrolysis yields obtained in the LBRs at different pHs on day 14. A hydrolysis yield of 718–729 g SCOD/kg VS_{added} was achieved in a pH range of 6.5–8.5 (LBR 6.5, LBR7.5 and LBR-8.5), which was 21–58% higher than those obtained at pH 5.5 (LBR-5.5; 577 g SCOD/kg VS_{added}) and the uncontrolled pH (LBR-UC; 462 g SCOD/kg VS_{added}). This result indicated that nearly neutral to alkaline pH ranges (pH 6.5–8.5) enhance the hydrolysis of food waste in LBRs. It has been reported that the hydrolysis of food waste in an LBR was improved when the pH was increased from an acidic pH (pH < 5.5) to a nearly neutral to alkaline pH [17,18,30,31]. Hussain [17] reported a 0.18–1.3-foldincrease in the hydrolysis yield obtained by increasing the pH from 4–5 (227–405 g SCOD/kg VS_{added}) to 6–7 (478–530 g SCOD/kg VS_{added}) during food waste fermentation in an LBR. Similarly, in another study treating food waste in an LBR, a 73%

higher hydrolysis yield was obtained at a pH of 6 (505 g SCOD/kg VS_{added}) compared to that obtained in uncontrolled pH conditions (292 g SCOD/kg VS_{added}) [31]. This enhanced hydrolysis of food waste at a nearly neutral to alkaline pH (i.e., 6.5–8.5 pH) can be attributed to better hydrolytic activity of the bacteria in these pH ranges, which results in the improved solubilization of particulate organic matter in the food waste [17,18].

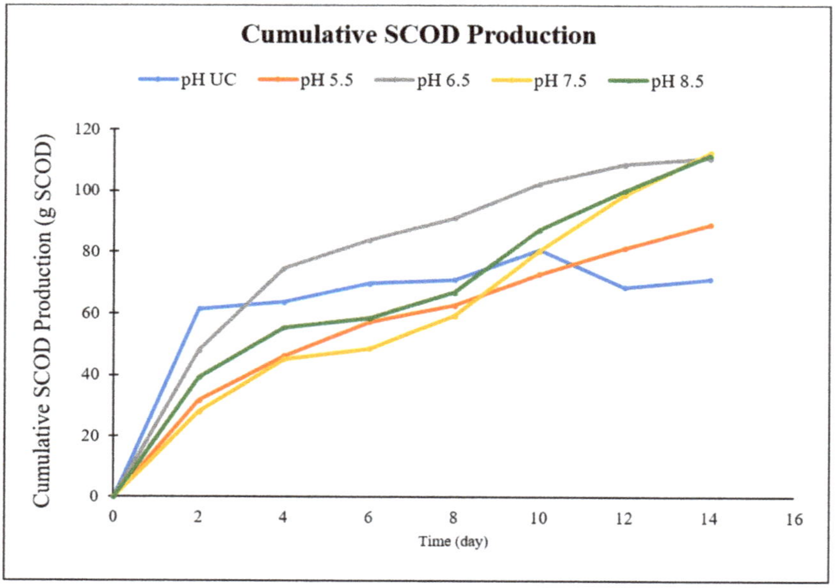

Figure 2. Cumulative SCOD production in LBRs at different operating pHs at room temperature (22 °C).

Table 1. Performance of LBRs at different operating pHs in this study.

Parameters	LBR-UC	LBR-5.5	LBR-6.5	LBR-7.5	LBR-8.5
Cumulative SCOD production (g SCOD)	71.3 ± 0.1	89.0 ± 3.4	110.8 ± 1.7	112.5 ± 2.6	111.7 ± 4.1
Hydrolysis yield (g SCOD/kg VS_{added})	462 ± 6.0	577 ± 22.1	718 ± 11.2	729 ± 17.1	724 ± 26.6
Acetate (g COD_{FA}/L)	3.85 ± 0.0	4.5 ± 0.42	6.15 ± 0.52	7.43 ± 0.3	10.7 ± 0.75
Propionate (g COD_{FA}/L)	0.57 ± 0.05	1.76 ± 0.08	2.57 ± 0.22	2.6 ± 0.2	4.1 ± 0.2
iso-Butyrate (g COD_{FA}/L)	-	0.08 ± 0.0	0.09 ± 0.00	0.24 ± 0.01	0.23 ± 0.05
n-Butyrate (g COD_{FA}/L)	5.72 ± 0.0	9.72 ± 0.13	16.3 ± 1.4	5.9 ± 0.3	3.63 ± 0.1
iso-Valerate (g COD_{FA}/L)	-	0.13 ±	0.15 ± 0.01	0.32 ± 0.02	0.31 ± 0.00
n-Valerate (g COD_{FA}/L)	-	0.76 ± 0.00	1.2 ± 0.09	1.3 ± 0.0	0.32 ± 0.01
iso-Caproate (g COD_{FA}/L)	-	0.09 ± 0.01	0.08 ± 0.00	0.1 ± 0.0	0.09 ± 0.0
n-Caproate (g COD_{FA}/L)	-	2.54 ± 0.01	2.83 ± 0.23	0.24 ± 0.1	0.11 ± 0.0
Heptanoate (g COD_{FA}/L)	-	0.29 ± 0.00	0	0.28 ± 0.0	0
TVFA production (C2-C7) (g COD_{FA})	25.5 ± 0.2	71.4 ± 0.6	99.2 ± 3.0	67.4 ± 3.4	65.3 ± 4.2
TVFA yield (g COD_{FA}/kg VS_{added})	165 ± 1.1	463 ± 4.3	643 ± 19.2	437 ± 22.1	423 ± 27.4

Table 1. *Cont.*

Parameters	LBR-UC	LBR-5.5	LBR-6.5	LBR-7.5	LBR-8.5
TVFA (C2-C7):SCOD (%)	36	80	90	60	58
Lactate (g $COD_{Lactate}$)	39.3 ± 0.3	-	-	-	-
Lactate yield (g $COD_{Lactate}$/kg VS_{added})	255 ± 2.1	-	-	-	-
Lactate: SCOD (%)	55	-	-	-	-
Acidification yield (g COD/kg VS_{added})	420 ± 5.8	463 ± 4.8	643 ± 19.2	437 ± 22.1	423 ± 27.4
Acidification (%)	91	80	90	60	58

Interestingly, the hydrolysis of the food waste was faster at the pH of 6.5 (LBR-6.5) than in the other pH conditions (LBR-UC, LBR-5.5, LBR-7.5, LBR-8.5). Notably, in LBR-6.5, the cumulative SCOD production on day 8 was 91 g SCOD, which was 82% of the cumulative SCOD produced on day 14 (Table 2). Comparatively, on the same day (day 8), the cumulative SCOD production for LBR-7.5 and LBR-8.5 was 53% and 60%, respectively, of the cumulative SCOD produced on day 14 in the reactor (Table 2). Additionally, the cumulative SCOD production on day 8 for LBR-6.5 was statistically ($p \leq 0.05$) higher (LBR-6.5) than that obtained in the other pH conditions on day 10 (LBR-UC, LBR-5.5, LBR-7.5, LBR-8.5). This faster hydrolysis in LBR-6.5 indicates that operating the reactor at a pH of 6.5 can significantly shorten the fermentation time to 10–12 days instead of the 14 days required at the pHs of 7.5 and 8.5 to achieve the same SCOD production/hydrolysis yield.

Table 2. Cumulative SCOD production in LBRs at different operating pHs in this study.

Time	Cumulative SCOD Production (g SCOD)									
Day	LBR-UC		LBR-5.5		LBR-6.5		LBR-7.5		LBR-8.5	
	g SCOD	%	g SCOD	%	g SCOD	%	g SCOD	%	g SCOD	%
0	0.0	0%	0	0%	0	0%	0	0%	0	0%
2	61.5	86%	31.8	36%	47.9	43%	28.1	25%	39.3	35%
4	63.8	89%	46.2	52%	74.6	67%	45.1	40%	55.5	50%
6	69.8	98%	57.3	64%	83.9	76%	48.5	43%	58.6	52%
8	71.1	100%	62.7	70%	91.1	82%	59.3	53%	66.9	60%
10	80.6	113%	72.8	82%	102.3	92%	80.3	71%	87.2	78%
12	68.6	96%	81.3	91%	108.6	98%	98.8	88%	100.1	90%
14	71.3	100%	89.0	100%	110.8	100%	112.5	100%	111.7	100%

3.2. TFA Production at Different pH

TFA production was calculated as the sum of SCFAs (i.e., acetate, propionate, n-butyrate, iso-butyrate, n-valerate, iso-valerate) and MCFAs (i.e., n-caproate, iso-caproate, heptanoate) produced in a particular LBR (Equation (2)). The TFA production also varied depending on the operating pH. On day 14, the maximum TFA production of 99.2 ± 3.0 g COD_{FA} was obtained in LBR-6.5, followed by 71.4 ± 0.6 g COD_{FA} in LBR-5.5, 67.4 ± 3.4 g COD_{FA} in LBR-7.5, 65.3 ± 4.2 g COD_{FA} in LBR-8.5 and 25.5± 0.2 g COD_{FA} in LBR-UC. These results showed that the TFA production in the LBR operated at pH 6.5 (LBR-6.5) was statistically ($p \leq 0.05$) higher than that obtained at the other pHs (LBR-UC, LBR-5.5, LBR-7.5, LBR-8.5) (Table 1). Moreover, the TFA production at pH 6.5 (LBR-6.5) was 47–52% higher than that obtained at pH 7.5–8.5 (LBR-7.5, LBR-8.5), even when the cumulative SCOD production was statistically the same at pH 6.5–8.5 (Table 1). The lower TFA production at pH 7.5–8.5 could be attributed to alcohol production under alkaline pH conditions [4]. Higher alcohol production (ethanol, butanol, etc.) at neutral and slightly alkaline pHs has been reported in previous studies [4,17,18]. It is due to enhanced hydrolysis in these alkaline pH ranges of 7.5–8.5 (Table 1), resulting in a greater availability of carbon as a source for alcohol production during the solventogenesis phase of the metabolic pathway in bacteria [4,17,18].

The positive impact of pH 6.5 on TFA production can be further determined from the TVFA yield and TVFA/SCOD ratio (Table 1). The highest TFA yield of 643 ± 19.2 g COD_{FA}/kg VS_{added} was obtained for LBR-6.5, followed by 463 ± 4.3 g COD_{FA}/kg VS_{added} for LBR-5.5, 437 ± 22.1 g COD_{FA}/kg VS_{added} for LBR-7.5, 423 ± 27.4 g COD_{FA}/kg VS_{added} for LBR-8.5 and 165 ± 1.1 g COD_{FA}/kg VS_{added} for LBR-UC. This implies that a nearly neutral pH (pH 6.5) resulted in higher TFA production, along with improved hydrolysis yields (Table 1). Yu [32] reported a 34% increase in the TFA yield during the acidogenic fermentation of food waste when the pH was increased from pH 5.5 to pH 6.5. Likewise, Cysneiros [30] reported a high TFA yield of 720 g COD_{FA}/kg VS_{added} at a pH of 6.5 in an LBR treating maize, which was 76% higher than that obtained under uncontrolled pH conditions (410 g COD_{VFA}/kg VS_{added}). This higher TFA production at pH 6.5 could be due to better acidogenic activity at pH 5.5–6.5 [33].

The maximum TFA/SCOD of 90% was achieved at pH 6.5 in LBR-6.5 (Table 1), followed by 80% at pH 5.5 (LBR-5.5) and between 58 and 60% for a pH of 7.5–8.5. A low TFA/SCOD of 36% was obtained in LBR-UC due to the transformation of solubilized matter into lactate rather than FAs at an uncontrolled pH (Table 1). Lactate production of 39.3 ± 0.3 g $COD_{Lactate}$, a lactate yield of 255 ± 2.1 g $COD_{Lactate}$/kg VS_{added} and a lactate/SCOD of 55% were obtained for LBR-UC. Notably, no lactate was produced at the controlled pH of 5.5–8.5. The higher lactate production at the uncontrolled pH was due to a shift in the microbial community from FA-producing bacteria to lactate-producing bacteria (discussed in Section 3.4). Lactate-producing bacteria such as *Lactobacillus* can thrive under acidic conditions (pH 3.5–4.5) [19,34]. High lactate production at an uncontrolled pH was also observed in other studies. For instance, Kim [35] reported significantly higher lactate production from food waste at a pH of 3.3–3.4 than a pH of 7.2–7.9. Similarly, Ye [36] obtained higher lactate production at an uncontrolled pH as compared to a controlled pH of 6–8 during the acidogenic fermentation of vegetable waste.

3.3. *TFA Composition at Different pH*

The TFA composition at different pHs is shown in Figure 3. Butyrate (56% of TFA) was the most dominant FA at the uncontrolled pH (LBR-UC), followed by acetate (38% of TFA). At the pH of 5.5 (LBR-5.5), butyrate (50% of TFA) and acetate (23% of TFA) were the prevalent FAs and, together, constituted 73% of the TFA. Similarly, the produced amounts of butyrate (56% of TFA) and acetate (21% of TFA) were higher than those of the other FAs at pH 6.5 in LBR-6.5. The higher production of acetate and butyrate at pH 5.5–6.5 implies that bacteria follow the acetate–butyrate metabolic pathway at pH 5.5–6.5 [18]. On the other hand, the main FA content shifted from butyrate to acetate, along with a comparatively higher production of propionate at pH 7.5 (LBR-7.5). Acetate constituted 36% of the TFA, being the most dominant FA, followed by butyrate (29% of TFA) and propionate (27% of TFA). At pH 8.5, acetate (57% of TFA) was still the dominant FA, but the production of propionate (22% of TFA) increased significantly. Butyrate constituted 19% of the TFA at pH 8.5. Overall, an alkaline pH of 7.5–8.5 promotes the production of acetate and propionate, which is in agreement with the findings of other research studies. Bacteria will conserve their energy by producing acetate and balance their intracellular reducing power by producing propionate at the alkaline pH of 7.5–8.5 [18].

Caproate was the main MCFA (Table 1). A relatively high composition of caproate was observed at pH of 5.5, forming 13% of the TFA produced (Figure 3). Other MCFAs, such as heptanoate, constituted a very low fraction of the TFA (0–1.5%). Overall, the results demonstrate that the operating pH is a key parameter that impacts the production and composition of SCFAs and MCFAs during the acidogenic fermentation of food waste.

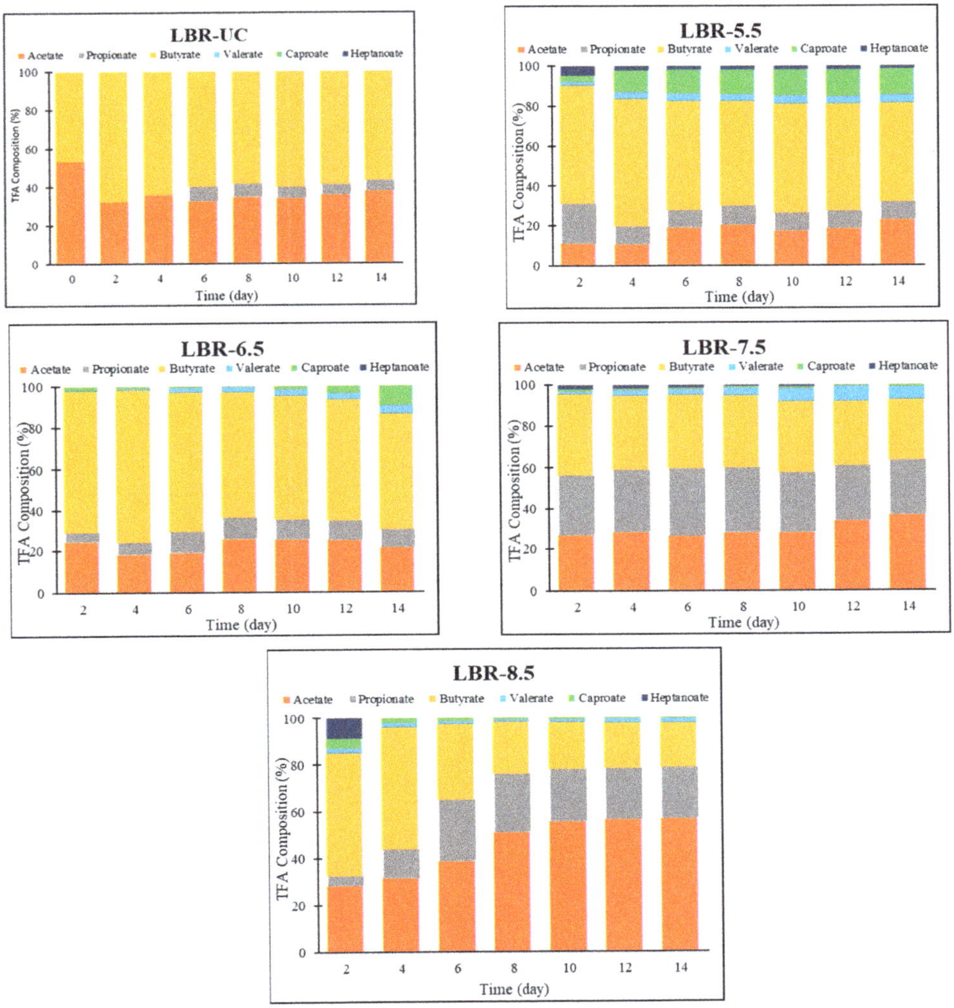

Figure 3. TFA compositions in the LBRs at different pHs. Butyrate refers to the sum of n-butyrate and iso-butyrate. Caproate refers to the sum of n-caproate and iso-caproate. Valerate refers to the sum of n-valerate and iso-valerate.

3.4. Microbial Community Composition

Figure 4 shows the microbial community at the genus level of the leachate and food waste for LBRs at different pHs. In the inoculum, *Clostridium* (43%) and *Marcellibacteroides* (34%) were mostly dominant. While the composition of *Marcellibacteroides* became significantly smaller throughout all the pH conditions, *Clostridium* was consistently found at all pHs in both the leachate and food waste. *Clostridium* is known for acetate, butyrate and hydrogen production through organic fermentation [18].

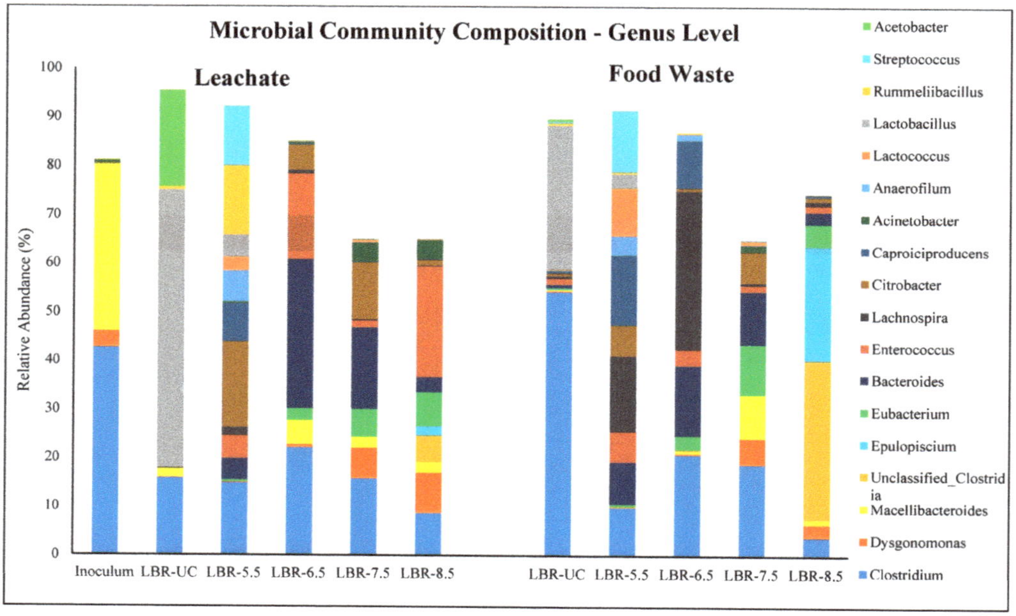

Figure 4. Microbial community compositions of food waste and leachate for LBRs operated at different pHs.

Different microbial communities were found in the leachate in response to different pHs. For the uncontrolled pH (LBR-UC), *Lactobacillus* was the most dominant (57%). *Lactobacillus* are lactate-producing bacteria and have been found under acidic conditions (pH 3.5–4.5) [34], which corresponds with the pH measured for LBR-UC. The high relative abundance of *Lactobacillus* supports high lactate production in LBR-UC. LBR-5.5 had the most diverse genera composition in its leachate, including *Clostridium* (15%), *Citrobacter* (18%) and *Rummeliibacillus* (14%). Among them, *Citrobacter* is a known fermentative bacteria, especially for hydrogen production through acetogenesis [37]. Recently *Rummeliibacillus suwonensis*, one of the species of *Rummeliibacillus* which is known for caproic acid production, was isolated [38]. Given that caproic acid was produced at a high rate at pH 5.5, this genus is suggested to be a major player in the production of caproic acid. In LBR-6.5 at pH 6.5, *Bacteroides* (31%) and *Enterococcus* (18%) were predominantly found with *Clostridium*. *Enterococcus* was reported as a fermentative bacteria producing butyrate and acetate, and it is also known to be a fermenter of carbohydrate and lignocellulose [39,40]. Bacteroides generates butyrate, acetate and propionate [41,42]. This genus was also found predominantly in LBR-7.5 at pH 7.5 (17%). In addition, the composition of *Dysgonomonas* became larger when the pH increased to 7.5 and 8.5. *Dysgonomonas* ferments glucose and produces propionate, acetate, lactate and succinate [43]. This result is consistent with our previous study that reported an abundance of *Dysgonomonas* at a high pH [18].

The bacterial communities in the food waste exhibited different compositions from those in the leachate, except for the uncontrolled conditions, in which *Lactobacillus* and *Clostridium* were dominant. At both pH 5.5 and 6.5, *Lachnospira* (16 and 33%, respectively) and *Caproiciproducens* (14 and 10%, respectively) were predominantly found, while these genera were less abundant in the leachate. *Lachnospira* is known as a type of pectin- and glucose-fermenting bacteria [44]. In our previous study, *Caproiciproducens* was mostly found in food waste and not in leachate [18]. Based on the literature, *Caproiciproducens* can hydrolyze cellulose using extracellular enzymes [45], and it also ferments fatty acids [46]. While LBR-7.5 at pH 7.5 showed a similar microbial composition, pH 8.5 exhibited a different composition, leading us to identify unclassified *Clostridia* (32.7%) and *Epulopiscium* (24%). Unclassified *Clostridia* were only identified in LBR-8.5 at pH 8.5. *Clostridia* is a class-level bacterium, and *Clostridium* also belongs to the *Clostridia* class. Thus, it is assumed that these populations would be involved in hydrolysis and fermentation at a specific high pH. However, it is challenging to fully comprehend their roles due to the limitation of identification to the lower phylogenetic level.

4. Conclusions

The operating pH significantly affects the solubilization and formation of fermentative products from food waste. High hydrolysis and acidification yields of 718 g SCOD/kg VS_{added} and 643 ± 19.2 g COD_{FA}/kg VS_{added}, respectively, were obtained at pH of 6.5. The acidification yield at pH 6.5 was 47–52% higher than that obtained at pH 7.5–8.5, even when the hydrolysis yields were statistically similar in a pH range of 6.5–8.5. A higher TFA/SCOD ratio of 90% was also achieved at pH 6.5. Butyrate was the dominant fermentative product at the pH of 5.5–6.5, whereas acetate formed the major proportion of the TFA composition at pH 7.5–8.5. Lactate-producing bacteria were the most prevalent at the uncontrolled pH, thus resulting in a high lactate production in LBR-UC. The pH of 5.5 (LBR-5.5) resulted in the highest level of MCFA production, constituting 13% of the TFA produced.

Author Contributions: Conceptualization, A.H. and X.F.; methodology, J.L. and A.L.; software, J.L.; investigation, P.R. and A.L.; resources, A.H. and X.F.; data curation, J.L.; writing—original draft preparation, P.R. and A.L.; writing—review and editing, A.H., X.F., S.M. and J.L.; supervision, A.H. and X.F.; funding acquisition, A.H. and X.F., All authors have read and agreed to the published version of the manuscript.

Funding: Natural Sciences and Engineering Research Council of Canada (NSERC) and Ministry of Education (MOE).

Data Availability Statement: Data available on request.

Acknowledgments: The authors are grateful to the Natural Sciences and Engineering Research Council of Canada (NSERC) and Ministry of Education (MOE), Singapore, for the financial support.

Conflicts of Interest: The authors declare that they have no known competing financial interests or personal relationships that could have appeared to influence the work reported in this paper.

References

1. Arras, W.; Hussain, A.; Hausler, R.; Guiot, S. Mesophilic, thermophilic and hyperthermophilic acidogenic fermentation of food waste in batch: Effect of inoculum source. *Waste Manag.* **2019**, *87*, 279–287. [CrossRef] [PubMed]
2. Silpa, K.; Lisa, Y.; Perinaz, B.T.; van Woerden, F. *What a Waste 2.0 A Global Snapshot of Solid Waste Management to 2050*; World Bank Group: Washington, DC, USA, 2018.
3. Xiong, Z.; Hussain, A.; Lee, H.S. Food waste treatment with a leachate bed reactor: Effects of inoculum to substrate ratio and reactor design. *Bioresour. Technol.* **2019**, *285*, 121350. [CrossRef] [PubMed]
4. Zhou, M.; Yan, B.; Wong, J.W.; Zhang, Y. Enhanced volatile fatty acids production from anaerobic fermentation of food waste: A mini-review focusing on acidogenic metabolic pathways. *Bioresour. Technol.* **2018**, *248*, 68–78. [CrossRef]
5. Agler, M.T.; Wrenn, B.A.; Zinder, S.H.; Angenent, L.T. Waste to bioproduct conversion with undefined mixed cultures: The carboxylate platform. *Trends Biotechnol.* **2011**, *29*, 70–78. [CrossRef]
6. Holtzapple, M.T.; Wu, H.; Weimer, P.J.; Dalke, R.; Granda, C.B.; Mai, J.; Urgun-Demirtas, M. Microbial communities for valorizing biomass using the carboxylate platform to produce volatile fatty acids: A review. *Bioresour. Technol.* **2022**, *344*, 126253. [CrossRef] [PubMed]
7. Lee, W.S.; Chua, A.S.M.; Yeoh, H.K.; Ngoh, G.C. A review of the production and applications of waste-derived volatile fatty acids. *Chem. Eng. J.* **2014**, *235*, 83–99. [CrossRef]
8. Bastidas-Oyanedel, J.R.; Bonk, F.; Thomsen, M.H.; Schmidt, J.E. Dark fermentation biorefinery in the present and future (bio)chemical industry. *Rev. Environ. Sci. Biotechnol.* **2015**, *14*, 473–498. [CrossRef]
9. Coma, M.; Martinez-Hernandez, E.; Abeln, F.; Raikova, S.; Donnelly, J.; Arnot, T.C.; Allen, M.J.; Honge, D.D.; Chuck, C.J. Organic waste as a sustainable feedstock for platform chemicals. *Faraday Discuss.* **2017**, *202*, 175–195. [CrossRef]
10. Chang, H.N.; Kim, N.-J.; Kang, J.; Jeong, C.M. Biomass-derived volatile fatty acid platform for fuels and chemicals. *Biotechnol. Bioprocess Eng.* **2010**, *15*, 1–10. [CrossRef]

11. Bhatt, A.H.; Ren, Z.; Tao, L. Value Proposition of Untapped Wet Wastes: Carboxylic Acid Production through Anaerobic Digestion. *iScience* **2020**, *23*, 101221. [CrossRef]
12. Moretto, G.; Valentino, F.; Pavan, P.; Majone, M.; Bolzonella, D. Optimization of urban waste fermentation for volatile fatty acids production. *Waste Manag.* **2019**, *92*, 21–29. [CrossRef] [PubMed]
13. Radadiya, P.; Lee, J.; Venkateshwaran, K.; Benn, N.; Lee, H.-S.; Hussain, A. Acidogenic fermentation of food waste in a leachate bed reactor (LBR) at high volumetric organic Loading: Effect of granular activated carbon (GAC) and sequential enrichment of inoculum. *Bioresour. Technol.* **2022**, *361*, 127705. [CrossRef] [PubMed]
14. Talan, A.; Pokhrel, S.; Tyagi, R.; Drogui, P. Biorefinery strategies for microbial bioplastics production: Sustainable pathway towards Circular Bioeconomy. *Bioresour. Technol. Rep.* **2022**, *17*, 100875. [CrossRef]
15. Uçkun Kiran, E.; Trzcinski, A.P.; Liu, Y. Platform chemical production from food wastes using a biorefinery concept. *J. Chem. Technol. Biotechnol.* **2015**, *90*, 1364–1379. [CrossRef]
16. Wu, S.-L.; Luo, G.; Sun, J.; Wei, W.; Song, L.; Ni, B.-J. Medium chain fatty acids production from anaerobic fermentation of waste activated sludge. *J. Clean. Prod.* **2021**, *279*, 123482. [CrossRef]
17. Hussain, A.; Filiatrault, M.; Guiot, S.R. Acidogenic digestion of food waste in a thermophilic leach bed reactor: Effect of pH and leachate recirculation rate on hydrolysis and volatile fatty acid production. *Bioresour. Technol.* **2017**, *245*, 1–9. [CrossRef]
18. Xiong, Z.; Hussain, A.; Lee, J. Food waste fermentation in a leach bed reactor: Reactor performance, and microbial ecology and dynamics. *Bioresour. Technol.* **2019**, *274*, 153–161. [CrossRef]
19. Wang, K.; Yin, J.; Shen, D.; Li, N. Anaerobic digestion of food waste for volatile fatty acids (VFAs) production with different types of inoculum: Effect of pH. *Bioresour. Technol.* **2014**, *161*, 395–401. [CrossRef]
20. Browne, J.D.; Allen, E.; Murphy, J.D. Improving hydrolysis of food waste in a leach bed reactor. *Waste Manag.* **2013**, *33*, 2470–2477. [CrossRef]
21. Chakraborty, D.; Karthikeyan, O.P.; Selvam, A.; Palani, S.G.; Ghangrekar, M.M.; Wong, J.W. Two-phase anaerobic digestion of food waste: Effect of semi-continuous feeding on acidogenesis and methane production. *Bioresour. Technol.* **2021**, *346*, 126396. [CrossRef]
22. Hussain, A.; Lee, J.; Xiong, Z.; Wang, Y.; Lee, H.-S. Butyrate production and purification by combining dry fermentation of food waste with a microbial fuel cell. *J. Environ. Manag.* **2021**, *300*, 113827. [CrossRef] [PubMed]
23. Jiang, J.; Zhang, Y.; Li, K.; Wang, Q.; Gong, G.; Li, M. Volatile fatty acids production from food waste: Effects of pH, temperature, and organic loading rate. *Bioresour. Technol.* **2013**, *143*, 525–530. [CrossRef] [PubMed]
24. Xu, S.Y.; Karthikeyan, O.P.; Selvam, A.; Wong, J.W. Effect of inoculum to substrate ratio on the hydrolysis and acidification of food waste in leach bed reactor. *Bioresour. Technol.* **2012**, *126*, 425–430. [CrossRef] [PubMed]
25. Lim, S.-J.; Kim, B.J.; Jeong, C.-M.; Choi, J.-D.; Ahn, Y.H.; Chang, H.N. Anaerobic organic acid production of food waste in once-a-day feeding and drawing-off bioreactor. *Bioresour. Technol.* **2008**, *99*, 7866–7874. [CrossRef]
26. Saha, S.; Lee, H.S. High-rate carboxylate production in dry fermentation of food waste at room temperature. *Sci. Total Environ.* **2020**, *714*, 136695. [CrossRef]
27. Takahashi, S.; Tomita, J.; Nishioka, K.; Hisada, T.; Nishijima, M. Development of a prokaryotic universal primer for simultaneous analysis of Bacteria and Archaea using next-generation sequencing. *PLoS ONE* **2014**, *9*, e105592. [CrossRef]
28. Callahan, B.J.; Mcmurdie, P.J.; Rosen, M.J.; Han, A.W.; Johnson, A.J.A.; Holmes, S.P. DADA2: High-resolution sample inference from Illumina amplicon data. *Nat. Methods* **2016**, *13*, 581–583. [CrossRef]
29. Caporaso, J.G.; Kuczynski, J.; Stombaugh, J.; Bittinger, K.; Bushman, F.D.; Costello, E.K.; Fierer, N.; Gonzalez Peña, A.; Goodrich, J.K.; Gordon, J.I.; et al. QIIME allows analysis of high-throughput community sequencing data. *Nat. Methods* **2010**, *7*, 335–336. [CrossRef]
30. Cysneiros, D.; Banks, C.J.; Heaven, S.; Karatzas, K.-A.G. The effect of pH control and "hydraulic flush" on hydrolysis and Volatile Fatty Acids (VFA) production and profile in anaerobic leach bed reactors digesting a high solids content substrate. *Bioresour. Technol.* **2012**, *123*, 263–271. [CrossRef]
31. Xu, S.Y.; Lam, H.P.; Karthikeyan, O.P.; Wong, J.W. Optimization of food waste hydrolysis in leach bed coupled with methanogenic reactor: Effect of pH and bulking agent. *Bioresour. Technol.* **2011**, *102*, 3702–3708. [CrossRef]
32. Yu, P.; Tu, W.; Wu, M.; Zhang, Z.; Wang, H. Pilot-scale fermentation of urban food waste for volatile fatty acids production: The importance of pH. *Bioresour. Technol.* **2021**, *332*, 125116. [CrossRef] [PubMed]
33. Strazzera, G.; Battista, F.; Garcia, N.H.; Frison, N.; Bolzonella, D. Volatile fatty acids production from food wastes for biorefinery platforms: A review. *J. Environ. Manag.* **2018**, *226*, 278–288. [CrossRef] [PubMed]
34. Gu, X.Y.; Liu, J.Z.; Wong, J.W.C. Control of lactic acid production during hydrolysis and acidogenesis of food waste. *Bioresouce Technol.* **2018**, *247*, 711–715. [CrossRef] [PubMed]
35. Kim, J.K.; Han, G.H.; Oh, B.R.; Chun, Y.N.; Eom, C.-Y.; Kim, S.W. Volumetric scale-up of a three stage fermentation system for food waste treatment. *Bioresour. Technol.* **2008**, *99*, 4394–4399. [CrossRef]
36. Ye, N.F.; Lü, F.; Shao, L.M.; Godon, J.J.; He, P.J. Bacterial community dynamics and product distribution during pH-adjusted fermentation of vegetable wastes. *J. Appled Microbiol.* **2007**, *103*, 1055–1065. [CrossRef] [PubMed]
37. Maru, B.T.; Lopez, F.; Medina, F.; Constantí, M. Improvement of biohydrogen and usable chemical products from glycerol by co-culture of Enterobacter spH1 and Citrobacter Freundii H3 using different supports as surface immobilization. *Fermentation* **2021**, *7*, 154. [CrossRef]

38. Liu, C.; Du, Y.; Zheng, J.; Qiao, Z.; Luo, H.; Zou, W. Production of caproic acid by Rummeliibacillus suwonensis 3B-1 isolated from the pit mud of strong-flavor baijiu. *J. Biotechnol.* **2022**, *35*, 33–40. [CrossRef]
39. Yin, Y.; Wang, J. Optimization of fermentative hydrogen production by Enterococcus faecium INET2 using response surface methodology. *Int. J. Hydrogen Energy* **2019**, *44*, 1483–1491. [CrossRef]
40. Esquivel-Elizondo, S.; Ilhan, Z.E.; Garcia-Peña, E.I.; Krajmalnik-Brown, R. Insights into Butyrate Production in a Controlled Fermentation System via Gene Predictions. *MSystems* **2017**, *2*, e00051-17. [CrossRef]
41. Yamada, H.; Watabe, Y.; Suzuki, Y.; Koike, S.; Shimamoto, S.; Kobayashi, Y. Chemical and microbial characterization for fermentation of water-soluble cellulose acetate in human stool cultures. *J. Sci. Food Agric.* **2021**, *101*, 2950–2960. [CrossRef]
42. Brame, J.E.; Liddicoat, C.; Abbott, C.A.; Breed, M.F. The potential of outdoor environments to supply beneficial butyrate-producing bacteria to humans. *Sci. Total Environ.* **2021**, *777*, 146063. [CrossRef] [PubMed]
43. Sun, X.; Yang, Y.; Zhang, N.; Shen, Y.; Ni, J. Draft Genome Sequence of Dysogonomonas Macrotermitis Strain JCM 19375^{T}, Isolated from the Gut of a Termite. *Genome Annoucements* **2015**, *3*, e00963-15.
44. Slováková, L.; Dušková, D.; Marounek, M. Fermentation of pectin and glucose, and activity of pectin-degrading enzymes in the rabbit caecal bacterium Bifidobacterium pseudolongum. *Lett. Appled Microbiol.* **2001**, *35*, 126–130. [CrossRef] [PubMed]
45. Opdahl, L. Identification of Candidate Cellulose Utilizing Bacteria from the Rumen of Beef Cattle, Using Bacterial Community Profiling and Metagenomics. 2017. Available online: https://openprairie.sdstate.edu/cgi/viewcontent.cgi?article=2688&context=etd (accessed on 24 January 2023).
46. Andersen, S.J.; De Groof, V.; Khor, W.C.; Roume, H.; Props, R.; Coma, M.; Rabaey, K. A Clostridium group IV species dominates and suppresses a mixed culture fermentation by tolerance to medium chain fatty acids products. *Front. Bioeng. Biotechnol.* **2017**, *5*, 8. [CrossRef] [PubMed]

MDPI AG
Grosspeteranlage 5
4052 Basel
Switzerland
Tel.: +41 61 683 77 34

Fermentation Editorial Office
E-mail: fermentation@mdpi.com
www.mdpi.com/journal/fermentation

www.ingramcontent.com/pod-product-compliance
Lightning Source LLC
Chambersburg PA
CBHW061619210926
51850CB00009B/2379
* 9 7 8 3 7 2 5 8 1 9 7 8 2 *